DESCARTES' METAPHYSICAL PHYSICS

SCIENCE AND ITS CONCEPTUAL FOUNDATIONS

David L. Hull, *Editor*

DESCARTES'
METAPHYSICAL
PHYSICS

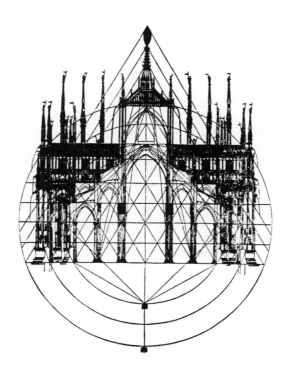

DANIEL GARBER

THE UNIVERSITY OF CHICAGO PRESS
Chicago and London

The University of Chicago Press, Chicago 60637
The University of Chicago Press, Ltd., London
© 1992 by The University of Chicago
All rights reserved. Published 1992
Printed in the United States of America

00 99 98 97 96 95 94 93   5 4 3 2

ISBN 0-226-28217-1 (cloth)
ISBN 0-226-28219-8 (paper)

Library of Congress Cataloging-in-Publication Data

Garber, Daniel, 1949-
    Descartes' metaphysical physics / Daniel Garber.
        p. cm.—(Science and its conceptual foundations)
    Includes bibliographical references and index.
    1. Physics—Philosophy. 2. Metaphysics. 3. Matter—Philosophy.
4. Motion—Philosophy. 5. Descartes, René, 1596–1650—Knowledge—
Physics. I. Title. II. Series.
QC6.G25   1992
530′.01—dc20   91–30190
                                                         CIP

TO MY FATHER AND TO THE
MEMORY OF MY MOTHER

Mens humana non potest cum Corpore absolutè destrui;
sed ejus aliquid remanet, quod aeternum est.
Spinoza, *Ethics* V, Prop. XXIII

CONTENTS

Preface
ix

Abbreviations
xiii

Prologue
1

1
DESCARTES' VOCATION
5

2
DESCARTES' PROJECT
30

3
BODY: ITS EXISTENCE AND NATURE
63

4
DESCARTES AGAINST HIS TEACHERS:
THE REFUTATION OF HYLOMORPHISM
94

5
DESCARTES AGAINST THE ATOMISTS:
INDIVISIBILITY, SPACE, AND VOID
117

CONTENTS

6
MOTION
156

7
MOTION AND ITS LAWS:
PART 1, PRELIMINARIES AND THE
LAWS OF PERSISTENCE
197

8
MOTION AND ITS LAWS:
PART 2, THE LAW OF IMPACT
231

APPENDIX
255

9
GOD AND THE GROUNDS OF THE LAWS OF MOTION:
IMMUTABILITY, FORCE, AND FINITE CAUSES
263

Afterword
307

Notes
309

Bibliography
369

Index
383

PREFACE

THIS BOOK GREW out of a curiosity about the larger programs of the figures I had been reading since my first years as a philosophy student; though Descartes, Leibniz, and others of their age now studied mainly in philosophy departments were acknowledged to have been deeply involved in the scientific world of their day, somehow or another the physical world seemed to have relatively little role in the version of the history of early modern philosophy that I was given as a student.

And so, following the familiar adage that those who cannot do, teach, I announced seminars on the physics of Descartes and Leibniz. I taught my first seminar on that topic in spring 1979 at the University of Minnesota, then the following year at the Johns Hopkins University, and then, finally, at my home institution, the University of Chicago. I would like to thank the kind audiences at all of those universities (especially John Earman, then at Minnesota, and Peter Achinstein at Johns Hopkins), as I stumbled through the material, learning as quickly as I could. While teaching those seminars, I boldly decided to write a book, the book that I had wished someone could have handed me when I first was interested in the area. Initially, I had hoped to write a single book on the natural philosophy of both Descartes and Leibniz. I soon realized that such a book, in essence a history of the development of physics in the seventeenth century was impossible, at least for me to write, and I retrenched, limiting myself to a book on Descartes' natural philosophy alone. This is the result.

I finally got the opportunity to put words to paper on this project in the academic year 1985–86, when I received a fellowship from the American Council of Learned Societies (under a program funded by the National Endowment for the Humanities) and an invitation to be a member of the School of Historical Studies at the Institute for Advanced Study in Princeton for the year. During that very happy year, and the summers that preceded and followed it, I wrote drafts of about three-quarters of the book. The book then sat for a long while, aging in

ix

a box in my office as other things took precedence. I was goaded again into action by meeting John Schuster at a conference on Descartes at San Jose in April of 1988. There, in a moment of somewhat inebriated bravado, I announced that I would have the manuscript ready to submit for publication a year later, on 15 April 1989, and would buy him dinner if I didn't. John, wherever you are right now, I owe you dinner. But I did take the bet seriously, and started work on the project, again. By a happy accident, the following summer, the summer of 1988, I was the director of the NEH/Council for Philosophical Studies Summer Institute on the History of Early Modern Philosophy. I had committed myself to give a series of sessions at the Institute, and that seemed like a perfect opportunity to do more thinking on the project. I would like to thank the many participants of that seminar for their kind reception of my presentations, and the NEH for their support to me and to all of the other participants. During that very busy but most rewarding of summers I did extensive reworking of the earlier material, and finished drafts of the remaining chapters. During the summer and fall of 1989 I did more reworking, and finally submitted the manuscript in December 1989.

Parts of this book formed the basis of lectures I gave at a number of places during the time I was working on it, and ideas have appeared in various publications. Some of the ideas about Cartesian method found in chapter 2 were first given at a lecture at the Université de Paris IV (Sorbonne), later at the Washington (D.C.) Philosophical Club and the Philosophy of Science Association, and published in "La méthode du *Discours*," which appeared in Grimaldi and Marion 1987. Parts of chapter 3 were delivered at the University of Massachusetts at Amherst, and the University of Georgia. I would especially like to thank Fred Feldman for the response he gave in Amherst; his careful probing led to some reformulations of my ideas. Chapter 4 was delivered to an audience at the Program for History of Science at Princeton University. Some themes from chapter 6 were delivered to a conference on Descartes and Newton at Hobart and William Smith. I would like to thank Scott Brophy for organizing that event as well as his active participation in the discussion; I would also like to thank the other participants, Alan Gabbey, Marjorie Grene, and Ted McGuire, who let nothing pass without challenging objections. Some of the material in that chapter will appear in the forthcoming proceedings of that conference, to be published in *Synthese*. Material on the ontology of force from chapter 9 was given at Emory University. Material on occasionalism from that chapter was also given as an invited address at the American Philosophical Association, Eastern Division, and appeared as Garber 1987. The book as a whole formed the basis of a chapter on Descartes' physics which will appear in the *Cambridge Companion to Descartes,* edited by John Cottingham.

There are many, many people to whom I owe thanks for the help and encouragement they have given me in connection with this project. First, let me thank the two reviewers for the University of Chicago Press, John Schuster and Anonymous. Both produced long and very valuable detailed comments, longer and more valuable than any reports that I can remember ever having written (or even read) before; I kept them both at my side while I did the final revisions on the manuscript. While I suspect that neither will be entirely satisfied with the changes I have made in the book in response to their remarks, the book is much better than it would have been otherwise. The shortcomings that remain are mine and mine alone. I would also like to thank Emily Grosholz for detailed comments that she sent on an earlier draft. Of the many colleagues who discussed topics in this book with me, I would like to single out Roger Ariew and Alan Gabbey for special thanks. Roger has been an ideal colleague and a swell friend over many years; he always has time to read a manuscript, and say a nice word, as well as a wise one, kindly put. I eagerly look forward to his help and active collaboration as we extend our work to the margins of the history of philosophy and the history of science in a forthcoming study of the social, political, and intellectual background to the Cartesian revolution. Alan Gabbey's writings on Descartes' physics and, more generally, on seventeenth-century philosophy and science have been particularly important to me; it is on his work, more than on that of anyone else I know, that I have modeled my own. Though we may disagree from time to time, and though he may (stubbornly) fail to understand an obvious point or two I try to make, I am honored to call him my friend and colleague. My colleagues in the Department of Philosophy, the Committee on the Conceptual Foundations of Science, and the Morris Fishbein Center for the History of Science and Medicine also deserve many thanks for their support and stimulation during the time this book was written. My Chicago colleague Howard Stein has been a particular inspiration to me; though he may strongly disapprove of Descartes, if not my interest in Descartes, I have tried to keep his high standards of rigor and scholarship in mind throughout this project. I would also like to thank my former and present students for giving me a hard time and keeping me honest, especially Stephen Menn and Heather Blair (Heather has also been an invaluable help as my sometimes research assistant over the last few years), as well as the members of my seminar at the University of Chicago in the fall of 1989, who patiently read and discussed the whole manuscript as it was nearing completion. Susan Abrams, my friend and editor at the University of Chicago Press, encouraged me time and again; her help in seeing this project through to the end was considerable.

Finally, some more personal acknowledgments. Ben Martinez, Jacquie Block, and their two charming daughters, Chloe and Nora, helped make these last many summers on the Massachusetts seashore so lovely for us, and helped to create the perfect atmosphere for doing the work I couldn't do in Chicago; they were there when the project began in the summer of 1985, and they are with us still.

Above all, I would like to thank my own family, my wife Susan Paul, and daughters Hannah and Elisabeth, who put up with more than anyone could be expected to, and then put up with more; I am deeply indebted to them for their love and their support over these years.

I dedicate this book to my parents, to my father, William Garber, and to the memory of my mother, Laura Coplon Garber; words, even these many pages of words, cannot express my love and gratitude.

ABBREVIATIONS

AT   The original language text of Descartes' writings used is the edition of Adam and Tannery (Descartes 1964–74 in the bibliography), abbreviated 'AT', followed by volume number and page number.

Pr   References to the *Principles of Philosophy* begin with 'Pr', followed by the part and section number. Where it is important to distinguish between the text in the Latin and French versions of the *Principles* (found in AT VIIIA and AT IXB, respectively), 'L' or 'F' is appended to the section number. So, for example, 'Pr II 37F' designates *Principles,* part II, section 37 in the French version.

PS   I also refer to Descartes' *Passions of the Soul* by section number, preceded by 'PS'. I cite original language editions first, with translations in parentheses following, if available.

K   For translations of Descartes' correspondence, references are given in parentheses to page numbers in Kenny's collection (Descartes 1970), abbreviated 'K', when available.

Ols   For translations from the *Meteors* and *Dioptrics,* references are given to page numbers in Olscamp's translation (Descartes 1965), abbreviated 'Ols'.

I cite original language editions first, with translations in parentheses following, if available. Since Cottingham, Stoothoff, and Murdoch give AT page numbers in their two-volume translation of Descartes' philosophical writings (Descartes 1985), I add no special citation for translations of standard works like the *Rules for the Direction of the Mind, Discourse on the Method,* and the *Meditations.*

Since the Miller and Miller translation of the complete *Principles of Philosophy* (Descartes 1983) is, of course, keyed to the part and section number, no special reference is made to that translation either.

Michael Mahoney's valuable translation of *The World* (Descartes 1979) is keyed to the pagination in the original publication. I have not given direct references to that translation, but they can easily enough be recovered, insofar as Adam and Tannery's text of *The World* is also keyed to the original pagination.

JB   References to Isaac Beeckman's *Journal* (de Waard 1939–53) are abbreviated 'JB', followed by volume number and page number. When a passage from JB is also found in AT, I indicate the location in AT following the citation in JB.

CM   The correspondence of Mersenne (Mersenne 1932–88) is abbreviated 'CM', followed by volume number and page; Descartes' correspondence with Mersenne is cited in AT only.

G   References to Leibniz in the Gerhardt editions of the philosophical and
GM   mathematical writings (Leibniz 1875–90 and Leibniz 1849–63) are abbreviated 'G' and 'GM', respectively, followed by volume number and page.

RM   The works of Malebranche are cited in the Robinet edition (Malebranche 1958–84), abbreviated by 'RM' and followed by volume number and page.

Other references are either to the short form of the bibilography entry, or to the standard scholarly abbreviations. Though I have consulted a number of available translations, and may have borrowed here and there, translations are my own unless otherwise indicated; the parenthetical reference to translations is generally given only as a convenience to the reader who would like to look up the full context.

PROLOGUE

THIS BOOK IS principally concerned with Descartes' accounts of matter and motion, the joint, as it were, between what we could call Descartes' philosophy and what we would consider his scientific interests, the point at which the metaphysical doctrines on God, the soul, and body developed in writings like the *Meditations* give rise to physical conclusions regarding atoms, vacua, and the laws that matter in motion must obey. To call the one group of interests philosophical and the other scientific is, of course, anachronistic. This is not to say that Descartes recognized no distinctions here; like most of his contemporaries, Descartes distinguished in some sense between metaphysics and physics, first philosophy and natural philosophy. But for Descartes, both metaphysics and physics are parts of philosophy, properly understood, and both metaphysics and physics, properly done, are sciences in the sense in which he understood the term *scientia,* knowledge in the strictest sense. As he characterized the notion in his early *Rules for the Direction of the Mind,* "all *scientia* is certain and evident knowledge *[cognitio]"* (AT XI 362), something that, for Descartes, characterizes our knowledge of the soul and God as well as it does our knowledge of the laws of motion and the nature of the material world. Furthermore, throughout his career, Descartes emphasized over and over again the unity of his thought; though he certainly drew distinctions between different domains of inquiry, there is a sense in which they are all of a piece, and fit together as integral parts of a larger enterprise. Descartes' project was not just metaphysics and epistemology, and not just physics, but a systematic attempt to deal with *all* knowledge. As we shall discuss in more detail below, Descartes was educated in a scholastic intellectual culture in which Aristotelian philosophy is joined with Christian theology in a grand synthesis. Descartes, in the first generation of seventeenth-century thinkers responsible for making the new mechanical philosophy respectable, saw as one of his principal goals the refutation of Aristotle and the establishment of a new philosophical system around the me-

1

chanical world view, a new philosophical synthesis to replace the then worn-out synthesis constructed around Aristotelian principles.

It is the wide scope of his program that separates Descartes most clearly from many modern conceptions of philosophical activity, and from the more strictly scientific and less metaphysical attitudes of Galileo and Newton, to choose two visible and influential thinkers of a different temperament. Descartes' intellectual program failed, of course; while pieces of the program may have proved important inspirations to later thinkers, as an approach toward understanding the natural world Descartes' program turned out to be a dead end. But while the design may have been faulty, and the edifice doomed from the start, it is fascinating to contemplate the entire structure as the architect planned it, an alternative way the world might have been, but wasn't.[1] And insofar as the details of this project are of interest to philosophers, historians of philosophy, and historians of science, it is important to see these details in the context of the larger program into which they fit. In tracing out the path Descartes followed from the metaphysical concerns of the works like the *Meditations* to the foundations of his mechanist physics, I hope to place both the metaphysics and the physics in their proper intellectual context, and, in so doing, illuminate both in ways that more specialized studies of individual arguments and doctrines cannot do.

Though I will have to say something about other areas of Descartes' thought, I shall concentrate on the theory of matter and motion and their metaphysical foundations as given in part II of the *Principles of Philosophy,* in parallel texts in *The World,* and in relevant passages from other works, correspondence, and notebooks. I shall assume that the reader is reasonably familiar with at least the standard themes in Descartes' metaphysics; given the enormous literature on that aspect of Descartes' thought, I feel under no obligation to give yet another detailed and systematic exposition of the *Meditations,* though, of course, more particular discussions will be appropriate from time to time. Also, the scientific essays of 1637—the *Dioptrics, Geometry,* and *Meteors*—will play a relatively minor role in my account. As I shall discuss below, these texts, while important in general for Descartes' thought, are explicit attempts to divorce Descartes' scientific conclusions from their metaphysical foundations, and so are less relevant to the theme I have chosen to emphasize than are other works, like the *Principles,* where Descartes attempts the full and proper development of his system.

My main interest in this book is in understanding Descartes' thought, and my main focus will be on Descartes' own writings. The book involves other figures, but only to the extent that I feel that discussing them can illuminate Descartes' thought. Full and systematic studies of late scholastic natural philosophy, and of other early modern attempts to deal with, for example, motion and relativity, the analysis of circular motion, and the analysis of bodies in collision have yet to be written; when they are, they will certainly illuminate Descartes' thought.[2] But such studies, done properly, would amount to a systematic history of sixteenth- and early seventeenth-century physics, something well worth doing, but something that I cannot do here.

Also important is the social and political context of Descartes' thought. Social studies of the history of science have become extremely fashionable these days, and a book like this, concentrating as it does on the texts and the interpretation of Descartes' ideas, may seem decidedly old-fashioned. It is, and I make no apologies for it. But at the same time, I think that understanding the social context of the Cartesian program is crucially important, too; in work in progress now, Roger Ariew and I hope to illuminate the larger social, cultural, and political background to the rejection of Aristotelianism and the rise of mechanist programs like Descartes' in Paris in the early seventeenth century. But important as such studies may be, much of Descartes' interest for us, as for his contemporaries, lies in his texts and his ideas, and though we may not want to limit our attention to these aspects of the history, it's not a bad place to begin.

This book does not have a single, overarching thesis, nor is it a grand reinterpretation of Descartes' thought. That isn't to say that there aren't important themes. One thread that runs through these chapters is the concern with the place Descartes' program occupies in the rejection of the broadly Aristotelian natural philosophy of his day, and the establishment of the so-called mechanical philosophy; another related theme is the mechanizing of the world by way of the replacement of scholastic matter and form with extended substance and God. And while I do attempt some significant reinterpretations of Descartes' thought, for instance the treatment of method in chapter 2 or of motion in chapter 6, one can only offer a grand reinterpretation if there is a standard interpretation of Descartes' natural philosophy against which to react, and in general, there isn't. What is needed, and what I hope to have provided, is a book that pulls together various aspects of Descartes' metaphysical approach to the world of body and presents them in a systematic and coherent way, a kind of handbook of Cartesian physics, a general introduction to the mechanical philosophy as Descartes or a sympathetic but not uncritical contemporary of his might have understood it. At the same time, I also hope to have sprinkled enough fresh insights and controversial claims in the text to keep the interest of those already familiar with the texts and issues.

My project begins in chapter 1 with a brief intellectual biography, an exposition of the aspects of Descartes' life and development most directly relevant to the task at hand. In chapter 2, I then discuss the organization of his overall project for philosophy, and some closely related aspects of his conception of method. With these preliminaries out of the way, I then turn to a systematic treatment of the main business of this book, the discussion of Descartes' doctrines of matter and motion and their laws. In chapter 3, I discuss Descartes' account of body, and in chapters 4 and 5, the relation between Descartes' views on body and those of both the schoolmen and the atomists. Chapter 6 deals with his account of motion and its supposed relativity. In chapters 7 and 8, I treat the laws of motion, concluding in chapter 9 with an account of how these laws are to be grounded in God, as well as the related issues of force and occasionalism.

DESCARTES' VOCATION

FIRST YEARS

RENÉ DESCARTES was born on 31 March 1596, in the town of La Haye, now known as Descartes. In 1606 or 1607, his father Joachim entrusted his education to the Jesuits, and sent the young René to the Collège Royal de la Flèche, founded in 1604 with the help and encouragement of Henri IV.[1] There he stayed until 1615 or so, when he went to the Université de Poitiers, where he received his *baccalauréat* and his *licence en droit* in late 1616.[2] With this he ended his formal education.

Little direct documentary evidence remains of Descartes' years in school.[3] But from his own report of his intellectual development in the *Discourse on the Method,*[4] and from the copious information we have about Jesuit education in the early seventeenth century,[5] it is clear that the education he received at La Flèche was an interesting combination of the old and the new, traditional scholasticism and the new learning. Instruction, of course, was in Latin, and his first three years were spent mastering Latin, together with some amount of Greek. As he reports in the account of his school years in the *Discourse:* "I knew that the languages that one acquires there are necessary for understanding the books of the ancients" (AT VI 5).[6] Also prominent in his account of his schooling in the *Discourse* are philosophy and theology. The philosophy curriculum which Descartes most likely followed in his last three years in school, under the direction of Étienne Noël, was grounded in Aristotle: "In Logica, et Philosophia naturali et morali, et Metaphysica, doctrina Aristotelis sequenda est," dictates the Jesuit manual of instruction.[7] In his first year he would have studied logic, followed by work based on Aristotle's *Physics, De Caelo,* and the first book of *De Generatione et Corruptione* in the second year, and the second book of *De Generatione et Corruptione, De Anima,* and the *Metaphysics* in the third.[8] While he did not follow the course of studies in theology, he was certainly exposed to theological issues by his teachers who, in accordance with Jesuit practice, would have studied theology through a course of studies grounded in St. Thomas.[9]

The core of the curriculum was, in a way, quite traditional. But it also shows the influence of later thought. The very fact that Greek was taught serves to differentiate Descartes' education from that a thirteenth- or fourteenth-century student would have received. Also significant are the two years he would have spent between the years mastering Latin and the three years devoted to studying philosophy. The fourth year was to be spent in the humanities, the *studia humanitatis.* There the student might have been exposed to the Latin essayists, poets, and historians that formed the core of the new humanist learning, including perhaps Cicero, Salust, Livy, Virgil, and Horace.[10] Rhetoric, the study of eloquence, which, Descartes admits, "has incomparable force and beauty" (AT VI 5), would have been taken up in the following year. Though Aristotle's *Rhetoric* would have been studied, the central figure would have been the favorite author of the humanists, Cicero, together with selections from other Roman writers who illustrated Cicero's points.[11] Morality, too, would have been taught from a humanist point of view, not through Aristotle or Thomas, but through literature, through lessons drawn from the ancient writers.[12] And finally, though not an official part of the curriculum, Descartes even seems to have been exposed to the sixteenth-century preoccupation with magic, alchemy, and the occult, though probably not in a sympathetic way.[13]

Even the teaching of Aristotle in the Jesuit College was influenced by more modern currents. In later years, twenty-five years or so after leaving La Flèche, Descartes recalled three textbooks he used at school: the Coimbrian Commentaries, Toletus, and Rubius (AT III 185 [K 78]). All three were late sixteenth-century authors of commentaries on Aristotle widely reprinted in the sixteenth and seventeenth centuries and widely used in colleges and universities.[14] In many ways these were traditional commentaries, the sort that could have been written centuries earlier. But in some ways they were quite different, particularly the Coimbrian commentaries. The Coimbrian commentaries, *Commentarii Collegii Conimbricensis,* were a complete course on Aristotelian philosophy, assembled by the Jesuit Fathers of the College of Coimbra, and published in nine volumes between 1592 and 1607. The volumes include both a complete text of the work under discussion and a commentary. The text chosen shows an obvious concern with the philological questions that occupied Renaissance scholars; the text is given in both Latin and Greek, unlike medieval commentaries, which use only Latin, and the Latin translation chosen is not a medieval version but a contemporary humanist translation. Furthermore, though standard medieval authorities are cited in the commentary, the commentary also discusses the views of more recent humanist thinkers and the views of quite non-scholastic sources such as Hermes and, significantly enough, given Descartes' later mechanical physics, Lucretius.[15] Though Aristotle was the core, it was not the Aristotle of the thirteenth- or fourteenth-century schools, but the Aristotle of the sixteenth century that students at La Flèche were exposed to.[16]

But most interesting is the treatment of mathematics and physics at La Flèche. There is a story, difficult to substantiate, of the young Descartes reading a poem to the assembled crowd at La Flèche celebrating the discovery of the moons of Jupiter by Galileo in 1610. Even if true, it is difficult to say how significant this is. While the Jesuits were certainly open to new scientific ideas, after a fashion, it is unlikely that the Copernican conclusions that Galileo was eager to draw from his observations got a very sympathetic treatment by Descartes' teachers.[17] But what he would have learned in the classroom is, perhaps, more significant.

Although Descartes was not altogether satisfied with his mathematical education at La Flèche, it was one of the subjects that did impress him most, at least according to the *Discourse*. He writes there that he "was above all pleased with mathematics because of the certainty and evidence of its reasoning" (AT VI 7). It was, in fact, one of the innovations of Jesuit education to introduce the serious study of mathematics into the curriculum at this time. Mathematics would have been taught in the second year of philosophical studies, at the same time as Father Noël taught Descartes his physics, but by a separate teacher, in Descartes' case a young man named Jean François.[18] He remarked in the *Discourse* that he "did not yet notice its true use, and thinking that it served only the mechanical arts, I was astonished that though its foundations were so firm and solid, nothing more elevated had been built on its foundations" (AT VI 7). This may reflect the way that mathematics was taught to the young Descartes at La Flèche. But there is another important strand in Jesuit pedagogy. Christopher Clavius, the important Jesuit mathematician of the late sixteenth and seventeenth century, was charged with drafting the mathematics component of the *Ratio Studiorum* and formulating a course of studies to train the teachers of mathematics in the Jesuit schools. One thing striking about Clavius' approach to mathematical education was his insistence on the importance of mathematics as giving philosophers examples of certain and solid demonstration.[19] In Clavius' writings one finds the contrast between the certainty of mathematics and the uncertainty of scholastic philosophy (including physics) drawn out in a way that echoes Descartes' own concerns in the period. Clavius wrote, for example, in the prolegomena to the first volume of his *Opera Mathematica,* published in 1611 and, no doubt, received at La Flèche soon after:

> Mathematics demonstrates and establishes by the firmest arguments everything about which disputation can be raised to such an extent that it truly brings forth knowledge *[scientia]* in the mind of the student and completely eliminates all doubt, something that we can scarcely ascribe to other branches of knowledge *[scientia],* since in other branches of knowledge, in judging the truth of conclusions, the intellect, undecided and uncertain, is repeatedly thrown into perplexity by the multitude of opinions and the variety of views. I am brought to this view of the matter by the many sects of Peripatetics (not to mention other philosophers, for the moment) which are derived from Aristotle like branches from a trunk and which disagree with one another and, sometimes, with Aristotle, their very source, to such an extent that one does not altogether know what Aristotle wants for himself, and whether the dispute is about names or about things.[20]

Clavius concludes by asserting that because of its certainty, we should grant mathematics "first place among all the other sciences." In this passage from Clavius we can see much that may well have impressed the young Descartes, the dissatisfaction with the uncertainty of scholastic philosophy, the attraction to mathematics for its very certainty, and the underlying suggestion that philosophy, too, should strive for the same standards as the mathematicians have, themes prominent in Part I of the *Discourse,* themes that recur in his work from his earliest years. His final views may be quite different than anything Clavius may have had in mind or Father François may have taught him. Indeed, we cannot be sure exactly how much of Clavius' work or attitudes Descartes may have known, either directly or through Father François.[21] But it is not implausible to suppose that the initial inspiration for Descartes' later project, the rejection of scholastic physics and the erection of a science on a firm foundation, may derive from his early years at La Flèche.

The *Discourse* presents a Descartes who, in a sense, rejected his education: "As soon as I reached the age at which I could leave the control of my teachers, I gave up the study of letters altogether. And I resolved to seek no other knowledge *[science]* but the one I could find in myself, or better, in the great book of the world" (AT VI 9; cf. p. 4). This, of course, is largely true of the historical Descartes. He certainly rejected much of the content of his education, as we shall later see in connection with his natural philosophy. And it will be twenty-five years or so before he will open another scholastic textbook, and then only to get a clearer idea of what his opponents are up to, a clearer idea of what it is that his arguments must refute (see AT III 185 [K 78]). He, in fact, is reported to have made a gift of his textbooks to the library at La Flèche upon his departure (AT X 646).[22] But one should not thereby underestimate the value he placed on his education and the value it had for his later work. Throughout his life he maintained a keen interest in how his work was regarded by his former teachers and the order to which they belonged; upon sending a copy of his newly published *Discourse, Dioptrics, Geometry,* and *Meteors,* he wrote to Father Noël, his teacher in physics: "I am very happy to present [this book] to you as a fruit that belongs to you, the first seeds of which you planted in my soul, just as I also owe those of your Order [i.e., the Jesuits] all of the scant knowledge I have of letters" (AT I 383).[23] And when in 1638 a friend asked Descartes for advice on where he should send his son for school, he recommended La Flèche in the highest terms: "there is no place in the world where I think they teach better than at La Flèche." Interestingly enough, he singled out the teaching of philosophy for special praise: "since it [philosophy] is the key to the other branches of knowledge *[sciences],* I think that it is very useful to have studied the entire course of it, in the way that it is taught at the Jesuits' schools" (AT II 378). Though he rejected much of what they taught him, the historical Descartes (as opposed to the mythical protagonist of the *Discourse*) seems to have appreciated his teachers and the disciplined education that they gave him.

DESCARTES AND BEECKMAN

According to the *Discourse,* after leaving school, Descartes wandered through Europe: "I devoted the rest of my youth to traveling, to seeing the courts and armies, to frequenting people of different sorts and circumstances, to collecting different experiences, to proving myself in the episodes that fortune sent my way, and above all, to reflecting on the things that presented themselves so that I could draw some benefit from them" (AT VI 9). Part II of the *Discourse,* which relates the next important step in the development of the protagonist of the *Discourse,* begins in a small room in Germany in 1619, we can infer, when Descartes presents what he later came to see as the start of his own philosophy. But in the *Discourse* Descartes neglects to mention an incident that preceded this in his own life, one of the most important incidents of his early years: his meeting with Isaac Beeckman.

Descartes, having chosen the life of a soldier, had gone to Breda as a volunteer ("miles voluntarius") in the army of Maurice of Nassau.[24] There, on 10 November 1618, he met, quite by chance, as he later emphasized, one Isaac Beeckman.[25] Beeckman was at that time a young man of thirty years, eight years Descartes' senior, who had been a devoted scientific amateur since his early twenties. His important scientific journals, rediscovered only in this century, show an inquiring mind ranging over a wide variety of scientific, medical, and mathematical questions.[26] They also provide the record of Beeckman's discussions with Descartes, variously identified as Renatus Picto, Renatus Descartes Picto, and Mr. du Peron. The two young men got on extremely well and saw one another regularly until 2 January 1619, when Beeckman left Breda. Letters passed between them for a few months (until late April, early May of 1619), and then they lost contact until 1628.[27]

Beeckman's journal shows evidence of discussion on a variety of issues.[28] Beeckman set problems for Descartes in music and acoustics, physics and mathematics. In addition to the record of the informal discussions, Beeckman's journal contains copies of three more formal writings Descartes presented to Beeckman. The longest is his *Compendium Musicae*, presented to Beeckman as a New Year's present in December 1618. In addition there is a shorter manuscript on hydrostatics and one on the motion of a stone in free fall.[29]

His short time with Beeckman was crucial to the young Descartes' development. Beeckman differed from Descartes in a number of important respects. It is wrong to see him as having derived his whole physics from Beeckman by any means, but the encounter with Beeckman probably represents Descartes' first serious (and sympathetic) introduction to some of the ideas that will characterize his own later physics.[30]

Basic to Beeckman's physics was the principle that "things once moved never come to rest unless interfered with" (JB I 24; AT X 60). Though Beeckman's understanding of the principle was not unproblematic,[31] it was central to his physics from 1613 onward. Furthermore, Beeckman carefully distinguished his principle, by which motion per se continues indefinitely, from the so-called impetus theory, widespread among Aristotelian physicists from at least the fourteenth century onward, by which a body in motion continues in motion by virtue of a force or impetus that constitutes a genuine cause of the body's continued motion, a cause without which the moving body would naturally come to rest.[32] Descartes was certainly introduced to the impetus theory in his years at La Flèche; it was a standard topic in late scholastic physics.[33] But it was almost certainly Beeckman who introduced Descartes to the somewhat different principle he had discovered for himself, a principle, suitably transformed and made his own, that was to become an important element of his later physics.[34] In the early writings connected with Beeckman we also find anticipations of the distinction Descartes later draws between motion and the tendency to motion, and the analysis of a motion into its momentary parts, a technique he will later use in his derivation of the laws of motion (JB IV 51 [AT X 77–78]; JB IV 54 [AT X 72]). Also found in entries Beeckman wrote in his journal before and during his discussions with Descartes are issues important to his later thought, the conservation of motion and the view that gravity results from the action of a particular ether on a heavy body, from which it follows that gravitational acceleration is not continuous, strictly speaking.[35] Though these latter views do not explicitly appear in his writings from these months or the record of his conversations with Beeckman, it seems likely that they were topics of conversation between the two friends.

In addition to these rather specific doctrines, Descartes, no doubt, absorbed many of the more general features of Beeckman's approach. Basic to Beeckman's thought was corpuscularianism, the idea that all qualities of bodies can be explained in terms of the size, shape, and motion of their tiny parts.[36] Now Descartes had surely come across ideas like these earlier, quite possibly even at La Flèche; they were certainly very much in the air. But his Jesuit teachers are unlikely to have taken such views very seriously; Beeckman is quite likely the first person with whom Descartes worked seriously who genuinely believed corpuscularianism as an account of the way the world is. While there is no record of an extensive conversation on the general principles of the mechanist and corpuscularian approach to natural philosophy between Descartes and Beeckman, it was, no doubt, in the background of all of their discussions, and appears as an unargued assumption in documents connected with their discussions.[37] Also important is Beeckman's program for "joining physics with mathematics" (JB I 244 [AT X 52]).[38] It is extremely difficult to say precisely what this meant to Beeckman, and a number of plausible but quite different interpretations have been offered.[39] But the very idea that mathematics and physics are connected no doubt helped put Descartes on the path he was later to follow.

Descartes was later to deny that Beeckman had any significant influence on him at all.[40] But this cannot be true. Though the contact lasted only a few months, the impact was profound. When he met Beeckman he was preparing to be a soldier, having set aside any former interests he may have had in matters philosophical. But, as he told Beeckman a few months after separating: "You are truly the only one who roused my inactivity, who recalled from my memory knowledge that had almost slipped away, and who led my mind, wandering away from serious undertakings, back to something better" (AT X 162–63). Furthermore, when he met Beeckman, he was still very much a student of La Flèche. Beeckman's earliest reports of Descartes show a young man "involved with many Jesuits and other studious and learned men" (JB I 244 [AT X 52]). Beeckman introduced him to new problems, new ideas, new approaches important to him later, and began the transformation from a (perhaps dissatisfied) Jesuit schoolboy of twenty-two years to a mature thinker.

But there was quite a ways to go. In a celebrated letter to Mersenne in 1638 Descartes complains of Galileo that he "continually makes digressions and does not stop to explain any single matter, which shows that he has not examined things in order and that he has only sought the explanation of certain particular effects without having considered the first causes of nature, and thus that he has built without foundation" (AT II 380). This, in a way, characterizes Descartes' work with Beeckman. There are hints that he may have had a treatise on mechanics in mind at the time, a general framework to underly the particular questions on which he was working (AT X 159, 162; JB IV 52, 54 [AT X 67–68, 72]). But, by and large, Descartes seems to have been quite happy to work out solutions to individual problems considered in isolation; there is yet no serious attempt to provide a unified foundation for his work in various branches of physics, what will become one of the hallmarks of his mature thought, I shall argue.

THE EARLY VISION

In the months after his contacts with Beeckman, Descartes traveled. His apprenticeship seems to have ended and his career proper seems to have begun in a small stove-heated room in a town on the banks of the Danube, where he had stopped in late autumn of 1619, planning to spend the winter there.[41] It is at this point that Part II of the *Discourse* opens:

> I was then in Germany, where I was called by the wars which had not yet ended. As I was returning to the army from the coronation of the Emperor, I was halted by the start of winter in a place where, finding no conversation to divert me, and, thankfully, having no other cares or passions to trouble me, I remained all day by myself, closed up in a small stove-heated room *[dans un poêle]* where I had the complete leisure to entertain my thoughts. (AT VI 11)

The *Discourse* then goes on to report something of what Descartes claims to have discovered in the winter of 1619–1620, the idea of formulating a unified scientific system and the method for carrying out such a project. He likened the sciences as he found them to old buildings, remodeled by several hands, to ancient cities that grew up over time in a haphazard way, and to codes of law that gradually grew as circumstances demanded (AT VI 11–13). Implicit in these metaphors is what I shall later argue is one of his key ideas, the idea of the interconnectedness of all knowledge; as with a house or a city or a legal code, the parts function not independently but as components of a larger whole. And so, he claimed, he decided then to reject his former beliefs, what he had been taught in school, and by himself, to rebuild the edifice of knowledge (AT VI 13–14). In order to execute this project, he claims, he formulated a method, four rules that enabled him "to use my reason on everything, if not perfectly, then at least as well as I could" (AT VI 21), rules that soon showed their fruitfulness in a series of mathematical discoveries he claims to have made with their help (AT VI 20–21). But realizing that the principles of all sciences "are borrowed from philosophy" (AT VI 21–22), and thus that philosophy must be the first inquiry to be undertaken, he thought himself too immature to undertake the project at that time in his life. And so, that winter he resolved to prepare himself for this great project by "uprooting from my mind all of the erroneous opinions that I had received before this time, in gathering many experiences that will later be the matter for my reasoning, and in constant exercise in the method I had prescribed for myself, so that I could get better and better at it" (AT VI 22). At the end of Part III of the *Discourse,* where he gives a provisional morality, a code of conduct that he also presents as having been formulated that winter, he claims that "nine years passed [i.e., until 1628 or 1629] before I took any sides concerning the difficult matters which are usually disputed among the learned, and began to seek the foundations of any philosophy more certain than the common one *[la vulgaire]*" (AT VI 30).

The documents that survive from the earliest of these years, the late autumn of 1619 and the period that immediately follows, to some extent corroborate the report of the *Discourse.*[42] It is highly unlikely that Descartes formulated his provisional morality at that time.[43] But it is highly likely that the idea of the interconnectedness of the sciences and the idea that there is some method for constructing them date from that period. The crucial day appears to have been 10 November 1619. Baillet quotes the *Olympica:* "November 10, 1619 when I was full of enthusiasm and discovered the foundations of a remarkable science."[44] There is little question but that this is the night of his famous three dreams.[45]

The dreams are complex, subtle, and very difficult to interpret.[46] What is important, though, is what Descartes drew from them. Much of the significance of the dreams concerned his life and the personal choices that he faced. Baillet reports that he saw the dreams as a call to action, a call to settle down and begin his life's work. In the third dream, Descartes reports having read a poem that begins: "quod vitae sectabor iter," "What course of life shall I follow?" (AT X 184).[47] Baillet reports that "M. Descartes . . . judged that the piece of verse on the uncertainty of the sort of life that one ought to choose . . . indicates the good counsel of a wise man, or that of moral theology itself" (AT X 184). Descartes was well aware of the conflict between his external life as a soldier and his developing interests as a mathematician, physicist, and philosopher since Beeckman reawakened his intellectual interests exactly one year earlier. It was time to choose, the dream told him. Furthermore, it told him which to choose: "he [i.e., Descartes] was bold enough to persuade himself that it was the Spirit of Truth that wanted to open to him through this dream the treasurehouse of all of the branches of knowledge [les sciences]" (AT X 185). Baillet reports that he interpreted "the thunder whose crash he heard [in the second dream]" to be "the sign of the Spirit of Truth, which descended upon him in order to possess him" (AT X 186). More significant for our immediate interests, though, are other features of the dreams. A dictionary, or, as has been suggested, an encyclopedia that appears in the third dream suggested to him "all of the branches of knowledge [les Sciences] gathered together," and the anthology of poetry, the Corpus Poëtarum that follows immediately is interpreted as "philosophy and wisdom [la Sagesse] joined together" (AT X 184). A melon given to Descartes in the first dream is interpreted as "the charms of solitude," a strange interpretation that puzzled his contemporaries (AT X 185).[48] More recent interpreters have seen this too as a symbol of the interconnectedness of knowledge, appealing to sixteenth- and seventeenth-century symbolism.[49]

Other contemporary documents show a similar interest in the unity of knowledge. A passage from the Praeambula, perhaps from mid-November 1619, reads as follows: "The sciences are now in masks; were the masks to be removed, they would appear most beautiful. To those who see the interconnectedness [catena] of the sciences, it will seem no more difficult to contain them in the mind than it is to contain the sequence of numbers" (AT X 215).[50] The Rules for the Direction of the Mind also begins with the doctrine of the interconnectedness of the sciences. In what may well be a passage written in the enthusiasm that possessed Descartes immediately after the dreams, he wrote: "We must believe that everything is interconnected in such a way that it is far easier to learn them all together and at the same time than it is to separate one from the others" (AT X 361). And, as I shall later argue, it is the order he sees in the body of knowledge, the fact that "things mutually cohere and are connected, in such a way that some follow from others" (AT X 204)[51] that allows him to formulate the method that he will later present in the *Discourse,* a method that may well have been first formulated in November 1619. For, if the sciences are connected in such a way that more complex sciences are derived from simpler ones, then we can hope to solve problems in the more complex science by tracing them back to the simpler science, as we shall later see.

The concerns with the interconnectedness of knowledge and the formulation of a method that held Descartes' attention in autumn 1619 were neither original nor well thought out. And while his vision may have been clear, it was hardly distinct; it is not obvious that he had anything but the vaguest idea of how knowledge was interconnected and how method was to unravel it.[52] But in the decade or so following the vision, the original enthusiasm was translated into a clear program. His whereabouts and activities in the years immediately following November 1619 are difficult to document. He traveled, certainly to Italy, likely in France, and perhaps in Germany and Prague, collecting experience to free himself from prejudice, as he reports in the *Discourse* (AT VI 22).[53] During those years he is thought to have worked out some of the central mathematical problems later to appear in the *Geometry,* and may have begun work on dioptrics.[54] Descartes may also have been working on formulating his method in more detail. Parts of the *Rules* may have been drafted then, as well as a treatise, now lost, called the *Studium bonae mentis.*[55]

The progress Descartes made in his project is obvious from his reception in Paris in 1625, where he was to remain, except for a few relatively short trips, until spring 1629.[56] In Paris he became associated with Marin Mersenne and the remarkable group of philosophers and mathematicians that surrounded him. He participated actively in the scientific and more generally intellectual life of the Mersenne circle, where mechanist and anti-Aristotelian ideas were being discussed.[57] By early 1626 Mersenne refers to Descartes as an "excellent mathematician," and a correspondent of Mersenne's asked for details of his accomplishments: "I will be greatly obligated to you and to Mr. des Chartes [sic] whenever you would share with me his fine method and his fine discoveries *[inventions]*."[58] Also indicative of Descartes' progress on his project is a public performance in 1627 or 1628, the so-called Bérulle affair. The occasion was a gathering including "the Papal Nuncio, Cardinal de Bérulle, Father Mersenne, and all that great and learned company assembled at the Nuncio's palace to hear M. de Chandoux lecture on his new philosophy." Descartes reports that there he presented results following from "my fine rule, or natural method," and goes on to say that "I made the whole company acknowledge what the art of reasoning well can do with respect to the mind of those who are only barely clever, and how my principles are better established, more true and more natural than any of the others which have been received up until now by the learned world."[59]

In this Paris period in the late 1620s, Descartes worked further on his mathematics, and is thought to have done some of his most important work in optics, solving the problem of refraction and related problems.[60] Also important from these years is work he did on the *Rules for the Direction of the Mind*. Though it is thought to have been started in the enthusiasm of 1619, it is only in the Paris period that it attained its final form. Abandoned by him in 1628 or so and never finished, it is the first systematic work of Descartes to survive, the early *Compendium Musicae* aside, and an important preface to the mature work of the 1630s and 1640s. The *Rules* will be central to unraveling his project in the next chapter, and we shall consider it in some detail there. But briefly, the *Rules* contain a full and detailed statement of just what the method Descartes was developing was, the fullest account in his corpus. Furthermore, parts of the *Rules* thought to have been written during the Paris years show hints of his metaphysical views in the period and show how his methodological interests led him to undertake the investigation into "what human knowledge *[cognitio]* is, and how far it extends," a question reminiscent of the later *Meditations* (AT X 397). The project that he envisioned in 1619 seems ready to be undertaken.

THE EARLY SYSTEM

The *Discourse* reports that it was nine years after the events of November 1619 and eight before the publication of the *Discourse* in 1637, that is, in 1628 or 1629, when Descartes "began to seek the foundations of a philosophy more certain than the common one" (AT VI 30). Aware that he was thought by many to have completed his philosophy and wanting to "make myself worthy of the reputation that I had been given," he decided to abandon Paris and move to the Low Countries to undertake his project in relative solitude (AT VI 30–31).[61]

Descartes' first years in the Low Countries were among the busiest and most productive of his life. He seems to have left his explicit concern with method behind when he left France; the unfinished *Rules* were never to be completed, and in his entire published corpus only a few lines are devoted to an exposition of the method that occupied so many pages in his early work.[62] Instead of dwelling on the preliminaries, he seems more concerned to get down to business on his philosophical system proper.

One of the first projects Descartes seems to have undertaken was his metaphysics. Writing to Mersenne on 15 April 1630, he noted:

I think that all those to whom God has given the use of this reason have an obligation to employ it principally in the endeavour to know him and to know themselves. That is the task with which I began my studies; and I can say that I would not have been able to discover the foundations of physics if I had not looked for them along this road. It is the topic which I have studied more than anything and in which, thank God, I have not altogether wasted my time. At least I think that I have found out how to prove metaphysical truths in a manner which is more evident than the proofs of geometry. . . . During my first nine months in this country I worked on nothing else. (AT I 144 [K 10–11])

This first metaphysics was composed, it seems, somewhere between the spring of 1629, when he first arrived in Holland, and April 1630.[63] Although by the time he told Mersenne about the metaphysics, he had definitely decided not to publish it (at least not until his physics was published; AT I 144 [K 11]), his work on metaphysics had taken the form of a treatise, in Latin (AT I 17, 182 [K 19], 350 [K 31]). It is to this work that he is apparently refering when he discusses "the first meditations I made" in the opening of Part IV of the *Discourse* (AT VI 31). The document itself does not survive; one can be reasonably certain that the metaphysics of *Discourse* IV, a preliminary sketch of the *Meditations,* represents a later stage in his thought, and it is difficult to say what precisely that early treatise may have contained. A slightly later letter to Mersenne fleshes the account out a bit; writing to him on 25 November 1630, he tells his friend that in the "little treatise on metaphysics" he "set out principally to prove the existence of God and of our souls when they are separated from the body, from which follows their immortality" (AT I 182 [K 19]). The 15 April letter, which begins a short but crucial series of letters on his claim that God created the eternal truths, suggests that this, too, may have been part of that original tract. But whatever precisely it may have contained, the metaphysics of 1629–30 is clearly connected with his physics. As he told Mersenne in that letter, he "would not have been able to discover the foundations of physics" without having worked out his metaphysics, and although he does not want to publish it as a whole, he does tell Mersenne that "in my treatise on physics I shall discuss a number of metaphysical topics" (AT I 144–45 [K 10–11]). Though the content of the metaphysics may have changed between 1630 and 1637, when the *Discourse* was published, it is plausible to suppose that the role of metaphysics did not; in 1630, as in the *Discourse,* Descartes seemed to think that the basic principles of sciences "are taken from philosophy," and that "one must try to establish it before anything else" (AT VI 21–22).[64]

17

But at the same time as he was working on the metaphysical foundations of the sciences, he was working on the sciences themselves. Letters from 1629 and 1630 show evidence of work on the theory of motion (AT I 71f), space and body (AT I 86), on the mechanist explanation of physical properties of bodies (AT I 109, 119–20), on optics and light (AT I 13, 53f, 179), and on the explanation of atmospheric and celestial phenomena (AT I 23, 127). On 18 December 1629, Descartes reports wanting to begin studying anatomy, apparently for the first time, a subject that occupied much of his attention in the years following (AT I 102).[65] Unlike the metaphysics, these studies were intended for publication in a work that came to be called *Le monde,* that is, *The World.*[66]

Originally, in October 1629, Descartes seems to have intended to publish only a small book, giving only a small portion of his thought, "a sample of my philosophy," an account of sublunar phenomena with special attention to his account of the rainbow; he intended to remain "hidden behind the scenes in order to hear what people say about it" (AT I 23). But by a month later, on 13 November 1629, Descartes had changed his mind. Telling Mersenne, somewhat overoptimistically, as it turned out, that the treatise would not be finished for more than a year, Descartes told him that "instead of explaining only one phenomenon, I have decided to explain all of the phenomena of nature, that is, all physics" (AT I 70). A year later, in November 1630, he is referring to this project as *"mon Monde,"* the title it was later to have (AT I 176, 179 [K 18]). A month later, in December 1630, he is struggling with the question of how God created the universe and how from the initial state all else follows, a crucial question that has a central place in the finished treatise (AT I 194; cf. *The World,* chap. 6). Less than a year later, in October or November 1631, he is promising Mersenne that the treatise will be ready by the following Easter (AT I 228). It was not (AT I 242 [K 22]). But by June 1632 he had finished the part on inanimate bodies, and, abandoning (for the moment, at least) a projected section on the origin of animals in the world, he turned to write what he originally intended to be a relatively brief section on human anatomy (AT I 254–55). This section grew beyond his original intentions, though, and it was not until more than a year later, in July 1633, that he could say that the treatise was almost done (AT I 263, 268).

But the work did not appear during Descartes' lifetime. Galileo had been condemned in Rome in June 1633, and by November of that year Descartes found out about the condemnation (AT I 270–72; cf. AT I 285–86 [K 25–26]). He withdrew the work from publication, and it did not appear until the 1660s, when Clerselier arranged for its publication. A summary of the work, though, was published in 1637 as Part V of the *Discourse,* following the account in Part IV of what Descartes represented as his first metaphysics.

The account in the *Discourse* contains somewhat more than the surviving text does, and it is not clear whether parts of the work were lost, suppressed, or never written.[67] When published, the work was presented in two parts, *The World, or Treatise on Light,* and *Treatise on Man,* though the account in the *Discourse* suggests that Descartes himself did not so divide his treatise. In the treatise of 1633 we have an incomplete but remarkably ambitious sketch of a mechanist physics and biology. He cautiously presents his world in an oblique manner, not as fact but as a fantasy; the world of the physics is presented not as the real world, but as an imaginary world that God created to Descartes' mechanistic specifications somewhere beyond ours, and the man of the biological sections is a machine God built to look and act like our bodies. But the intention is clear. After a few preliminary chapters, *The World* begins with an account of the geometrical nature of the matter in Descartes' new world, and the laws of motion it obeys (chaps. 6–7). The project then is to show how everything in the world can be explained in terms of the size, shape, and motion of this geometrical matter, one part acting on another through collision alone in a world without empty space. He begins with a conjecture about creation (chap. 8),[68] and shows how planetary systems, vortices of etherial matter carrying planets around central suns, would naturally evolve from initial chaos as bodies moving in the plenum would produce circular streams; he shows how gravity would arise, what light is and its central properties, and how this world would appear to the inhabitants of others.[69] The *Discourse* goes on to mention a number of other topics, including mountains, seas, springs, rivers, metals, and the natural origin of plants, topics not discussed in *The World* as we have it (AT VI 44). He had originally hoped that this sort of natural evolutionary account could be extended from planets, mountains, and plants to animals as well. But he couldn't solve the problem (AT VI 45; AT I 254), and so in the *Treatise on Man* we begin with the hypothesis that God created a human body from the same geometrical matter that composes the rest of the world. The *Treatise* then goes on to examine the behavior of this remarkable machine, the human body considered without its soul, discussing functions such as digestion and circulation, nonvoluntary motion, and the operation of the sense organs, all from a purely mechanistic point of view, explaining these functions in terms of size, shape, and motion. The *Discourse* goes on to report that the work of 1633 contained a discussion of the human being proper, the rational soul joined to the machine of the body, but only hints of this discussion are found in the text that survives (AT VI 46; AT XI 143); like the account of the genesis of animal bodies through natural processes, one suspects that it was never written.

With *The World* and the lost metaphysical manuscript of 1629–30 that he had hoped to publish after he had tested the waters with *The World,* Descartes had completed an extensive sketch of his system, from its metaphysical foundations to its physics and biology. His remaining years were to be spent revising this initial sketch, filling it out, and making it public.

THE MATURE SYSTEM

When Galileo was condemned, Descartes withdrew his book from publication, and, in fact, expressed the desire never to publish at all. Writing to Mersenne in late November 1633 upon just having heard about Galileo, he said: "There are already enough opinions in philosophy which are plausible and which can be maintained in debate, so that if mine are no more certain and cannot be approved without controversy, I never want to publish them" (AT I 271–72)[70] There are many speculations as to why precisely he withdrew his system at this time, explanations that go from cowardice to genuine faith in the wisdom of the Church and genuine doubt about his own physics.[71] But it is indisputable that the physics of *The World* was Copernican, and, he thought, necessarily so: "it [i.e., Copernicanism] is so connected with all of the parts of my treatise that I cannot detach it without rendering the remainder completely defective," Descartes wrote to Mersenne in that same letter (AT I 271; cf. AT I 285 [K 25–26]). The world he built in 1633 was a world of vortices, whirlpools of extended stuff that carry planets around central suns.[72] And since light for Descartes is just the pressure of celestial matter, the illumination of a central sun follows in a relatively straightforward way from what later came to be called the centrifugal force of the vortex, the (apparent) tendency of bodies in circular motion to recede from that center.[73] The view seems wedded to Copernicanism; it is difficult to see how a sun turning around a central earth could give it light, or, for that matter, how a central earth could fail to be a center of illumination. Abandoning Copernicanism would, it seems, force him to abandon either his theory of light or his entire cosmology. Even worse, Descartes thought that the world of vortices was a direct consequence of his laws of motion together with virtually any initial chaos; in fact, he thought, however God created the world, the laws of motion operating in a plenum would insure that matter would eventually sort itself out into vortices.[74] And so, any attack on the vortex theory threatened the very most basic features of the Cartesian world.

But, as it turned out, Descartes' despair did not last. In April 1634, five months after he first got word of Galileo's fate, he was still complaining to Mersenne (AT I 285–86 [K 25–26]). However, by September or October of that year, he seems to have been back to work on his projects (AT I 314).[75] And a year and a half later, in March 1636, a new work was finished, a collection of essays, all in French. Descartes' original title for it was "The Plan [project] of a Universal Science to raise our nature to its highest degree of pefection, with the Dioptrics, the Meteors and the Geometry; in which the most curious topics which the author has been able to choose in order to give proof of his Universal Science are explained in such a way that even those who have never studied them can understand them" (AT I 339 [K 28]). When it appeared a little over a year later it was titled: "Discourse on the Method for conducting one's reason well and seeking truth in the sciences, together with the Dioptrics, the Meteors, and the Geometry, which are essays in this method." The preliminary essay, the *Discourse*, was originally described as the first of four treatises (AT I 339 [K 28]).[76] Structured as a history of the author's intellectual development, it contains "a part of my method" (AT I 339 [K 28]), as well as "a certain amount of metaphysics, physics, and medicine . . . in order to show that my method extends to topics of all kinds" (AT I 349 [K 30]; cf. AT I 370). The summary of Descartes' philosophical system, a summary of *The World* (without mentioning its Copernicanism) preceded by what purports to be a brief exposition of the metaphysics of 1629–30, also gives the reader a sense of the Cartesian program as a whole, the "plan of a universal science" he may be referring to in the original title. But he resisted including anything more than a brief sketch of his physics. Instead he returned to his original idea of October 1629 to give only "a sample of my philosophy" (AT I 23). Three essays were chosen to illustrate his program; one, the *Geometry*, was intended as pure mathematics, one, the *Meteors*, as pure "philosophy," and one, the *Dioptrics*, as a mixture [meslé] of philosophy and mathematics (AT I 370).

Much of the material was not so much new as it was newly drafted and arranged. Some of the central parts of the *Dioptrics*, like the law of refraction, go back to the Paris years in the late 1620s. Work on the *Dioptrics* was closely intertwined with the work on *The World*, both in Descartes' letters, and in the surviving text of *The World*, where it is referred to as a separate treatise (AT XI 102, 106); in January 1632, he also sent one correspondent what he called "the first part" of "my Dioptrics" (AT I 234–35). Work seems to have been resumed on it in autumn 1634, and by October 1635, it was finished (AT I 314, 322, 325–27; cf. 315, 585–86). Descartes' original inclination seems to have been to publish it alone (AT I 322).[77] But by early November 1635, his plan seems to have changed, and he then intended to join to the *Dioptrics* the *Meteors,* and publish them together with a preface (AT I 329–30). While preparing *The World* he had collected a great deal of material on meteors that had not found its way into the abandoned treatise,[78] and throughout the period he seems to have maintained an interest in meteoric phenomena.[79] By the time the essay on meteors turns up in the correspondence in November 1635, it is practically finished and ready to be copied out (AT I 329–30). A letter from 1638, after the project had been published, confides that "I only, as it were, composed [the *Geometry*] while my *Meteors* were being printed" (AT I 458).[80] This is probably something of an exaggeration. But even though the *Geometry* was composed of elements that had been worked out at earlier times, it is likely that the decision to include it was a late thought, between November 1635 and March 1636, when the project appears to be complete (cf. AT I 329–30 with AT I 339–40 [K 28–29]). The *Discourse* was also probably drafted during these busy months, perhaps drawing on earlier material; it appears as a projected preface in November and then, the following March, as one of four treatises ("traittez").[81] The work, largely finished by March 1636, was printed on 8 June 1637, and copies were sent to friends shortly thereafter, including a copy to his philosophy teacher at La Flèche, Father Noël.[82]

A central feature of the *Discourse* and the *Essays* that accompanied it is the lack of the full framework of physics and metaphysics that, Descartes admitted, lay under the samples of work that he presented; not only is there no mention of Copernicanism, but there is no serious presentation of the basic principles of his system. And so, in the *Dioptrics* and *Meteors* he proceded hypothetically, laying down hypotheses *(suppositions),* and showing what wonderful conclusions follow from them. But though he argues hypothetically, he is clear that this is only an expository device, satisfactory as such, but that the arguments of the *Dioptrics* and *Meteors* do not give the full and proper account of his system: "I have called them hypotheses so that people will know that I think I can deduce them from first truths which I have already explicated, but that I explicitly desired not to give them" (AT VI 76).[83] As in 1629, when he began to think about composing a work for publication, he is hiding behind the scenery, waiting to see the reaction to his work, waiting for the right moment to publish his full system (AT I 23). Writing to a correspondent in May 1637, just before the *Discourse* and *Essays* appeared, he confided:

> As to the treatise on Physics which you paid me the compliment of requesting me to publish, I would not have been so imprudent as to talk about it in the way in which I did unless I had the hope

of bringing it to light should the world desire it and should I find it worthwhile and prudent. But I want to tell you that my entire plan as far as what I am publishing this time is only to prepare the path for it and to see how the land lies *[sonder le guay]*. (AT I 370)

And to Vatier in February 1638 Descartes wrote: "The only thing left to answer you concerns the publication of my Physics and Metaphysics, on which I can tell you, in a word, that I desire it as much or more than anyone, but only under certain conditions, without which I would be imprudent to desire it" (AT I 564 [K 49]). Those conditions seem to have obtained, for shortly afterward, he began to make plans to publish the full system that the publications of 1637 promised.

The years immediately following the publication of the *Discourse* and *Essays* were very busy; a survey of the correspondence shows that one of Descartes' main preoccupations was simply answering criticisms of and inquiries about his published views. In this period the Cartesian philosophy was beginning to be taught with the aid of the new textbook Descartes had provided, if not in Catholic schools like La Flèche, then at least in the more liberal Protestant universities in the Low Countries.[84] At Utrecht, his friend of some years, Henricus Reneri (Henri Rénier) started teaching from Descartes' newly published work, though he died in 1639.[85] The tradition was continued by Henricus Regius (Henri de Roy), a man just slightly Descartes' junior who was later to prove one of Descartes' most unfortunate disciples. Regius, lecturing on what he represented as Descartes' doctrine, was at least partly responsible for calling down the wrath of the important Dutch Reformed theologian, Gisbertus Voëtius, an influential member of the theology faculty at Utrecht, who accused Descartes of grave impiety and involved him in a war of pamphlets that led even to legal actions, a messy affair that occupied all too much of his time in the early and mid-1640s.[86] There was also interest at Leiden, where Descartes had both disciples (François du Bon, Adriaan Heereboord) and an influential enemy (Revius, the director of the College of Theology).[87] But despite the controversy, Descartes decided to begin publishing his full system of philosophy.

First published was the metaphysics. As early as April 1637, before the *Discourse* and *Essays* ever appeared, Descartes showed some interest in reviving, revising, and publishing the metaphysical treatise of 1629–30, to be bound together with a projected Latin translation of the *Discourse* (AT I 350 [K 31]).[88] But he makes no clear mention of it in his letters until two and a half years later, 13 November 1639, by which time the project seems well along (AT II 622 [K 68]).[89] By 11 March 1640, it seems to have been finished, and copies were sent to Regius in May and

Mersenne in November, for distribution to a variety of philosophers and theologians for their comment (AT III 35–36, 61, 63 [K 73], 230, 235 [K 83]). Responses began to come in by December 1640, and the first part of 1641 was occupied with answering objections.[90] The *Meditations on First Philosophy,* as it was called, was published in Latin in Paris on 28 August 1641, together with a selection of objections and Descartes' replies. A second, somewhat altered version of the text was published in Amsterdam in May 1642.

The *Meditations* is, of course, Descartes' central philosophical work, when philosophy is understood in the modern sense; it is generally accepted as the canonical text for his first philosophy, his account of the soul, of God, of knowledge, and as such there are no lack of commentaries on it. However, from the point of view of the themes that concern us here, the *Meditations* is largely of interest not in and of itself, but as it relates to another work of the period of the early 1640s, the *Principles of Philosophy.*

When Descartes was on the verge of publishing the *Meditations,* in January 1641, he confided to Mersenne:

> I may tell you, between ourselves, that these six *Meditations* contain all the foundations of my physics. But please do not tell people, for that might make it harder for supporters of Aristotle to approve them. I hope that readers will gradually get used to my principles, and recognize their truth, before they notice that they destroy the principles of Aristotle. (AT III 297–98 [K 94]; see also AT III 233 [K 82])

The *Meditations,* then, are at least in part to prepare readers for the full physics; Descartes is still waiting in the wings and sending out trial balloons.

The decision to publish the full physics, though, was a difficult one. In 1639, for example, two years after publishing the *Discourse* and *Essays,* his friend Constantijn Huygens tried to convince Descartes to "give *The World* to the world," a request that he gracefully declined to honor (AT II 679; see also 680–81).[91] On 11 March 1640, even after the *Meditations* were completed, he still hesitated to publish his physics (AT III 39 [K 70–71]).[92] But his mind began to change in autumn of 1640. In September 1640 he wrote to Mersenne asking him to suggest some scholastic textbooks he might peruse, presumably to prepare himself for the remarks on the *Meditations* he expected to receive from the schoolmen, remarks that he intended to answer in preparation for the publication of the work the following year. By 11 November he had received Mersenne's sugestions, and reports having purchased the *Summa philosophica* of Father Eustachius a Sancto Paulo, which he judged to be "the best book of its kind ever made" (AT III 232 [K 82]). It is in this same letter that he first announces the publication of his system:

> I would willingly answer your question about the flame of a candle and similar things; but I see that I can never really satisfy you on this until you have seen all of the principles of my philosophy. So I must tell you that I have resolved to write them before leaving this country, and to publish them perhaps within a year.[93] My plan is to write in order a complete course in my philosophy in the form of theses. I will not waste any words, but simply put down all my conclusions, with the true premises from which I derive them. I think I could do this without many words. In the same volume I plan to have printed an ordinary course in philosophy, perhaps Father Eustachius', with notes by me at the end of each question. In the notes I will add the different opinions of others, and what one should think of them all, and perhaps at the end I will make a comparison between the two philosophies. (AT III 232–33 (K 82)[94]

The plan to reprint Eustachius, a fat tome by itself, was abandoned by January, in part because of Eustachius' death in December, which prevented Descartes from getting his personal permission, and in part because he came to think that such an explicit attack was unnecessary (AT III 260, 286, 470, 491–92 [K 126–27]). But the idea of a short course in the full Cartesian philosophy, giving both metaphysical and physical foundations of the *Dioptrics* and *Meteors,* had taken firm root. By 31 December 1640, while waiting for the objections to the *Meditations* to be submitted, Descartes reported being at work on Part I of a work to be used in teaching his philosophy, a characterization he gives of the *Principles* on at least one other occasion (AT III 276 [K 92]; cf. AT VII 577).[95] More than a year later, in January 1642, deeply involved in the affairs at Utrecht, he wrote Huygens: "Perhaps these scholastic wars will result in my *World* being brought into the world. It would be out already, I think, were it not that I want to teach it to speak Latin first. I shall call it the *Summa Philosophiae* to make it more welcome to the scholastics" (AT III 523 [K 131]; cf. AT III 465). And in February 1642, writing to Regius in connection with the dispute with Voëtius, he told Regius that "I have decided to finish [my philosophy] this year" (AT III 529).

Actually, it turned out to be harder than he thought to teach *The World* to speak the language of the learned, and the Voëtius affair turned out to be more troublesome than he could have anticipated. Work continued on the project for more than two years longer, until summer 1644. The first substantive public announcement of the work was in the Amsterdam printing of the *Meditations* in May 1642, in the letter to Father Dinet printed there (AT VII 574–77).[96] In February 1643, probably working only on the cosmology in part III of the book, Descartes planned to begin having the *Principles* printed in the summer, a process he anticipated might take as long as a year (AT III 615, 646–47).[97] But in April he had to put down the cosmological sections he was working on, "leave the heavens for some days," and "shuffle some papers to try to redress some wrongs done to me on earth" (AT III 647).[98] The events connected with the dispute with Voëtius must have been time-consuming throughout the year, and in January 1644 he was still working on the final sections of part IV while the earlier parts of the work were being set.[99] The *Principles of Philosophy*, the first comprehensive statement, in proper order, of his philosophical system, finally appeared on 10 July 1644, while Descartes was visiting France for the first time in over fifteen years.

Though he represented the project to Huygens as teaching *The World* to speak Latin, the *Principles* are, in reality, something somewhat different. Unlike *The World* he finished some ten years or so earlier, the *Principles* contain an account of Descartes' first philosophy, his metaphysics; part I, as he later put it, is "an abrégé of what I wrote in my *Meditations*" (AT V 291 [K 246]).[100] Parts II-IV correspond more closely to *The World*. But even here the correspondence is not exact. Part II, on space, body, and motion, a central text for the study that follows, corresponds in content to chapters 6 and 7 of *The World*. But, as we shall later see when we discuss these questions in more detail, Descartes has given the issues considerable thought since their first presentation. Parts III and IV correspond roughly to chapters 8–15 of *The World*. As in *The World*, Descartes in the *Principles* presents a vortex theory of the planetary system that is unmistakably Copernican, despite the apparent attempts to argue that, on his view, the Earth is more truly at rest than it is in other theories (Pr III 19ff).[101] But there are also significant differences. While there are many discussions of light in the *Principles*, there light does not have the central organizing role that it does in *The World*. Also the *Principles* contain extensive discussions of magnetism not found in the early writings. Clearly, the composition of the *Principles* entailed extensive reworking and expansion of his earlier thought.

Descartes was not to complete another project of the same magnitude and comprehensiveness during his lifetime; the *Principles* is the final statement on his program in physics, and it will be one of the main sources we shall use in explicating that program. But it is by no means the end of his career.

The correspondence from Descartes' last years, from the publication of the *Principles* in the summer of 1644 to his death five and a half years later, shows continued active interest in metaphysics, physics, biology, psychology, and morals. Particularly important on these last topics is a set of letters between him and the young Princess Elisabeth of Palatine, to whom the *Principles* were dedicated in 1644. Starting in May 1643 and extending to his death, the correspondence is particularly illuminating about parts of his thought that had not (yet) found their way into his publications. The discussions begin with the celebrated series of letters on mind-body union and interaction, and go on to discuss topics in psychology and morality. Out of these discussions grew the *Passions of the Soul,* his final publication, which appeared in autumn 1649, only months before death. More directly relevant to the topics we shall be taking up is a fascinating set of letters that passed between Descartes and the Cambridge Platonist Henry More from December 1648 to Descartes' death. More had recently developed an interest in his thought, and the careful and penetrating questions he asked about metaphysics and physics will help to clarify some of the material in the *Principles* we shall later examine, as will other letters of the period.[102]

The controversy at Utrecht continued through the period. More important for our purposes than the theological and, increasingly, personal debates was the break with his erstwhile disciple, Regius. Descartes had embraced Regius as a follower, and in the letter to Dinet that accompanied the Amsterdam edition of the *Meditations* in 1642 and in the *Letter to Voëtius* of 1643 he had publicly endorsed the man who was teaching his doctrines at Utrecht (AT VII 582–83; AT VIIIB, 162–63). But Regius undertook to write his own Cartesian textbook, which he sent to Descartes in June 1645. Descartes was not pleased, and urged Regius not to publish (AT IV 239–40, 248–50, 256–58). Part of it was, no doubt, the fact that Regius' work extended the Cartesian system to biology and anatomy, drawing on unpublished notes that Descartes himself intended to work into publishable form (AT XI 673). But he had other substantive criticisms, and felt that the work was badly argued, and could lead to readers unfairly rejecting his own ideas (AT IV 510 [K 204–5]). When Regius went ahead with his project and in August 1646 published the work as his *Fundamenta Physices,* Descartes broke with him and publicly denounced the book in 1647 in the preface to the French translation of the *Principles* (AT IXB 19–20). Regius responded by publishing a summary of his main theses, the program to which Descartes replied in the *Notes against a Program,* published in early 1648 (AT VIII B 335–69).[103]

Beside the *Passions* and the *Notes against a Program,* there were two other publications of some importance: the French translations of the *Meditations* and the *Principles,* both of which appeared in 1647. There is much dispute over relative value of the French and Latin editions of the *Meditations*. More important for our purposes, though, is the question of the French version of the *Principles*. As we shall later see, the translation published in 1647 shows a number of significant differences from the Latin of 1644, differences that could only have come from Descartes' hand.[104] The French translation also contains a "Letter by the Author to the Translator [Abbé Claude Picot] that can serve as a Preface," a crucial document for understanding Descartes' mature views on his system of philosophy (AT IXB 1–20).[105]

But perhaps most interesting is a project that Descartes only began in this period, the completion of the *Principles* with the addition of two books on living things, plants, animals, and human beings. Referred to obliquely in Pr II 2, named as *De homine (traitté . . . de l'homme)* in Pr II 40, the additions are outlined in Pr IV 188 of both the Latin and French editions: "I should not add more to this fourth part of the *Principles of Philosophy* if (as I had previously intended) I were still going to write two other parts, namely, a fifth concerning living things, or animals and plants, and a sixth concerning man. But because all the things which I would wish to discuss in those parts have not yet been perfectly examined by me, and because I do not know whether I shall ever have sufficient leisure to complete them . . . I shall add here some few things concerning the objects of the senses" (Pr IV 188L).[106] He then goes on to discuss some of the topics that, he thinks, should find place in the projected fifth and sixth parts.

Descartes is certainly not optimistic in this passage about actually completing the final parts, not nearly as optimistic as the earlier references might suggest.[107] But he almost certainly began the project. The closing sections indicate some of what he intended to include, material on the mechanism of sense perception. The *Passions* is probably also connected with this project, and a careful study of Regius' *Fundamenta Physices* may possibly yield the contents of some of the unpublished notes he accused Regius of publishing without his consent. But the largest piece of the work he did toward that project is probably contained in an unfinished essay, written after the publication of the French *Principles* and published after his death, the *Description of the Human Body* (AT XI 223–86).[108] It contains, as the title suggests, an account of human anatomy, at least selected aspects. But it also contains a long "digression" on sexual reproduction, the generation of the animal body (AT XI 252–86).[109] This was the missing link, as Descartes recognized in 1632, between the genetic account of the visible world that evolved from chaos through natural means, and the appearance of life in the world; Descartes' apparent hope was that if he understood how animals arose mechanistically through other animals, he might be able to understand how they could arise mechanistically from inanimate matter.[110]

It would have been interesting to see how all of these strands of Descartes' last work would have come together, and how he would have completed the system he outlined and began to execute in the writings that survive. But Descartes died before he had a chance. In Sweden at the Court of Queen Christina from early October 1649, the climate as well as the rigors of life in Christina's court caused Descartes to fall ill in early February 1650, and on 11 February, he died.

DESCARTES' PROJECT

IN THE PREVIOUS chapter we saw something of the development of Descartes' thought, the stages that he passed through in working out his initial vision and presenting it to the public. In the course of that account we touched superficially upon some of the principal elements of his program. Now that we have some of the more prominent landmarks in view, we can begin the process of probing more deeply. Eventually we shall narrow our focus and examine Descartes' account of the physical world. But before we can do that we must understand the comprehensive program for the sciences he put forward. One of the most characteristic features of Descartes' thought, what for him differentiated his approach to problems in physics and optics, medicine and meteors, from that followed by most of his contemporaries was scope; from the sketchbooks of 1619–20, to the *Discourse* of 1637, and on to his last writings, Descartes emphasized the interconnectedness of the different branches of knowledge he pursued, the importance of grounding the sciences on one another in the appropriate way. It is the structure of this program that I will examine in this chapter.

This inquiry divides itself neatly into two principal parts. We know little for certain about the exact details of the program Descartes dreamed up in November 1619. But the *Rules for the Direction of the Mind,* probably begun in that month and abandoned roughly ten years later, gives us a reasonably detailed picture of the program as worked out by the late 1620s, a picture closely linked to his doctrine of method. We shall begin by examining method and order in the *Rules.* We shall then examine how the view evolved in Descartes' later writings, the *Discourse, Meditations, Principles,* and other documents from the early 1630s to the end of Descartes' life, looking first at the question of method in the later works, and then at the doctrine of order. We shall end with a brief attempt to place the Cartesian program in the context of early seventeenth-century thought.

DESCARTES' PROJECT:
METHOD AND ORDER IN THE RULES

The *Rules for the Direction of the Mind,* started as early as 1619 and abandoned in 1628, is a very difficult work; despite its superficial organization, a series of brief rules and commentaries on those rules, it is often strikingly unmethodical and disorderly for a work that is generally taken to be Descartes' most systematic exposition of his method. It is quite obviously a work in progress that never progressed to anything like a finished draft, and the text we have shows both contradictions and what can only be interpreted as successive drafts of the same material, obvious signs of having been picked up and put down at different times throughout the period of composition.[1] It is important to examine the *Rules,* if only briefly, to get a picture of the shape of Descartes' intellectual project in the period that immediately preceded the first metaphysics of 1629–30 and the first exposition of Descartes' physical system in *The World.*

The order of knowledge is a basic theme in the *Rules.* But to understand the conception of the order of knowledge that underlies the *Rules,* it will be helpful to begin with an account of the method. And to understand method in the *Rules,* we must understand something of what Descartes takes the goal of inquiry to be.

From the earliest portions of the *Rules,* portions thought to date from as early as mid-November 1619, shortly after the dreams of November 10, Descartes is clear that the goal of method is certainty. Thus he wrote in the very first rule: "The goal *[finis]* of studies ought to be the direction of one's mind *[ingenium]* toward making solid and true judgments about everything which comes before it" (AT X 359).[2] And in the second, probably from the same period, he wrote: "We should concern ourselves only with those objects for which our minds seem capable of certain and indubitable cognition" (AT X 362).

Descartes' conception of certainty in the *Rules* has two principal elements, the mental processes that result in certainty, intuition, and deduction, and the primary objects of certainty, what Descartes calls simple natures. In Rule 3 he announces that it is only intuition and deduction that give us the sort of knowledge which he seeks: "Concerning things proposed, one ought to seek not what others have thought, nor what we conjecture, but what we can clearly and evidently intuit or deduce with certainty; for in no other way is knowledge *[scientia]* acquired" (AT X 366). Similarly, he says that intuition and deduction are the only acts of intellect that lead to a knowledge of things "without any fear of being deceived" (AT X 368).[3] He defines intuition as follows:

By *intuition [intuitus]* I understand, not the fluctuating faith in the senses, nor the deceitful judgment of a poorly composed imagination; but a conception of a pure and attentive mind *[mens]*, so easy and distinct that concerning that which we understand *[intelligere]* no further doubt remains; or, what is the same, the undoubted conception of a pure and attentive mind, which arises from the light of reason alone. (AT X 368)

Deduction is characterized as a chain of successive intuitions. Descartes writes:

Many things are known with certainty, even though they are not themselves evident, only because they are deduced from true and known principles through a continuous and uninterrupted movement of thought, perspicuously intuiting each individual thing. . . . We can therefore distinguish an intuition of the mind from a certain deduction, by the fact that in the one, we conceive a certain motion or succession, but not in the other. (AT X 369–70)

Or, as he puts it a bit more clearly in Rule 11: "Deduction . . . involves a certain movement of our mind, inferring one thing from another" (AT X 407).

Certainty as characterized in terms of intuition and deduction has a number of salient properties. Certainty, first of all, seems to entail indubitability in at least some weak sense; intuition, the ground of certainty, is, after all, the "undoubted conception of a pure and attentive mind." Furthermore, certainty involves a particular faculty of the mind; intuition, Descartes tells us, is not derived from the senses or the imagination, but "arises from the light of reason alone" (AT X 368). And finally, he claims, for the mind to have an intuition, and thus, for it to perform a deduction, it must be in a special state, it must be both "pure" and "attentive."

Descartes also thinks certain knowledge has its own special set of objects, a particular collection of notions or concepts; certain knowledge, strictly speaking, is knowledge of these simple natures and their combinations. Simple natures are first introduced in Rule 6, which most likely dates from mid-November 1619, the same time he drafted the passages on intuition and deduction that we have been discussing. When introduced in Rule 6, they are connected not so much with the characterization of certainty as with the method as sketched in Rule 5, a question to which we shall shortly turn. In Rule 6 he explains how "things" (the Latin *res*, noncommittal in the extreme) can be arranged in a series, "not, indeed, insofar as they are referred to some genus of being, as the philosophers divide them up into their categories, but insofar as some can be known from others" (AT X 381). At the beginning of the series are those things he calls complete *(absolutus)*, things that "contain in themselves a pure and simple nature" (AT X 381). The "pure and simple natures" these complete things contain are, according to Rule 6, few in number and can be grasped *(intueri,* i.e., intuited) "first and through themselves, not dependently on the grasping of others, but through experience itself or through a certain light inherent in us" (AT X 383). Later in the series come relative things, things understood through the simple natures or absolute things (AT X 382–83). Descartes' examples here suggest that simple natures include cause, the one, and the equal, for example, and it is through these notions that we comprehend the effect, the many, and the unequal (AT X 381–83).

Simple natures reappear in Rule 12, in a passage written some years later, probably between 1626 and 1628. There the list has changed; instead of the grab bag of diverse and supposedly basic notions or concepts found in Rule 6, Descartes recognizes three kinds: those purely mental *[intellectualis]* (thought, doubt, will, etc.), those purely material (shape, extension, motion, etc.), and those that pertain to both (existence, unity, duration, etc.; see AT X 419). According to Rule 12, these simple natures are the primary objects of intuition and, indeed, the primary objects of all knowledge: "We can never understand anything but these simple natures and a certain mixture or composition of them"; "all human knowledge *[scientia]* consists in this one thing, that we distinctly see how these simple natures come together at the same time in the composition of other things" (AT X 422, 427).

This conception of certainty constitutes a kind of epistemology that underlies the methodological project of the *Rules.* Though unargued and unjustified until the very latest strata of the *Rules,* as I shall later discuss, it tells us what knowledge is, where it derives from, and what it comprehends. And insofar as it defines knowledge, it sets a goal for method: the construction of a body of knowledge grounded in intuition and deduction, derived from the intellect's acquaintance with simple natures, derived from the light of reason.

Descartes declares in Rule 4 that "method is necessary for seeking after the truth of things" (AT X 371). He continues:

> By method, moreover, I understand certain and easy rules which are such that whoever follows them exactly will never take that which is false to be true, and without consuming any mental effort uselessly, but always step by step increasing knowledge *[scientia],* will arrive at the true knowledge *[vera cognitio]* of everything of which he is capable. (AT X 371–72)

It is important to understand here that Descartes' idea is not that method will somehow help us in actually performing intuitions or deductions. Rather, method is supposed to be a tool by virtue of which the mind is led toward the *discovery* of genuine knowledge, knowledge based in intuitions and deductions, knowledge that the mind is in principle, though not perhaps in practice, capable of having independently of the method itself (AT X 372).

But what are the "certain and easy rules" that Descartes promises? Descartes summarizes them as follows in Rule 5:

> The whole of method consists in the order and disposition of those things toward which the mental insight *[mentis acies]* is to be directed so that we discover some truth. And this [rule] is observed exactly if we reduce involved and obscure propositions step by step to simpler ones, and thus from an intuition of the simplest we try to ascend by those same steps to a knowledge of all the rest. (AT X 379)

Descartes' rule of method has two steps: a *reductive* step, in which "involved and obscure propositions" are reduced to simpler ones, and a *constructive* step, in which we proceed from the simpler propositions back to the more complex.[4] But the rule makes little sense, nor does it connect very clearly with the account of knowledge and certainty in terms of intuition and deduction unless we know what he means here by the reduction to simples, and the construction, or *re*construction of the complex from the simple.

The account of the workings of the method given in the earliest strata of the *Rules* is obscure and very difficult to translate into concrete terms. It is quite possible that Descartes' vision in those texts is still somewhat cloaked in poetic enthusiasm and that he himself may not have had a clear and distinct idea of precisely how the method was to work in actual practice.[5] But matters are clarified considerably in an example of methodological investigation he gave late in the composition of the *Rules*. The example I have in mind is that of the anaclastic line, which Descartes gives in the commentary to Rule 8. The example he chooses is closely connected with the optical investigations that he undertook between 1626 and 1628, and it is highly probable that the example was added to the *Rules* sometime in that period (see Schuster 1980, pp. 55, 88, n. 68). As such it may well represent an understanding of the method that he simply did not have when he gave his first exposition of the method in 1619. Also, it is difficult to connect the account of method in Rule 8 with the account of method he gives in some other passages of the *Rules*, particularly in Rule 4, where the method seems closely connected with the *mathesis universalis,* or in Rule 14, where it seems more obviously mathematical.[6] But this example is still useful for understanding one central theme in Descartes' thought on method, as it is developed in the Paris years of the late 1620s, when he was doing some of his most important scientific work, and just before he began to compose the first version of his system.

The problem that Descartes poses for himself is that of finding the anaclastic line, that is, the shape of a surface "in which parallel rays are refracted in such a way that they all intersect in a single point after refraction."[7] Now, he notices—and this seems to be the first step in the reduction—that "the determination of this [anaclastic] line depends on the relation between the angle of incidence and the angle of refraction." But, he notes, this question is still "composite and relative," that is, not sufficiently simple, and we must proceed further in the reduction. He suggests that we must next ask how the relation between angles of incidence and refraction is caused by the difference between the two media (e.g., air and glass) which in turn raises the question as to "how the ray penetrates the whole transparent thing, and the knowledge of this penetration presupposes that the nature of illumination is also known." But, he claims, in order to understand what illumination is, we must know what a natural power *(potentia naturalis)* is. This is where the reductive step ends. At this point, Descartes seems to think that we can "clearly see through an intuition of the mind" what a natural power is. While Rule 8 is not entirely clear on this, it is plausible to connect this intuition to the conception of simple natures he had come to hold in the late 1620s. As I noted earlier, in Rule 12, written shortly after the completion of Rule 8, Descartes lists the simple natures pertaining to body as "shape, extension, motion, etc." (AT X 419).[8] Rule 9, again dated from the same stage of composition, tells us, in turn, that in order to understand the notion of a natural power, "I will reflect on the local motions of bodies" (AT X 402).[9] This suggests that the understanding of illumination is, somehow, an intuitive judgment about the simple nature, motion, though it is not clear how exactly he thought this would work. Once we have an intuition, we can begin the constructive step, and follow, in order, through the questions raised until we have answered the original question. That would involve understanding the nature of illumination from the nature of a natural power, understanding the way rays penetrate transparent bodies from the nature of illumination, and the relation between angle of incidence and angle of refraction from all that precedes. And finally, once we know how the angle of incidence and angle of refraction are related, we can solve the problem of the anaclastic line.[10]

This example develops the programmatic statement of the method as given in Rule 5 in a fairly concrete way. If we take the anaclastic line example as our guide, then methodical investigation begins with a question, a question which, in turn, is reduced to questions whose answers are presupposed for the resolution of the original question posed (i.e., q_1 is reduced to q_2 if we must answer q_2 before we can answer q_1). The reductive step of the method thus involves, as Descartes suggested earlier in Rule 6, ordering things "insofar as some can be known from others, so that whenever some difficulty arises, we will immediately be able to perceive whether it will be helpful to examine some other [questions] first, and what, and in what order" (AT X 381). And so, in a sense, the reduction leads us to more basic and fundamental questions, from the anaclastic line, to the law of refraction, and back eventually to the nature of a natural power and to the motion of bodies. Ultimately, he thinks, when we follow out this series of questions, from the one that first interests us, to the "simpler" and more basic questions on which it depends, we will eventually reach an intuition, presumably an intuition about simple natures. When the reductive stage is taken to this point, then we can begin the constructive stage. Having intuited the answer to the last question in the reductive series, we can turn the procedure on its head, and begin answering the questions that we have successively raised, in an order the reverse of the order in which we have raised them. What this should involve is starting with the intuition that we have attained through the reductive step, and deducing down from there, until we have answered the question originally raised. Should everything work out as Descartes hopes it will, when we are finished, it is evident that we will have certain knowledge as he understands it; the answer arrived at in this way will constitute a conclusion deduced ultimately from an initial intuition.

Descartes' strategy here is extremely ingenious. The stated goal of the method is certain knowledge, a science deduced from intuitively known premises. So, the principal goal of the reduction seems to be an intuition from which the original question posed can be answered deductively. But, he says, to simply launch off looking for such an intuition is folly; it is like searching the streets randomly for lost treasure.[11] What the method gives us, at least as he understood it by the late 1620s, is a *workable procedure* for *discovering* an appropriate path between intuition and the answer to the question posed. This workable procedure is the reduction of a question to more and more basic questions, questions that we can identify as questions whose answers are presupposed for answering the question originally posed.

There are numerous problems with the method as Descartes seems to have understood it by the late 1620s. As the example shows, the intuitive step remains profoundly obscure; it is not at all clear how one is supposed to pass from a question about the nature of illumination and the nature of a natural power to an intuition about simple natures. Nor is it obvious that once we have such an intuition, we will be able actually to do the deductions necessary to solve the original problem posed, as Descartes seems to assume we can. It is, for example, a non-trivial problem to pass from the law of refraction to the actual shape of the anaclastic curve. But the most basic, the most substantive assumption that the method makes is that of the structure of knowledge itself, the idea that from any question (at least, any question we are capable of resolving)[12] there is a string of questions that lead us back to an intuition. It is precisely because knowledge is ordered and interconnected in this way that the method is possible at all. Were knowledge to lack such order and connection, then method as Descartes conceives it would be impossible; and once we recognize such an order among things we seek to know, then the method for acquiring knowledge becomes relatively straightforward. It is here that the question of method links up with the question of the structure of Cartesian science in this period.

As discussed in chapter 1, the idea of order and the interconnectedness of knowledge is basic to Descartes' thought, and was probably one of the central components of his vision in November 1619. This vision of the interconnectedness of knowledge appears prominently in Rule 1, written almost certainly in the very earliest stratum of the composition of the *Rules*. The main theme in the commentary to Rule 1 is the claim that one must not limit one's attention to one branch of knowledge to the exclusion of others, that we must aim for "solid and true judgments concerning *everything* which comes before [the mind]" (AT X 359). Descartes argues, against Aristotle, that knowledge of the individual branches of knowledge do not interfere with the acquisition of one another, as plowing might make one's hands unsuitable for playing the lute (AT X 359–60); rather, he argues, "Since all sciences are nothing but human wisdom *[sapientia]*, which always remains one and the same, however much it may be applied to different subjects . . . , it is not necessary to confine minds *[ingenia]* within any limits" (AT X 360).[13] It is, then, possible to cultivate all of the sciences, and not confine ourselves to one. But, he argues, not only is this possible; it is also *necessary*. Because of the interconnection of the parts of knowledge, knowledge of any part is facilitated by knowledge of the whole, he claims:

> We often omit many things which are necessary for the understanding of others, either because they seem, at first glance, scarcely useful or scarcely interesting. But one must believe that everything is so interconnected that it is far easier to learn all

things together than it is to separate one from the others. Therefore, if one wants seriously to investigate the truth of things, one
should not select any individual science, for they are all connected with one another and depend upon one another. (AT X
361)

Though Descartes clearly saw in 1619 that the sciences are interconnected, it is by no means clear how he thought them to be interlinked.
Rule 6 suggests that statements about effects are linked to statements
about causes, statements about unequals linked to statements about
equals, and so on.[14] But this seems hardly worth the excitement. It
seems likely that as with the method to which it ultimately gave rise, the
initial picture of the order and connection of things is more program
than it is precise doctrine.

But by the later 1620s, the anaclastic line example of Rule 8 and the
discussion of simple natures in Rule 12 suggest that Descartes has a
much more precise idea of that structure of knowledge that got the
program going in 1619. The list of simple natures he gave in Rule 12
suggests that all knowledge is grounded in certain immediately intuited
propositions about a relatively few number of simple natures, mental,
material, and common (AT X 419). It is to these intuitions, intuitions
about the very most general features of the world—thought, extension,
shape, motion, existence, duration—that all questions are to be reduced and from which all answers are to be deduced. On these intuitions are grounded layers of successively less general propositions.
There are few grounds for speculating on what a systematic science of
the mental may have looked like for Descartes at this time. While Rule
12 contains a discussion of the intellect, which is clearly distinct from
body, there is no hint of how a systematic science of mind can be constituted.[15] But the anaclastic line example suggests a rather concrete
picture for physics, general propositions about motion (and, presumably, shape) grounding propositions about natural powers in general
and illumination in particular, and from there on to the specific properties of illumination and the laws it satisfies, the general and universal
grounding and giving rise to the more particular. One can see in this
structure the clear marks of his early vision of the interconnectedness
of knowledge.

In the end it was probably overstatement to suggest as Descartes did
in 1619 that *all* of the sciences are so interconnected that it is impossible to learn one without all the others. It would appear that a science
of mind, grounded in its own set of intuitions, might be pursued rather
independently of the science of body. But it does seem correct to see
everything in an account of body and the physical world connected if

not by direct deductive relations then at least by common ancestry, derivation from the same set of primitive and general intuitions. And were we to grasp those intuitions, as is necessary for any genuine, certain knowledge of any branch of physics, then to the pure and attentive mind, the mind capable of grasping truth at all, everything would seem to follow in all branches of the science of body, he might plausibly have held. At least in this qualified sense, in the late 1620s he is entitled to claim that it is "no more difficult to hold [the sciences] in the mind than it is to contain the sequence of numbers," and that one cannot "possess one [science] completely without the others following spontaneously" (AT X 215, 255).[16]

It must be emphasized that the order and connection of knowledge at issue here is strictly epistemological, strictly with relation to us and our ability to see truths through intuition and perform deductions.[17] As such, the doctrine of the order and connection of knowledge is intertwined with the underlying epistemology of the *Rules,* the account of knowledge in terms of intuition and deduction; the order is an order of propositions, deductively connected with one another and grounded in intuition. It is this order that allows there to be a method, as I emphasized earlier; it is because everything is structured the way it is that we can trace a question back along a chain of deductions to the intuition on which they all depend and from that intuition work down to a deductive answer to the question posed. In that way the order and connection of knowledge gives rise to a rather specific procedure for settling questions, a method in the narrow sense. But in a much more general sense, the picture of the structure of knowledge also determines what it is for a problem to be solved, for a question to be answered. Because all knowledge for Descartes must derive from intuition and deduction, and because more specific propositions are deductively connected to more general propositions in a chain that leads back to an intuition of the very most general truths, there is no real solution to any problem until more fundamental questions are confronted and until we confront the very most general questions and resolve them through intuition. In that way the order and connection of things determine not only the method, in the strict sense, but the very shape of Cartesian science considered as a body of knowledge. While Descartes never says so explicitly at this point in his development, it is not too much to suggest that by 1628 or so he had come to see all natural philosophy as grounded in general knowledge of the corporeal properties of extension, shape, motion, and all of the more specialized branches, dioptrics, meteors, even, perhaps, the biology of plants and animals, and large portions of medicine resting on top of those foundations in the precise sense that in order to know the higher sciences in the strictest sense one must pass through the more foundational disciplines.

This is the picture as of 1628 or so. But there is a particular problem that comes to interest Descartes in the course of writing the *Rules,* a problem whose solution will ultimately lead him to revise quite radically the carefully structured program for the sciences outlined there and abandon the method he spent so many years developing.

As I pointed out earlier in this chapter, the *Rules* is grounded in an underlying epistemology; the starting place of the *Rules* is the claim that all real knowledge is derived from either intuition or deduction. In the earliest strata of the *Rules* there is little concern with systematically justifying this claim; while he offers considerations in its favor and against the more formal scholastic logic and the appeal to authority, for example, there is no systematic defense of the claim.[18] But a few years later in Paris, Descartes faced the question head on. In the commentary to Rule 8, in a passage immediately following his exposition of the anaclastic line example we discussed earlier in this section, Descartes raises what he calls the "noblest example" one can give of his method, the project of "examining all the truths for the knowledge of which human reason suffices" (AT X 395). Most of the remainder of Rule 8 is concerned with sharpening this question, with reformulating it and clarifying the motivation for undertaking it. In the text that survives, we find three separate reformulations of this same question, what appear to be successive drafts of the same idea.

The project evolves and changes in its various reformulations.[19] But it is the last reformulation of this project that is particularly interesting to Descartes; it is the one he develops at greatest length, and the one that is directly linked to later rules, most notably Rule 12. The actual statement of the project is preceded by an intriguing analogy between the method developed in previous rules and blacksmithery. He writes:

> This method somewhat resembles those of the mechanical arts which do not require the assistance of others, but themselves show how their tools must be made. For if anyone wished to ply one of these trades, that of the blacksmith, for example, and lacked all tools, he would be forced at first to use a hard stone or some rough lump of iron as an anvil, to take a rock instead of a hammer, to prepare wood as tongs, and to arrange other things as they may be necessary. Then when these things have been done, he will not immediately attempt to forge swords or helmets or any of those things which are made of iron for others to use, but before anything else will make hammers, an anvil, tongs, and other tools useful to himself. We are taught by this example that

since in these first attempts we have been able to discover nothing but rather rough precepts which seem innate in our minds rather than derived by art, we should not immediately try by the aid of these to settle the disputes of philosophers or to solve the difficulties of mathematicians, but they should first be used in searching diligently for whatever other things are more necessary for the examination of truth. This is particularly true since there is no reason why it would seem more difficult to discover these things than it would be to solve any of the other problems usually propounded in geometry or physics or in other disciplines. (AT X 397)

One lesson Descartes draws from this analogy is that, like the provisional tools that the would-be blacksmith has fashioned out of sticks and stones, things immediately at hand, the rules of method previously laid down must be regarded as being merely provisional. The claim is that we should not try immediately to solve philosophical and scientific problems with them. Rather, what we should try to do is "search diligently for whatever other things are more necessary for the examination of truth." That is, we must use the provisional rules of method already laid down as provisional tools to build a proper set of tools for seeking truth. And, at this point, an epistemological project is announced:

In fact, there is nothing more useful to inquire about than what human knowledge might be and how far it extends [quid sit humana cognitio et quousque extendatur]. And therefore this very thing is now encompassed in the single question which we evaluate as the one which should be examined before all others by way of the rules previously laid down. This inquiry should be undertaken once in life [semel in vita] by any person who has the slightest regard for the truth, because in this investigation is contained the true instrument and the whole method of acquiring knowledge. (AT X 397–98)

The claim seems to be that a central part of building tools for catching truth is the investigation of a particular epistemological question.

In these passages Descartes argues that the investigation of knowledge, its scope and its nature, constitutes a necessary preliminary to any application of the method to specific scientific (and philosophical) questions, a prelude to genuine scientific investigation. But, one might ask, why does science need such an epistemological prelude? What is it that makes the method, like the would-be blacksmith's sticks and stones, provisional and unsuitable for investigating scientific questions, and how is it that an investigation of the scope and nature of knowledge is supposed to help us out? One suggestion seems obvious. As I noted earlier, the method that Descartes outlines in the *Rules* depends essentially on a certain conception of knowledge, a (there) unexamined epistemology according to which genuine knowledge is grounded in intuition and deduction. The method proper, I argued, is a device for generating such knowledge, a mechanism for constructing intuitive and deductive answers to questions posed. But there is an obvious problem here: where does this epistemology, the picture of knowledge that generates the whole method, come from? And how can we be sure that this epistemology is a good and appropriate ground on which to base our method? These, I think, are the questions that Descartes is raising in this passage from Rule 8; they are, I would conjecture, later reflections on the method that he had sketched a few years before with naive enthusiasm. It is interesting to note that there is no suggestion of any abstract skeptical worry here.[20] The epistemological question in this context seems to be a common-sense, almost a practical question: just as the blacksmith must make his tools before engaging in any serious projects, so must the scientist. Before we can enter into genuine scientific activity, we must (with the aid of our provisional method, based on a provisional epistemology) first discover what tools one has with which to forge truths, which are most dependable, what kinds of truth each tool is suited to, and so on; this is the question of the nature of knowledge.

The epistemological investigation that Descartes proposes is sketched twice in the *Rules*. At the very end of Rule 8, he gives a brief sketch of the project, immediately following on the third draft of its proposal that we have been examining (AT X 398–99).[21] The sketch is greatly expanded and appears as the commentary to Rule 12.[22] In order to explore the nature and extent of knowledge, Descartes tells us in both Rule 8 and Rule 12, we must deal with two sorts of things, "we who know, and the things themselves which are known" (AT X 411; see also 398). This, presumably, is the first step in the reduction of the question at hand to simpler questions. (It is striking that even the question of the justification of the foundations of method is a question that he attacks methodically.) But here, unfortunately, the treatment of the question becomes sketchier. Rather than continue the reduction and give a full answer, Descartes chooses merely to outline the kind of answer that, he thinks, would result from a proper investigation. He lays down a series of assumptions *(suppositiones)* which present the answers to the two simpler questions that, he thinks, he would ultimately arrive at, were the investigation carried out properly (AT X 411–12).[23]

The picture Descartes paints of the knowing subject is similar to the mechanistic account of the physiology and psychology of sensation that is given in the later works, from the *Treatise on Man* of 1633 onward, though he had not at this point undertaken any serious study of physiology. The senses transmit shapes *[figurae]* through other bodily parts to a common seat, the common sense *[sensus communis]*, perhaps the brain or the pineal gland (he does not say in the *Rules;* see AT X 412–14). Imagination, too, is for Descartes a bodily phenomenon (see AT X 414–15). At the end of the process is something nonmaterial, he hypothesizes, something purely spiritual *[pure spiritualis]*, "no less distinct from the whole body than blood is from bone, or the hand from the eye" (AT X 415). This spiritual thing receives the impressions of the senses, imagination, and memory, and is capable of initiating actions (AT X 415–16).[24] Following this hypothetical account of the knowing subject, he gives a similarly hypothetical account of the proper objects of knowledge, what he calls the simple natures and their combinations, as we discussed earlier (see AT X 417–25). There Descartes gives special attention to the roles that the cognitive faculties discussed earlier—intellect (i.e., intuition and deduction), senses, imagination, and memory—play in the perception of simple natures and their relations (see AT X 422–25).

Descartes' first conclusion on the basis of this survey is that "we can show distinctly and, I believe, by a sufficient enumeration that which we could only show in a confused and rough way, namely, that there are no ways open to human beings for the certain knowledge of truth, except evident intuition and necessary deduction" (AT X 425). He does not fill the reader in as to how, precisely, this conclusion is drawn. It looks to me like a conclusion drawn from a simple examination of the cognitive faculties and their objects; it seemed, perhaps, simply *evident* to the Descartes of the *Rules* that since all knowledge involves simple natures and their combinations, that the intellect and *only* the intellect, through the inner vision of intuition and deduction, is capable of knowledge.

By the last stages in the composition of the *Rules,* Descartes seems to have been quite taken with this problem of the scope and nature of knowledge. But it is important to point out that it is by no means clear just how this investigation is connected to the main project for the sciences as he conceived it in the *Rules,* the construction of a hierarchically organized body of knowledge, ultimately grounded in the immediate intuition of the very most simple and general features of the world. Though he seems clear that this epistemological investigation is a project that should be undertaken early on in the course of investigation, he is not at all clear whether it is a mere preliminary to real investigation, or part of the system of knowledge itself, whether it is essential in order for us to have any knowledge, or whether it is simply a practical suggestion about where we might begin.

DESCARTES' PROJECT: METHOD IN THE MATURE WRITINGS

In the *Rules for the Direction of the Mind* we have one conception of Descartes' project. Following the abandonment of the *Rules,* and his abandonment of Paris for the Low Countries, Descartes began the actual construction of the system of knowledge that he had contemplated and outlined in his earlier years. But the system as actually constructed turned out to be significantly different than what he imagined it to be in the years immediately following his dream. In this section we shall discuss what became of the method in the 1630s and 1640s; in the following section we shall turn to the question of the organization of knowledge in this period.

In the *Discourse,* the first of Descartes' published writings, the method is praised, and given almost magical powers. It is his method, he tells us, not any special acuteness of his mind that has led him to the wonderful discoveries which he made, discoveries some of which he will reveal to the world in the *Essays* that accompany the *Discourse* (see AT VI 2–3). The reader is led to expect that he will be shown the method Descartes formulated so that he, too, can progress, as Descartes did, to knowledge, truth, and, ultimately, happiness.

The reader is not entirely disappointed in his expectations. In Part II of the *Discourse* Descartes gives four "precepts" of method, covering in a few lines much of the ground covered by the lengthier and infinitely more complicated *Rules.* Briefly, the first precept of the *Discourse* is "never to accept anything as true that I did not know evidently to be such"; the second is "to divide each of the difficulties under examination into as many parts as possible"; the third is to "conduct my thoughts in order beginning with the most simple and easiest to understand, and advancing little by little, as by degrees to the knowledge of the most composite"; and finally, the fourth precept is "to make enumerations so complete and reviews so general that I could be sure I omitted nothing" (AT VI 18–19). While we must be very careful about reading the earlier view into the later text, it looks as if the method of the *Discourse* is meant to be a simplified exposition of that of the *Rules.*[25] As in the *Rules,* the goal is certainty, and as in the *Rules,* the procedure for attaining certainty is twofold, first a reduction of problems to simpler ones, and then a solution of complex problems in terms of simpler ones. But Descartes' exposition in Part II of the *Discourse* is not nearly detailed enough to be able to recover the full method of the *Rules* in all of its glory. These precepts are sketchy, thin, and not very impressive; it is not surprising that many of Descartes' contemporaries were convinced that he was hiding his secret. If we did not have the text of the *Rules,* it is not clear what possible sense we could make of them.

But, Descartes claims, he had not intended for people to learn the method from the *Discourse,* at least, not from the *Discourse* alone. Explaining his choice of titles for the *Discourse,* Descartes tells Mersenne shortly before the work appeared, "I did not give it as *Treatise on the Method* but as *Discourse on the Method,* which is the same as *Preface or Advice* [Aduis] *concerning the Method* to show that I did not intend to teach it but only to discuss it, for, as one can see from what I say in it, it [i.e., the method] consists more in practice than in theory" (AT I 349 [K 30]). This message, that the *Discourse* discusses the method without teaching it, strictly speaking, is something Descartes tells other correspondents as well (see. e.g., AT I 370; AT I 559 [K 45]). For the method, he tells Mersenne, we must turn to his practice. But what practice is relevant here?

Interestingly enough the three essays that accompany the *Discourse* are not, by and large, what Descartes had in mind.[26] The *Dioptrics, Meteors,* and *Geometry* are, of course, not unconnected to the method. As he wrote to Mersenne, continuing the explanation he gave of his title, "I call the treatises following [the *Discourse] Essays in this method* because I claim that the things they contain could not have been found without it and that through them one can understand its value" (AT I 349 [K 30]; cf. AT I 370, AT I 559 [K 45])." But though the *Essays* contain results obtained by applying the method, they do not themselves contain examples of reasoning in accordance with the method, by and large. As Descartes wrote to Vatier in February 1638, not long after the publication of the work:

> My intention was not to teach the whole of my method in the discourse where I propound it, but only to say enough about it to allow one to judge that the new opinions that one will see in the *Dioptrics* and *Meteors* were not conceived lightly, and that, perhaps, they are worth the trouble to examine. I could not show the use of this method in the three treatises which I presented because it prescribes an order for investigating things very different from that which I thought I had to use to explain them. (AT I 559 [K 45–46])

The mode of exposition Descartes chose for the *Dioptrics* and *Meteors* was, of course, hypothetical; both works begin with appropriate assumptions or hypotheses about the nature of light and its basic laws, in the case of the *Dioptrics,* and about the makeup of the things most commonly found on earth in the case of the *Meteors,* hypotheses that while plausible are not proved (see AT VI 83ff, 233ff [Ols. 66ff, 264ff]). By arguing from hypotheses, he thought that he could show some of his results without having to divulge the first principles of his physics, for which, he believed, the public was not ready. For this purpose he thought the hypothetical mode of argument quite well suited and he thought that he had succeeded quite well in the limited goals he set for himself in the *Essays.* But, at the same time, while pleased with his *Essays,* he was clear that they represent not the definitive treatment of his thought, in accordance with his method of inquiry, but, rather, interesting experiments in exposition.[27]

The only exception Descartes seems to make to this view of the *Essays* is the discussion of the rainbow in discourse 8 of the *Meteors.* This section contains the only explicit reference to the method in any of the three essays, and in the letter to Vatier from February 1638, it is singled out as an example of the method of the *Discourse* (see AT VI 325 [Ols. 332]; AT I 559 [K 46]). It is, I think, possible to make good on his claim here, and show how the account of the rainbow can be set out in accordance with the method.[28] But the actual presentation of the case in the *Meteors* is difficult and technical, and Descartes doesn't make any special attempt to relate the argument there to the method as presented in the preliminary *Discourse;* it is certainly not easy for the reader to see there the outlines of Descartes' simple method. Furthermore, it is interesting to note here that the solution to the problem of the rainbow that he calls attention to in 1637 and 1638 dates from some years earlier. While it is impossible to be certain, it is likely that that portion of the *Meteors* dates from late 1629, not long after the *Rules* were set aside.[29] When the account of the rainbow appears eight years later in the *Meteors,* it appears as a kind of ghost from an earlier period.

This leads us to a somewhat surprising conclusion. Though the method is apparently central to the rhetoric of the *Discourse,* it seems to play a minor role, if any at all, in the science presented in the accompanying *Essays.* Indeed, one is hard-pressed to find any clear and unambiguous example of the method being applied to nonmathematical questions in *any* text that follows the work on the rainbow in 1629.[30]

In *The World,* for example, Descartes divulges more of the foundations of his physics than he will do later in the *Essays;* though certain metaphysical issues that Descartes was concerned with at the time are hidden, he is forthcoming about the nature of matter, the nature of light, the role God plays in maintaining the world and determining the laws of motion, and so on. But even so it is difficult to discern the formal method of the *Rules* in *The World.* Descartes seems to leave aside any reductive stage, and, after some introductory chapters, he begins directly with the construction of the world, with God's creation of a plenum of material substance in chaos, which because of the laws it must obey, forms itself into the world as Descartes conceives of it, vortex after vortex, with central suns and planets.

And when in the 1640s Descartes sets aside his scruples and presents his system in its full and proper form, there is as little of the method as there was in *The World* and the *Essays*. This is evident in the *Principles of Philosophy*. Unlike *The World*, the *Principles of Philosophy* does contain an exposition of the metaphysics on which the physics presented is based. But like *The World,* the *Principles* lacks any apparent reductive step; after a few brief sections, exorting the reader to reject his former beliefs, Descartes begins at the beginning, with the *Cogito* in this case, and deduces on down, showing step by step how the Cartesian world, vortices, planets, and all, can be deduced from this initial intuition.

Evidences of the method of the *Rules* are also difficult to discern in the *Meditations,* a work whose origin, Descartes tells the Doctors of the Sorbonne, was in part a response to a request for him to apply his celebrated method to God and the soul (see AT VII 3), a work written in the analytic mode, he tells the second objectors, a work that is intended to follow "the true way through which a thing was methodically . . . discovered" (AT VII 155). In the *Meditations,* the intuition that constitutes the starting place of the deduction, the *Cogito,* is carefully prepared in the First Meditation. But the preparation does not seem to be a reduction in the precise sense of the term. The First Meditation does many things; it clears away prejudice, establishes a standard for certainty, introduces the problem of knowing our creator as the essential preliminary for any further knowledge.[31] But so far as I can see, it does not clearly pose a single question to be answered, and it does not sketch out the sequence of steps to be followed in deducing the answer to a question posed, the way a proper reduction is supposed to do. Nor is it followed by the sort of deduction mandated by the method. There is, to be sure, a thread of argument that leads one from the *Cogito* through God to knowledge and the external world, a kind of deduction from foundations to superstructure. But from the *Cogito* of Meditation II to the end of Meditation VI there are numerous places where the meditator tries to lead the argument into a dead end, where the meditator begins an argument that simply does not pan out. For example, at the beginning of Meditation III, the meditator tries unsuccessfully to demonstrate the existence of the external world, before giving the proof for the existence of God (AT VII 37–40).[32] This and other digressions are very important to the structure of the *Meditations.* The *Meditations* are addressed, in part, at a very specific audience that Descartes knows quite well: the unconverted, readers full of prejudice in favor of their senses and for the material world. And these digressions are very important to convince them that certain arguments that they are inclined to accept, arguments that take for granted a faith in the senses and a priority in belief in the external world, are mistaken.[33] Descartes' meditator is clearly feeling his way along from *Cogito* through God to the external world, and not following out a path sketched out in a reductive stage of the inquiry. There is method in this procedure, to be sure, but the method is not the method of the *Rules.*

This is a point worth emphasizing. Descartes' later writings are quite clearly governed by a relatively strict conception of the proper order of inquiry; as we shall see in more detail later in this chapter, Descartes never gives up the idea of a hierarchically ordered science that he had had since his earliest years. But, I claim, what he does seem to give up is the precise conception of the method as given in the *Rules* and the *Discourse.* One could, perhaps, express this by saying that the method is *changed* or *transformed* in the later writings. But given Descartes' own apparent avoidance of all talk of method after the *Discourse,* it does not seem too bold to say that he has given up the method.

In claiming that Descartes' later works do not display his earlier method I am making a controversial claim, one that would be challenged by other scholars, who have claimed to find the method of the *Rules* and *Discourse* in the *Meditations,* at very least.[34] But even if they are right (and I don't think they are), it is beyond dispute that he himself hardly mentions his method after the *Discourse* and the letters that immediately follow its publication; indeed, outside of the *Discourse* and correspondence directly connected with it, the method hardly comes up at all in any of the writings that follow the abandonment of the *Rules.* If method is the key to knowledge and the key to the later Cartesian system (as it seemed to be in the *Discourse*), Descartes himself does not call attention to that fact anywhere outside of that document. When the earlier method does come up in his later writings, it has a decidedly subordinate role in his thought. In the Letter to Picot that serves as a preface to the French *Principles of Philosophy* of 1647, he recommends that the student of philosophy "ought to study logic, not that of the schools, . . . but that which teaches one how to conduct his reason well in order to discover truths that one does not know" (AT IX B 13–14). It would be good, Descartes says, for him to "practice the rules concerning easy and simple questions for a long time" until "one acquires a certain habitude for finding truth in these questions" (AT IX B 14). But in this respect, the method has roughly the status of the provisional morality (which immediately precedes it in the Letter), one of those preliminaries that should be undertaken by the student of nature before undertaking the serious business of philosophy; it is an exercise useful primarily in sharpening the mind and helping us to recognize truth, an exercise that has in 1647 roughly the same role that Descartes earlier gave the scholastic logic he otherwise rejected in the *Rules* (see, e.g., AT X 363–64). Whatever it is, it is clearly not nearly as important to Descartes as it appeared to be in the *Rules* or in the *Discourse.*

This leaves us with some puzzles. First of all, why should Descartes have abandoned the method when he did? And if Descartes did abandon the method in his own practice by 1630, then why then did he write the *Discourse on the Method* in 1637, a work that seems to emphasize the centrality of method in his thought?

Let me address the second question first. It seems to me quite possible that when he wrote the *Discourse,* Descartes simply did not fully realize that he had in practice set aside the method that was so important to him in the 1620s. And so, when he sat down to write the introduction to his three *Essays,* out came the *Discourse on the Method,* a work that reflects a philosophical conception of scientific inquiry that had ceased to fit his actual practice for at least five or six years. Only when faced with puzzled readers, demanding further elucidation of the method might he have realized how little relevant it was to his current practice. Alternatively, even if he realized that the method was no longer relevant to his practice, he might have seen method, which was historically important to his own intellectual development, as a neat hook around which to organize the diverse observations collected in the preliminary *Discourse,* just as light had been so used, somewhat artificially, in *The World* or *Treatise on Light.* When pressed for further explanations of his marvelous method, Descartes may have wished that he had stuck with his original title for the *Discourse, The Plan of a Universal Science . . .* (a better description of what much of the *Discourse* is actually concerned to do), and represented the method merely as a stage in his intellectual biography.[35]

There are a number of factors that might explain why Descartes abandoned the method by 1630 or so. The case of the *Meditations* is instructive, I think. As I pointed out, the *Meditations* is carefully structured to address the reader who comes to the project of knowledge full of prejudice and bias. While the method of the *Rules* might work for an inquirer who is clear of mind, something else is called for in this real world, where common experience, reinforced by education, leads one to begin the search after truth with a clouded mind. Another factor might concern the relation between the method and the doctrine of the interconnection of knowledge on which it is grounded. This doctrine suggests that what we should be doing is not solving individual problems, but constructing the complete system of knowledge, the interconnected body of knowledge that starts from intuition and comes to encompass everything capable of being known. This, indeed, is the project of his mature writings. Starting from the metaphysics of 1629–30 and *The World,* Descartes undertook the construction of the whole body of knowledge, from foundations on up, considered as a unified system of knowledge. But the method of the *Rules* is not appropriate to this style of science. To make use of the method, we must first set a specific question for ourselves, what is the shape of a lens with such-and-such properties, or, what causes the rainbow, or whatever. Once we have a specific question, we can then apply the method, reduce the question to simpler questions until we reach an intuition, and deduce back up to an answer to the question originally posed. The method is a method for doing science as, say, Beeckman and the young Descartes conceived of it, as a series of discrete questions about the natural world. When Descartes ceased to be a problem-solver and became a system-builder, it is not surprising that the method, central to his earlier thought, would become obsolete.

Both of these factors contributed to the demise of the method, no doubt. But also important, indeed crucial, was a basic transformation in the very conception of the structure of knowledge that one finds in Descartes' mature thought, a transformation that one can trace back, at least in part, to the "noblest example" of the method, as presented in the *Rules.* In order to appreciate this last factor contributing to the demise of the method, we must understand Descartes' new conception of the organization of knowledge. Let us, then, turn to the question of structure in Descartes' later system.

DESCARTES' PROJECT: ORDER IN THE MATURE WRITINGS

Earlier, in connection with the *Rules,* I emphasized the importance of order and the interconnectedness of all knowledge. In the mature writings, Descartes' obsession with order in its various manifestations— order of investigation, order of inquiry, the interconnectedness of knowledge—continues, and continues to shape Descartes' thought in fundamental ways; it is, as one celebrated commentator has emphasized, the key to understanding his mature thought.[36]

But of more importance than the abstract ideas of order and interconnectedness themselves is the precise structure that Descartes thought knowledge had in his mature writings. Though the emphasis on order and structure is by no means new, the details of his conception of order differs considerably from the conception that he held in early 1628 when he set the *Rules* aside; in actually working out the system of knowledge he dreamed of and planned through the 1620s, the original sketches were altered in significant ways. But by 1636 at the latest, when the *Discourse* was completed, the rather rough order from general intuitions about simple natures to more particular propositions had evolved into a doctrine on order both broader and more precise.

The order of knowledge is exemplified in the very structure of the *Discourse,* an order that is supposed to represent the order that Descartes, the apparent protagonist of the fable, followed in actually constructing his philosophy. The exposition of the philosophical system begins in Part IV with his "first meditations," meditations that he characterizes as "metaphysical" (AT VI 31).[37] While we shall have to examine the specific contents of this metaphysics in more detail later, briefly, it includes an abbreviated exposition of one version of Descartes' doctrine of the soul, its existence, its nature, and its distinction from the body, his doctrine of God, and his derivation of the principle of clear and distinct perception from these. These are the "first truths," he claims, those on which other truths rest (see AT VI 40). In Part V some of these other truths are sketched in a brief (and cautious) exposition of the contents of *The World.*[38] As in *The World,* he declines to draw the connections with the metaphysics he has just given, and speaks hypothetically, not about this world, but "only about what would happen in a new one," should God create it "somewhere in imaginary spaces" (AT VI 42).[39] But next in the order of exposition, and presumably, following out of the metaphysics sketched in Part IV, is the very most general portion of physics, the doctrine of matter and the laws of nature. This is followed by a sketch of the world; following a hypothesis about creation Descartes shows how the basic features of our world would arise out of matter following the laws of motion (see AT VI 43–45). Though he admits that at the point of writing the *Discourse* he had not worked out the details, he implies that one ought to be able to show how even plants and animals can be derived in this way, that in a proper system of philosophy, our knowledge of living things would come from an understanding of how they derive by natural processes from the physical state of the world, like minerals and meteors (see AT VI 45). Put briefly, in the *Discourse,* knowledge seems to be ordered in a hierarchical way, with his metaphysical "first meditations" on the bottom, and followed in order by the general parts of physics, the physics of inanimate bodies, starting with light, the stars, and the planets and working down to the more detailed features of the world, ending with biology and the study of animals, those special and complex machines that are man's companions and competitors.

This picture of the order of knowledge is expanded somewhat and neatly summarized roughly ten years later, after the metaphysical part of the program had been published in the *Meditations* of 1641, followed three years later by an outline of the physics in the *Principles of Philosophy* of 1644. When in 1647 he published his *Principles* in French, Descartes appended a letter to the Abbé Claude Picot, the translator, to serve as a preface. In this short essay, like the *Discourse* the preface to a scientific work and in many ways an update of that earlier preface, he set out the order one should follow in instructing oneself. In the course of giving that advice, he makes clear how he thinks knowledge is ordered. Descartes begins with the preparations, strictly speaking, for undertaking the acquisition of knowledge. As in the *Discourse,* he recommends that the reader in search of knowledge begin by formulating "a code of morals that can serve to regulate actions in life" while his beliefs are undergoing reform, and by studying and practicing logic, not the logic of the schools, but what Descartes earlier called his method, the logic "that teaches how to conduct one's reason well in order to discover truths that one does not know" (AT IXB 13–14).[40] After these preliminaries he says:

> One ought to begin applying oneself to the true philosophy in earnest. The first part of the true philosophy is metaphysics, which contains the principles of knowledge, among which are the explication of the principal attributes of God, the explication of the immateriality of our souls, and the explication of all the clear and simple notions which are in us. The second is the physics in which, after having found the true principles of material things, one examines in general how the whole universe is composed, and then in particular, one examines the nature of this Earth and that of all the bodies one finds most commonly on it, like air, water, fire, the magnet [i.e., lodestone], and other minerals. After that it is also necessary to examine in particular the nature of plants, that of animals, and above all, that of man, so that after that one will be able to discover the other sciences which are useful to him. Thus all philosophy is like a tree, whose roots are metaphysics, whose trunk is physics, and whose branches, which grow from this trunk, are all of the other sciences, which reduce to three principal sciences, namely medicine, mechanics, and morals. I mean the highest and most perfect morals which, presupposing a complete knowledge of the other sciences is the highest degree of wisdom. (AT IXB 14)

Matters are somewhat more complex than Descartes represents them as being in his neat and simple tree of knowledge. The structure cannot be so neatly hierarchical as he represents it as being. The science of body that constitutes the trunk of the tree includes the biology of non-thinking things, and gives rise to medicine, insofar as medicine deals with the body as a complex physical object. But as Descartes fully recognized when, in the *Passions of the Soul*, he treated topics relating to morals, in order to treat morals we must understand the mind both taken by itself and as united to the human body, a topic that fits poorly into his neat schema. Furthermore, though each part of the true philosophy is supposed to grow out of the previous part, it is not entirely clear how this metaphor is to be cashed out; much of our task later in this chapter and, indeed, later in the book, will be to try to make precise how at least some parts of the true philosophy that Descartes is outlining in this preface and in the mature writings connected to it are related to one another, how the lower parts are supposed to support (to use the architectural metaphor) and give rise (to use the tree metaphor) to the later parts of the program. And finally, there seems to be no place for mathematics in this structure, very odd considering the importance mathematics had for Descartes.[41] But despite the problems and the obscurity, this gives us a tolerably clear idea as to how in general terms he thought knowledge to be ordered in his mature system.

There are a number of salient differences between the order of knowledge in the *Rules* and the order in the mature writings. In the *Rules* there is clearly the idea that knowledge forms a system, and that the system is ordered, for the most part, from general to particular, at least in the portions that relate to the material world. In the mature writings, though, Descartes is clearer about what subject matters belong at the various levels, what is foundational and what rests on what. In a sense, the order of reasons in the *Rules* has sorted itself out into an order of subject matters, metaphysics grounding physics, which in turn supports branches of knowledge like medicine, mechanics, and morals, those that yield the fruit. Furthermore, as I shall explore in more detail later in this chapter, the notion of support in play in the mature writings is somewhat more complex than the one to which Descartes appeals in the *Rules*, though this is by no means obvious at first glance. But the most obvious difference between the structure in the Rules and the structure found in the *Discourse* and the preface to the *Principles* is what it is that is taken to be foundational; unlike the *Rules*, where the structure of knowledge seems to be grounded directly in the immediate intuition of the very most general features of the mental and material world, in the mature works, the system is grounded in what he calls metaphysics or, alternatively, first philosophy. The unified science of the *Rules*, first germinated from the seeds of truth in November 1619, has grown metaphysical roots.

The metaphysics or first philosophy that is supposed to ground the

Cartesian program contains a variety of elements, if we take the *Meditations* as representative of what Descartes had in mind.

The metaphysical foundations include, first of all, an epistemological inquiry into the sources of knowledge, in particular, an investigation of the extent to which knowledge can be derived from sensation, imagination, and pure intellection. This part of the new metaphysical foundations derives in an obvious way from the "noblest example of all," the epistemological reflections that appear in the very latest stages in the composition of the *Rules.* As in the *Rules,* the project is to investigate "what human knowledge might be and how far it extends" (AT X 397). Furthermore, the conclusion at which Descartes arrives in the *Meditations* is very similar to the principal conclusion of the investigation as sketched in Rule 12, that knowledge derives principally from the intellect, though the senses and the imagination can be of some help. But there are also some significant differences. For one, there is the question of skepticism; the epistemological inquiry of the *Meditations* is, at least in part, intended to address the skeptic in a way in which the project of Rules 8 and 12 do not (see Curley 1978, pp. 44–45). But more important than this, I think, is a fundamental clarification in the place that the epistemology occupies in the enterprise. As I noted earlier in this chapter, the epistemological investigation suggested in Rule 8 and sketched in Rule 12 occupies a somewhat ambiguous position. Descartes is not at all clear whether it is part of the hierarchical system of knowledge, or whether it is a mere preliminary to the construction of that system, a system which properly speaking rests on the bedrock of intuitions of the most general features of the world and deductions from those intuitions. But in the mature writings the epistemological investigation of the sources of knowledge has become integrated into the system as a whole as a prominent component of the foundations of all knowledge; no mere practical preliminary to methodical investigation, it is unambiguously a component of the first philosophy on which the rest of knowledge stands, and from which the rest of knowledge grows.

A second element of the new first philosophy is the account of God and the soul. God and the soul certainly make their appearances in the *Rules;* Descartes presents as examples of immediate intuitions the propositions "I am, therefore God is; also, I understand, therefore I have a mind distinct from body" (AT X 421-22). But as suggestive as these might be of his later position, there is no reason to think that he gave these intuitions the special foundational status they have in the *Meditations,* even if we could figure out precisely what he meant by them in the *Rules,* and it is their status as foundational that is important in this context.[42] The central role of God in the later writings shows a striking difference with the *Rules.* In Rule 12 the suggestion is that intuition and deduction are to be certified through an examination of the knowing subject, the soul, and its objects, the so-called simple natures. But in the *Meditations,* it is the knowing subject and its creator, God, that ground knowledge. Furthermore, the same God whose veracity helped ground knowledge will, through his immutability, ground the laws of motion, as we shall see below in chapters 7–9, a position that Descartes is unlikely to have held before the metaphysics of 1629–30.[43] The doctrine of the soul developed in the first philosophy of the *Meditations* also grounds a number of other aspects of the mature system. As we shall see below in chapter 3, our knowledge of the nature of the soul, established in the metaphysics, is an important step in establishing the nature of body as an extended thing, a substance whose essence is extension. The doctrine of the soul developed in the metaphysics will also prove important to the elimination of substantial forms, as we shall discuss in chapter 4. So far as I can see, there is no indication in the *Rules* of such uses for the soul.

There is a third element that pertains to the *Meditations* and may, properly speaking, pertain to what Descartes has in mind among the metaphysical foundations of the sciences. That is the elimination of prejudice through radical doubt. The general motivation for the practice goes far back, to November 1619 and the very earliest strata of the *Rules.* There, in Rule 4, he notes that the seeds of truth implanted in us by God can be "suffocated by opposing studies," and that "nothing can be can be added to the pure light of reason without obscuring it in some way" (AT X 373).[44] The idea that we must reject our past beliefs in order properly to ground knowledge is, of course, one of the central themes of the *Discourse.* In the *Discourse* it is experience in the very most general sense that is supposed to free us from prejudice; travel and study of the book of nature is supposed to loosen the bias of youth and education and prepare us for undertaking the reconstruction of knowledge Descartes proposes there (see, e.g., AT VI 9–10, 22, 28–29). But in the *Meditations* he advances a new device for eliminating prejudice. Though the idea certainly precedes the *Meditations,*[45] it is for the first time in the *Meditations* that we find the contemplation of skeptical arguments in order to clear prejudice presented as a necessary first step in first philosophy. The skeptical arguments of Meditation I, Descartes tells the reader in the synopsis, "deliver us from all sorts of prejudices, and lay out a very easy way to detach our mind from the senses" (AT VII 12). The elimination of prejudice through radical doubt occupies a somewhat ambiguous position in the first philosophy of the *Meditations.* As practice rather than doctrine it cannot, strictly speaking, be a part of metaphysics, even though it does appear as a "postulate" in the Euclidean presentation he gives some of his arguments following the *Second Replies* (see AT VII 162). But while it is not metaphysical doctrine in any strict sense, the elimination of prejudice through radical doubt is, in Descartes' mind, absolutely inseparable from the proper conduct of metaphysics; one is simply incapable of grasping metaphysical truths without performing such an exercise, something that, it would seem, distinguishes this from other exercises he represents as preliminaries to investigation, like practicing the method and formulating a provisional morality.

In adding this particular metaphysical sub-basement to the system of knowledge originally envisaged in the *Rules,* Descartes has fundamentally transformed his conception of the project, his conception of the order and connection of knowledge. In the *Rules* the order seems to proceed from general to particular, from intuitions concerning the generalmost features of the world to more particular facts. In the mature writings, though, the system is grounded in particulars, in the self and in God; the starting place for knowledge, the ultimate ground is no longer a general proposition, like the nature of a natural power, but a radically particular proposition, the *Cogito.* But perhaps more important than this is another change. In the *Rules* what grounds knowledge is intuition *simpliciter;* a problem is reduced only to the point to which we can intuitively grasp the propositions that, ultimately, are necessary for its solution. While Rules 8 and 12 do, indeed, suggest an epistemological investigation of intuition and deduction, in the *Rules* this investigation occupies an ambiguous place in the enterprise, and may only be a preliminary to the system of knowledge, and not itself part of that system. But in the mature writings Descartes is absolutely clear that it is no longer sufficient merely to begin the construction of knowledge with intuitions or, as they later come to be called, clear and distinct perceptions.[46] In the mature system we do, indeed, begin with intuitions; intuitions after all are what ground the arguments concerning the self and God (see *Second Replies,* AT VII 140ff). But the first stage of the program essentially involves *validation* of those intuitions with which we begin, something not required in the earlier method.

In this last transformation we find, I think, the most important reason why Descartes came to set aside the celebrated method of the *Rules* and the *Discourse.* With this change, that is, with the new metaphysical foundations Descartes adds to his program, the method of the *Rules* by itself can no longer lead us to genuine knowledge. The reductive stage of that method starts with a question, and then takes us back to questions presupposed, until we finally reach an intuition. But when the reduction has reached an intuition it goes no farther. Thus the method of the *Rules* can at best give us imperfect knowledge, the moral certainty we get when we take intuitions for granted, rather than the metaphysical certainty that comes from knowing that our clear and distinct perceptions are the creation of a God who does not deceive.[47] In this way, what first appears as the noblest example of the method leads eventually to the demise of the method itself.

But it is not only the method that is undermined in this transformation. In placing this metaphysical program at the beginning of the system, Descartes radically altered the very meaning of order in his system, whether he was aware of it or not. In the *Rules* the notion of order is rather straightforward: A is prior to B means, in an obvious sense, A is a *premise* for B, and the body of knowledge grounded in intuition is grounded in the ultimate premises. But in the mature writings, this seems not to be enough. There we must go beyond the ultimate premises of physics, say, premises that were the stopping place of the reductive step in the method of the *Rules,* and ask further questions about what entitles us to take those intuitions *as* the starting place for science; Descartes has passed, it seems, from wanting a science *as* certain as geometry to wanting one *more* certain than geometry.[48] The epistemological project that now, in part, grounds the program is epistemologically prior to the rest of the project in a fairly plausible sense; the epistemological project Descartes envisions is one that, it would seem, should be undertaken before engaging in the construction of a system of positive knowledge. But it is not epistemologically prior in the precise sense that the first intuitions were epistemologically prior in the *Rules,* as genuine *premises* for further conclusions. Adding this metaphysics as the foundations of the system changes the very conception of order itself, of what it means for one thing to precede another in the order of knowledge. What this new sense of order comes to is by no means clear. It will become obvious later why Descartes thinks knowledge of God and the soul must precede knowledge of physics and why clearing the mind by radical doubt is, in a clear sense, a necessary condition for any further knowledge. But it is never clear the precise sense in which we are *obligated* to undertake the epistemological programs associated with the first philosophy first, before undertaking anything else. Ironically enough, as the details of the order of the sciences expands and comes clearer into definition, the principles on which it is constructed appear to lose the sharp outlines they had in the *Rules*.

Even though method seems to lose its centrality in the mature writings, the picture of scientific activity that dominates the other writings is very much of a piece with the picture of the *Rules*. The notion of order and interconnection, originally important, perhaps, largely insofar as it gave rise to the method, becomes itself the central organizing principle and inspiration of Descartes' project in later years. Though significantly transformed, the idea of order remains at the center of Descartes' thought in his later writings, just as it had from November 1619 onward.

With this the preliminary account of Descartes' program is complete. In later chapters we shall have to make more precise the way in which metaphysics serves to ground the physics by examining how Descartes actually argues for some of the central doctrines of his physics. But before turning to those questions, I would like to examine briefly some aspects of the larger context in which we must understand the project Descartes proposes to undertake.

THE PROJECT IN CONTEXT

Approaching Descartes in ignorance of the still-thriving scholastic intellectual culture of the sixteenth and early seventeenth centuries (the culture to whose extinction Descartes contributed in an important way), it is easy to overestimate the novelty of his approach, or, at least, to misidentify what precisely it is about his thought that is genuinely novel. Many elements of the program we have outlined in this chapter do, in fact, fit quite conveniently into the general intellectual framework in which Descartes was first nourished. His definition of knowledge in the strict sense, the goal of inquiry, and his demand for certainty, are very close to the traditional Aristotelian conceptions; his definition of knowledge in the *Rules* ("Omnis scientia est cognitio certa et evidens") is almost word for word the same as the definition Eustachius cites in his *Summa Philosophica,* an almost contemporary textbook of scholastic philosophy ("scientia appellatur omnis certa ac evidens notitia").[49] Furthermore, Descartes' demand for knowledge derived from an intuitive grasp of first principles fits nicely with the scholastics' preference for demonstrations *propter quid* over demonstrations *quia.*[50] Though method and order of inquiry and exposition were not common concerns of high scholasticism, in the sixteenth century method became all the rage, and lengthy discussions of method are common, both among adherents of the traditional school philosophy and among its opponents.[51] And so, when Descartes announced his method in 1637, while the method he claims to have discovered and followed may have been original, the idea of method itself was not. And finally, the categorization of the sciences and the claims Descartes makes about their interdependencies and the order in which they should be learned (claims, I have argued, that in an important sense define Descartes' very project) fit nicely into a tradition of such schemes that extends back into classical antiquity. Such questions were frequently treated in the introduction to scholastic commentaries and were the principle subject of a number of treatises.[52]

In pointing up these important similarities, though, it is worth noting the differences between Descartes and the tradition that he in important respects resembles. Although he shares with the schoolmen a strong preference for certain knowledge derived from first causes, he is considerably more optimistic than any of his teachers were about how far such knowledge extends. Eustachius, for example, recognized that most of physics is known only with probability. However, he argued that it still qualifies as a science (science, *scientia* = knowledge) because "for some whole discipline to be truly called a science, it is enough for some things to be known with certainty, even though not all are."[53] Descartes, on the other hand, often expressed the view that most if not all physics can be known with certainty. Addressing the scholastics' pessimism he says in the *Rules:*

> And although perhaps the learned have persuaded themselves that little such [knowledge] exists . . . nevertheless, I advise them that there is much more than they think, and that it is sufficient for demonstrating with certainty innumerable propositions which until now could only be treated with probability. (AT X 362)

In the mature writings this is true despite the fact that Descartes has come to adopt a standard of certainty, resistance to hyperbolic doubt, that is considerably higher than any standard that the scholastics were prepared to adopt.[54] And while Descartes shares the contemporary enthusiasm for method, it should be noted that his concern is in important ways different from those of many other writers. For scholastic writers, and for antischolastics like Peter Ramus who is probably responsible for introducing the notion in the sixteenth century, method is primarily conceived as a part of logic, one that follows the study of terms, propositions, and syllogisms, the three parts of the standard Aristotelian logic, and deals with questions relating to larger scale argumentation and organization of bodies of material for teaching or debate. And so in Eustachius' manual method is treated as part of logic, and is primarily defined as "the order or series of all of those things which are treated or arranged in knowledge taken as a whole *[universa doctrina]* or in one of its parts."[55] For these writers, the discovery of new knowledge is simply not part of the task of method. But while Descartes does seem to identify method with logic, method is primarily concerned with discovery. As he writes in the *Rules:*

> By method, moreover, I understand certain and easy rules which

are such that whoever follows them exactly will never take that which is false to be true, and without consuming any mental effort uselessly, but always step by step increasing knowledge *[scientia]*, will arrive at the true knowledge *[vera cognitio]* of everything of which he is capable. (AT X 371–72)[56]

These differences are important. But for our purposes, the most important difference concerns Descartes' conception of the categorization and interconnection of the various branches of knowledge. Comparing his conception of the tree of knowledge with that of his scholastic predecessors is a very tricky business. First of all one must take account of the differences between Descartes' conception of the branches of knowledge in question, for us principally metaphysics and physics, and those held among the scholastics. While there is no universally agreed upon definition of metaphysics among the schoolmen, there is a certain range of topics generally considered metaphysical. In his manual, for example, Eustachius lists God, separated substances (i.e., substances existing apart from matter), substance in general, finite being, and "being taken in the broadest sense."[57] Descartes' first philosophy contains an account of the soul and God, entities he identifies as "les choses immatérielles ou Metaphysiques" on at least one occasion, and he does discuss the notion of substance.[58] But the study of being as such has no apparent place in his metaphysics. Furthermore, Descartes' first philosophy includes a discussion of knowledge foreign to any scholastic conception of the subject.[59] His conception of physics is also somewhat different than that of his scholastic contemporaries. Physics for Descartes is the study of body (material substance), what it is, how it behaves, and how the cosmos and its contents are constituted from material substance. This, in a way, is not too different from what the scholastic physicist was trying to do. But scholastic philosophers defined the domain of physics in terms foreign to Cartesian thought. Characteristically physics was defined as the study of mobile (i.e., changeable) or natural things.[60] But what scholastics meant by mobility and motion was change in a very general sense, a notion that Descartes claimed to find unintelligible, as we shall later see.[61] And for scholastic philosophers, a natural thing is a thing that contains in itself a principle of motion and rest, something Descartes denied to any body, as again we shall later see.[62]

Over and above the problems that arise from attempting to link the branches of knowledge Descartes is ordering with those that the scholastics were concerned to categorize, there is the fact that many scholastic discussions of the order of the sciences are concerned with very different issues than concern Descartes. For him, the question is, in some sense, the epistemological interconnection of the branches of knowledge, what rests on what, and what sequence must be followed in order to discover and learn the sciences. But matters are not so clear with the scholastics, who seem to be concerned with categorization in a number of different dimensions, theoretical versus practical, abstract versus concrete, less certain versus more certain. Even when the proper order of teaching is under discussion, one must be careful to remember that considerations relating to practical pedagogy are often intermixed with more theoretical considerations relating to the epistemological (and metaphysical) interdependencies of some sciences on others.[63]

But realizing the difficulties that abound in making any such comparison between Descartes' own considered view and the views he was exposed to at La Flèche, it is fair to say that his view of the order of knowledge may well have represented a significant departure from the mainstream of the scholastic tradition. Though there are many differences between different scholastic writers,[64] there was wide agreement that knowledge of physics is largely independent of knowledge of metaphysics, however precisely either discipline is defined. And so, they claimed, one can (and, in fact, ought to) study physics before undertaking the more elevated studies of God and being as such that pertain to first philosophy. In demanding that metaphysics must *precede* physics, in arguing that physics as Descartes understands that discipline must be grounded in some sense in metaphysics, in knowledge of God and the soul, Descartes is stepping clearly outside of the tradition.[65] And so when by the mid 1630s Descartes came to hold to the priority of metaphysics in the strong sense he held it, his view would likely have been recognized as a clear departure from the received view.

But however important and revolutionary this view may have been for Descartes, and however important it may be to our understanding of his position in the history of philosophy,[66] it is interesting to point out that if it was a revolution, it went largely unnoticed by those against whom it was directed. While the phenomenon of the scholastic reaction to Descartes is still largely unstudied, a preliminary glance at the scholastic writers of the seventeenth century who reacted against Cartesian thought suggests that they were much more concerned with the details of his program, the radical doubt, the denial of substantial forms and final causes, than they were with the perhaps more radical conceptual shift that underlies those doctrines.[67] Furthermore, the doctrine of the priority of metaphysics was unnoticed or unwelcomed by many of the party of the moderns, both Cartesians and non-Cartesians, whom it was supposed to place in the ascendancy. Descartes certainly had followers in seeing the necessity of grounding physics in some way in metaphysics; Spinoza, Malebranche, and Leibniz can all, in their very different ways, be seen as continuing this particular line he started. But others who considered themselves his followers took a different view. Henricus Regius, once Descartes' disciple, published in 1646 his *Fundamenta Physices,* a work clearly of Cartesian inspiration; Descartes appears prominently in the dedicatory letter, and, he claims, the book contains nothing that is not derived from his own thought. But, he wrote in the preface to the French *Principles,* "because he copied [my thoughts] badly, changed the order, and denied certain metaphysical truths on which the whole of physics ought to rest, I am obligated to disavow it entirely" (AT IXB 19; see also AT IV 625–27). And, indeed, from Descartes' point of view the *Fundamenta* is badly organized. Regius begins with the foundations of physics, and it is only much later, after discussions of cosmology, astronomy, meteors, minerals, and animals, that the mind and its ability to discover the world are taken up.[68] God is never treated, indeed, is explicitly excluded from consideration, and plays no substantive role in the development of the physics, in radical contrast to what we shall find to be the case in Descartes' own system.[69] Jacques Rohault, an influential later Cartesian physicist and author of the *Traité de physique* (1671), a textbook of Cartesian physics, also goes against Cartesian orthodoxy on this point. Although he does begin with a very brief account of knowledge of the self and knowledge in general (see Rohault 1969, 1:3–12), God appears nowhere. Furthermore, the preface contains a diatribe against metaphysics and in the body of the work, Rohault explicitly endorses a hypothetical mode of reasoning, collecting phenomena, making conjectures, and testing them against experience.[70] The content is clearly Cartesian, but the dependence of physics on metaphysics Descartes himself took such pains to emphasize is all but ignored. And contemporary mechanist system builders outside the Cartesian tradition, like Gassendi and Hobbes, for example, felt no need to ground their physics in metaphysics; Gassendi treats God long after the foundations of his physics, and Hobbes seems to leave God out altogether. Both delay consideration of the nature of the mind until after the physics of inanimate bodies.[71]

Though unnoticed by some and unappreciated by others, the program and the conception of order on which it is grounded was of central importance to Descartes himself, and in order to understand his metaphysical physics we must see it in this context.

BODY: ITS EXISTENCE AND NATURE

IN THE PREVIOUS chapter we outlined Descartes' project, a complete body of knowledge, with knowledge understood in the strictest sense, organized hierarchically, beginning with metaphysics and progressing through physics to all other branches of knowledge, medicine, mechanics, and morals. Now it is time to turn to the foundations of Descartes' physics proper, and how that physics grows from and is nourished by its metaphysical roots.

Properly speaking the project begins roughly where the *Meditations* do, with the elimination of prejudice and the establishment of the existence of the self as a thinking thing through the *Cogito* and related arguments, and continues with the proof for the existence of God and the validation of clear and distinct perception through God's veracity. Some elements of these portions of Descartes' metaphysical project will, indeed, enter into our account of his physics. But given the general familiarity of this part of the Cartesian corpus, we shall begin at the ground-line of the tree of knowledge, with Descartes' views concerning the nature of body and its existence. Then, in chapters 4 and 5 we shall examine how his view of the makeup of the physical world relates to that of his two principal seventeenth-century adversaries, the scholastics and the atomists.

BODY AND EXTENSION: SUBSTANCE, ATTRIBUTE AND MODE

Descartes' view, of course, is that the bodies that exist in the world are extended things and extended things alone, the objects of geometry made real; while they can properly be said to have broadly geometrical properties like size, shape, relative position, and motion, they lack all of the sensory qualities like heat, cold, taste, and color that we are often inclined to attribute to them.

It is impossible to say precisely when Descartes came to hold this view. While it may go back as far as his days with Beeckman, it is first clearly suggested only in Rules 12 and 14, rules probably composed in the last stage of composition of the *Rules for the Direction of the Mind* in the late 1620s. There he claims that "shape, extension, motion, etc." comprise the simple natures in terms of which bodies are to be understood, and that "body" and "extension" signify the same idea for us, at least in the imagination *(phantasia)* (AT X 419, 444).[1] The geometrical view of body is clearly what he has in mind in *The World* he built in the early 1630s. The matter Descartes imagines God to create in the imaginary spaces is "a true body, perfectly solid, which equally well fills all of the length, breadth, and depth of this great space in the midst of which we have halted our thought" (AT XI 33). He goes on to say, as in Rule 14, that he does not want "to distinguish [body] from its own quantity and from its outward extension, that is, from the property it has of occupying place," and that "the quantity pertaining to the matter I describe differs no more from its substance than number from things numbered" (AT XI 35, 36; cf. Pr II 8). He concludes with a characterization of this "perfectly solid body" whose existence he posits: "I conceive of its extension, or the property it has to occupy space not as an accident, but as its true form and its essence" (AT XI 36). Though he is cautious about expressing this view in the *Discourse,* this view of body clearly persists throughout the rest of his career.

Descartes uses a number of terms interchangeably when discussing body: beside body, he uses the terms corporeal or extended substance, material substance, and matter, among others. In his early writings, he seems relatively casual about his metaphysical terminology. In the *Rules,* for example 'substance' appears only twice, both in contexts where he is not concerned to characterize body (or mind) in any general terms, and the terms 'essence,' 'attribute,' and 'accident' don't appear at all.[2] The term 'mode,' later to be central to Descartes' metaphysical vocabulary appears often. But it is almost always used in its nontechnical sense, meaning 'way' or 'fashion'; only once does he talk about something as a 'mode of body,' hardly grounds sufficient for attributing to him a fixed metaphysical vocabulary (AT X 446 l. 26). Matters are a bit more settled in *The World.* There he gives the following general characterization of the matter whose existence he posits: "I conceive of its extension, or the property it has to occupy space not as an accident, but as its true form and its essence" (AT XI 36)[3] The idea that extension is the essence or nature of body is found also in the *Meditations* (AT VII 63; cf. AT III 423 [K 111]); this, indeed, becomes one of the standard ways that Descartes expresses his view of body as geometrical object. But even in the *Meditations* it is not the only terminology he uses. Responding to the *Sixth Objections* he characterizes extension as the 'ratio' of body, using another scholastic term equivalent to essence (AT VII 440).[4] And even at the time of the *Meditations* his terminology for notions like specific sizes, shapes, and motions that genuinely pertain to body without being essential to it is entirely unsettled. He often uses the standard scholastic term 'accident' to characterize shapes, sizes, and motions (AT III 502; AT VII 48, 161, 175–76, 359; AT IXA 216), but he also feels free to use the terms 'quality' (AT II 367 [K 61]; AT VII 161), 'property' (AT VII 161), 'attribute' (AT VII 161, 222, 223, 360), and, with increasing frequency, 'mode' (AT VII 45, 433, 440, 444; see also AT III 503–4, 508).[5] Descartes seems at this time uninterested in adopting a standard terminology.

But some time in 1641 or 1642, while drafting Part I of the *Principles of Philosophy,* the situation changes. As part of the project of making his philosophy appropriate for teaching in the schools, Descartes attempted to give a clear exposition of the metaphysical framework into which his conception of body (and mind) fits, clear definitions and explanations not unlike the sorts that a student was likely to find in a manual like Eustachius', the manual he had at hand. This involved a clearer definition of substance than he had given thus far, and the singling out of the terms 'attribute' and 'mode,' once virtually interchangeable in his philosophical vocabulary, to play special roles in the metaphysical framework.[6]

While there is more than one strain of thought in Descartes' conception of substance, the view he seems to settle on by the time of the *Principles* is one in accordance with which substance is characterized in terms of the capacity for independent existence: "By substance we can understand nothing but a thing that exists in such a way that it requires nothing else in order to exist" (Pr I 51).[7] Substances, of course, are not recognized directly: "But yet we cannot initially notice a substance from the fact that it is a thing that exists, since substance alone and through itself does not affect us. But we can easily become acquainted with it through any of its attributes, by way of that common notion that that which is nothing has no attributes, or no properties or qualities" (Pr I 52).

Now, among the attributes (used here in a loose sense), properties, or qualities a thing has, one is special: "And indeed a substance can be known from any of its attributes. But yet there is one special property of any substance, which constitutes its nature and essence, and to which all others are referred" (Pr I 53L).[8] This special property is extension in body, and thought in mind. All other notions "are referred" to this special property insofar as it is through the notion of extension that we understand size, shape, motion, etc., and it is through the notion of thought that we understand the particular thoughts we have, he claims.

This special property is what Descartes in the 1640s comes to call the 'principal attribute' of a substance, or just its 'attribute' (Pr I 53, 56, 62; AT III 567; AT IV 348–49 [K 186–87]; AT VIIIB 348). Descartes claims that there is only what he calls a distinction of reason between a substance and its principal attribute, and that we cannot have a clear idea of a substance apart from its principal attribute:

> Thought and extension can be regarded as constituting the natures of intelligent and of corporeal substance, and then they must not be conceived otherwise than as thinking substance itself and extended substance itself, that is, as mind and body; in this way they will be understod most clearly and most distinctly. Indeed it is easier for us to understand extended substance or thinking substance than substance alone, omitting the fact that it thinks or that it is extended. For there is some difficulty in abstracting the notion of substance from the notions of thought or of extension, which indeed differ from it only by reason [i.e., through a distinction of reason]. (Pr I 63; see also Pr I 62, Pr II 8–9)

But though substance cannot be understood without an attribute, body without being extended, and mind without thought, body and mind can be understood without certain particular sizes or shapes or thoughts; while a body cannot fail to be extended, it need not be spherical or five feet long or in motion from New York to London. These are what Descartes comes to term 'modes'. Writing somewhat later than the *Principles,* he explains to a correspondent: "And so, properly speaking shape and motion are modes of corporeal substance, since the same body can exist now with this shape, now with another, now with motion, now without, though, on the other hand, this figure or this motion cannot be without this body. And so love, hate, affirmation, doubt, etc., are true modes in mind" (AT IV 349 [K 186–87]; see also Pr I 56, 61). While he does not abandon other terminology altogether,[9] 'mode' becomes his favored term for talking about shape, size, motion—things that pertain to material substance in nonessential ways.

The picture that Descartes sometimes paints is rather simple: all that is in body is simply comprehended through extension, just as all that is in mind is comprehended through thought; these properties constitute the "nature and essence" of the substance in question, and all other properties are to be "referred" to those principal attributes (see Pr I 53). But in reality, the picture is a bit more complex. In Rule 12 of the *Rules,* Descartes divides the simple natures we can comprehend into three classes, those that are purely intellectual (thought, doubt, ignorance, etc.), those that are purely material (shape, extension, motion, etc.), and those that are common to both the intellectual and the material (duration, existence, unity, etc.; see AT X 419). His conception of these common notions is amplified and developed in more detail during the 1640s, when he attempted to set his thought out in a systematic way. In the *Principles,* for example, "substance, duration, order, number" are listed among the most general notions, notions "that apply to all kinds of things" (Pr I 48).[10]

To appreciate the significance of Descartes' views on these common notions, let us examine his treatment of the notion of duration more carefully. Descartes closely links duration to the notion of substance itself. Duration is, he claims, an attribute of finite substance (Pr I 56),[11] and as with the attribute of extension in bodies and thought in mind, there is only a distinction of reason between substance and duration: "Since any substance that ceases to endure also ceases to be, there is only a distinction of reason between it [i.e., a substance] and duration" (Pr I 62).[12] And just as there would be no space or place were there no bodies, as we shall see in more detail below in chapter 5, there would be no duration were there no finite substances.[13] More suggested to Descartes on 5 March 1649 that were God to destroy the world and later create another one, "that period which was between worlds *[intermundium]* . . . would have its duration, whose measure would be in so many days, years, or centuries" (AT V 302). Descartes replied: "I think that it implies a contradiction for us to conceive of any duration coming between the destruction of the prior world and the creation of the new one. For if we relate this duration to the succession of divine thoughts or the like, we will commit an intellectual error, and not have a true perception of anything" (AT V 343 [K 249–50]).[14] The connection between substance and duration seems to hold in a sense even for geometrical objects. In Meditation V, discussing the geometrical ideas that constitute the bare idea I have of body, Descartes attributes to them motion, and through motion, duration, in addition to continuity, extension, number, size, shape, and situation (see AT VII 63). This, of course, is not to say that the geometrical objects he is concerned with in that context have actual duration in the way a table or chair or rational soul does. But still, for Descartes, the idea of a sphere, an object of geometrical reasoning, is the idea of an enduring thing, even if it may not, as a matter of fact exist and endure formally.[15]

Interpreted in the strictest sense, then, Descartes' claim that all accidents of body are "referred" to its principal attribute alone, extension, is false; enduring through a certain period of time may be an accident of a particular body, though it is not comprehended through extension at all, in the way that shape or size are.[16] And if we take this to imply, as he is often interpreted, that all real properties of body are modes of extension in the narrowest sense, then this would also seem to be false. But even though the letter of his philosophy may be violated, the spirit remains intact. Though not all accidents of body are modes of bare extension, ways of being extended *simpliciter,* they must all be ways of being an *extended substance.* That is to say, all accidents must be the sort of thing that could pertain to a purely geometrical object. But as an object, even the geometrical object must involve, at very least objectively, notions like duration, existence, number, and unity.

The terminological framework of substance, attribute, and mode, the framework that Descartes came to adopt to express his views in the 1640s, was an interesting and significant departure from the standard logical and metaphysical terminology common to a wide variety of scholastic writers. Put briefly, the distinction basic to the scholastic metaphysical framework is the distinction between substance and accident. As Eustachius put it, "substance is defined as a being in and of itself; an accident though is a being in another. . . . Moreover the subject of an accident is a substance" (1648, part 4, p. 52; cf. part 1, pp. 41f, 47). Now, a substance is intimately linked with certain accidents, those that constitute its form or nature or essence.[17] These characterize the substance, and should the substance lose any of those accidents, it would not be the same substance.[18] But those accidents that are not part of the essence of a substance bear a different relation to the substance. Nonessential accidents are, as it were, glued onto substances which are, in and of themselves, complete. And so, Thomas writes, "that to which an accident is added is a being complete in itself, subsisting in its own being, which being naturally precedes the added accident."[19] And so Eustachius can say that "a subject can be altogether deprived of its accidental form, even if none takes its place, as when air is deprived of light, since it does not cease to be a complete thing without it."[20]

Thus far there is nothing in the framework that Descartes could not have used for his own purposes. But what makes the scholastic metaphysical framework not altogether congenial to Descartes' purposes is the rather loose connection between the substance and its essence on the one hand, and its nonessential accidents on the other. Scholastics recognized certain accidents which while nonessential, are connected in important ways to the essence of a substance. For example, risibility, the capacity for laughing, while not essential to humans, properly speaking, is found in all humans as a consequence of their nature, it was thought; such accidents were called *propria,* or properties.[21] Other accidents, like the very act of laughing, while not found in all humans at all times, are found only in things having the nature of a human being; such things are called *accidentia propria,* or proper accidents.[22] But other run-of-the-mill accidents need bear no special relation to the essence; while such accidents must be understood as being in *some* substance, they are not necessarily conceived through the *essence* of that substance. But this is quite inimical to Descartes' way of thinking about body and substance. Descartes wants to make *all* of the properties of body geometrical. And so for him, modes, the term he comes to choose in preference to accident, are to be the expression of the very essence or attribute, in his terminology. For him, no accidents are to be merely tacked onto a substratum; for him all accidents are to be either *propria* or *accidentia propria,* intimately linked to the essence of the substance; as Descartes put it, all of the modes of body must be understood through its essence, its principal attribute, extension (see Pr I 53). And thus, when he claims that the essence of body is extension, he is not merely saying that all bodies have extension and necessarily so, as his scholastic contemporaries might have meant such a claim; he is saying something stronger, that everything that can really be attributed to body as such must be some way or another of being an extended thing.

Given this vision, it is not surprising that Descartes chooses "mode" over "accident." The term 'mode,' *modus* in Latin, means quite simply "way"; it is a perfect term to choose to express the fact that all accidents are *ways* of something being extended or being a thinking thing, *ways* of expressing the essence of a thing. In emphasizing the necessity of a link between essence and accident, attribute and mode, Descartes is saying something his teachers would not have, and so departs from their terminology. But while his language is a departure, it is not altogether original. I can find no precedent for his use of 'attribute' in the precise sense in which he uses it in his philosophy.[23] But 'mode' was a term already in general use, as Descartes knew full well.[24] However, one use of the term is of particular importance here. The term 'mode' had, by the late scholastic period, become a term of art within the theory of distinctions, an increasingly elaborate set of categories to deal with metaphysical and theological problems connected with such questions as the relation between being and essence, the attributes of God, and the three persons of the Trinity. At the two extremes were the real distinction, the distinction between two substances capable of existing apart from one another that Descartes appeals to in connection with the distinction between mind and body, and the distinction of reason, the distinction between two things that can neither exist apart from one another nor be conceived clearly apart from one another, such as body and extension (Pr I 60, 62).[25] While a number of proposals were made for distinctions intermediate between the two extremes, the one Descartes favored was the modal distinction, the distinction between a substance and its modes, or between different modes inhering in the same substance (Pr I 61).[26] In this context, Suarez, perhaps one of

Descartes' sources for the notion, characterizes a mode of quantity as "something bestowed upon *[afficiens]* it that, as it were, ultimately fixes its state, and its way *[ratio]* of existing, without adding a genuinely new entity, but only modifies a preexisting one."[27]

Descartes seems first to have worried about the theory of distinctions in any explicit way in the *Objections and Replies* to the *Meditations,* after Caterus raises a question about the proof for the distinction between mind and body in terms of the traditional theory of distinctions (see AT VII 100). Descartes' somewhat confused response to this question, a response which he finds he must correct later in the *Principles,* shows that it is a matter to which he had not given much thought, up until then.[28] My suspicion is that Descartes, dealing in the early 1640s with the theory of distinctions for the first time since his student days, saw in the notion of a mode as explicated in the context of the modal distinction an ideal notion around which to systematize his own metaphysics, a notion better suited to express his general view of the relation between a thing and its affectations and his rather more specific view of the relation between body and its properties than the more standard framework of substance and accident.[29] By adding to the notion of a mode the correlative notion of an attribute, Descartes was able to express his view of the metaphysics of mind and body in terms which, though not altogether traditional, had, at least, a familiar ring to those who taught and studied at the schools.

THE EXISTENCE OF BODY

Now that it is reasonably clear what Descartes' view is, we must turn to an examination of the grounds on which he defended it. It is only in the 1640s with the appearance of the *Meditations* and *Principles,* when Descartes turns to the canonical development of his philosophical system, that his view of body gets any serious and systematic argument. Consequently, it is on these relatively late texts that our account must center.

Our treatment of Descartes' arguments for a geometrical conception of body must begin with his argument for the real existence of body. Although he talked at great length about bodies throughout his career, it is only fairly late, in the *Meditations* of 1641, that he attempted to prove that bodies exist as real things, outside of minds and outside of God. As one might expect, given the complex structure of the *Meditations,* where later moves are based on arguments and positions carefully set up in earlier passages, the argument he gives in Meditation VI is difficult to disentangle from the rest of the work; in presenting the argument in isolation we shall have to make frequent reference to points Descartes makes earlier in the *Meditations*. But, in bare outline, the argument proceeds as follows.[30]

"Now there is in me a certain passive faculty for sensing, that is, a faculty for receiving and knowing the ideas of sensible things. But I could make no use of it unless a certain active faculty for producing or bringing about those ideas were either in me or in something else." So the argument begins. Descartes' strategy is to show that the active faculty in question is not in me (i.e., my mind), or in God, or in anything but bodies. "This [active faculty] cannot be in me, since it plainly presupposes no intellect, and these ideas are produced without my cooperation, and, indeed, often involuntarily," he writes. This step depends on the doctrines established earlier in the *Meditations*. First of all, it depends on the claim that the mind has only two faculties, intellection and volition, a doctrine suggested in the account of judgment in Meditation IV, and argued explicitly elsewhere in the corpus, in the *Principles* and in the *Passions of the Soul* (see AT VII 56f; Pr I 32; PS 17f). Second of all, it depends on the claim that all modes of mind must be understood either through intellection or through volition (see Pr I 32). Now, since the active faculty in question "plainly presupposes no intellect," it cannot be a mode of intellect, and since ideas of sensation are produced involuntarily, the active faculty cannot be in my volition. Thus, it cannot be in me at all. "Therefore it remains that it is in some substance different from me. . . . This substance is either body, or corporeal nature, namely, that which contains formally everything which is in the ideas [of bodies] objectively, or it is, indeed, in God, or some other creature nobler than body in which it [i.e., corporeal nature] is contained eminently." To show that bodies really exist, Descartes will eliminate the latter two possibilities, and show that the active faculty must be in bodies themselves. The argument proceeds as follows:

> And since God is not a deceiver, it is completely obvious that he did not give me these ideas immediately and through himself, nor even with the help of some creature in which their objective reality is contained not formally but only eminently. For since he plainly gave me no faculty for knowing this, but, on the contrary, gave me a great propensity for believing that they derive *[emittere]* from corporeal things, I do not see how he could be understood not to be deceitful if they derived from anything but corporeal things. Thus, corporeal things exist.

This final step depends, of course, on the veracity of God, the claim that God is not a deceiver as established in earlier *Meditations*.[31] It is interesting, though, how it is used in this argument. Descartes does *not* say that we clearly and distinctly perceive that bodies as such are the

71

cause of our sensory ideas, appealing to the principle of clearness and distinctness established in Meditation IV through God's veracity. Our sensory ideas must be caused by bodies because we have a "great propensity" for believing that they are, and God has given us no faculty to discern otherwise. And so, in this case at least, where God has given us no means to correct our natural inclination, we must trust it, and in this case, we are entitled to believe that our sensory ideas are caused by an active faculty in bodies.

This, in essence, is the argument Descartes offers. It is a complex argument, one that connects up with many elements of his program, and I do not pretend to have given anything like a complete account of its intricacies. For my purposes, though, the argument is mainly of interest as the point of departure into Descartes' physics, and it is to those aspects of the argument that we shall soon turn. But before doing so I would like to digress for a bit and examine some variants of this argument found in slightly later texts, the Latin *Principles* of 1644 and the French *Principles* of 1647.

In the *Principles of Philosophy* the proof for the existence of bodies comes at the very beginning of the second part, entitled "on the principles of material things," the part that begins the physics proper with a discussion of body, vacuum, motion, and the laws of motion. In the Latin version the argument proceeds as follows:

> Now, it can scarcely be doubted that whatever we sense comes to us from some thing which is distinct from our mind. For it is not in our power to bring it about that we sense one thing rather than another; rather, this [i.e., what we sense] plainly depends upon the very thing that effects our senses. But we can ask whether that thing is God, or whether it is different from God. But since we sense, or rather, impelled by sense, we clearly and distinctly perceive a certain matter extended in length, breadth, and depth, various parts of which are endowed with various shapes . . . , if God directly and through himself displayed to our mind the idea of this extended matter, or even if he brought it about that it were displayed by some thing in which there were no extension, shape, or motion, one could not come up with a reason why he should not be thought to be a deceiver. For we clearly understand that thing as something plainly different from God and from us (that is, different from our mind) and also we seem to ourselves clearly to see that its idea comes from things placed outside of us, things to which it [i.e., the idea] is altogether similar, and, as we have already observed, it is plainly repugnant to the nature of God that he be a deceiver. And thus, in all, we must conclude that a certain

thing, extended in length, breadth, and depth, having all of the properties which we clearly perceive to pertain to an extended thing, exists. And it is this extended thing that we call body or matter. (Pr II 1L)

The argument Descartes offers here is, in its basic strategy, very close to the argument offered in Meditation VI. As in the earlier argument, bodies are established as the cause of my sensations through eliminating first myself, then God, then something distinct from God or body as possible causes of my idea. And as in the Meditation VI argument, a major role is played by the belief I have that my sensory idea of body proceeds from a body itself. But there are also some interesting differences. One obvious difference is the role my beliefs about the origin of my sensory ideas play in the argument. In the Meditation VI argument, the belief in question is described as a "great propensity for believing," and it is justified by appeal to the consideration that God, who is not a deceiver, has given me no means to correct my propensity toward belief; the validation of this propensity points directly toward the more general validation of at least certain teachings of nature that Descartes will undertake later in Meditation VI.[32] In the version from the *Principles,* though, the belief that bodies cause my sensations is characterized as a *clear* belief, and the appeal to God the nondeceiver is an apparent appeal to the validation of clear and distinct perceptions.[33] It is not obvious whether Descartes has changed his view, or whether (as seems more likely to me) he simply does not want to raise the question in the *Principles,* oriented as it is toward more scientific questions. But the general validation of propensities, inclinations, and the teachings of nature that, in the *Meditations,* supplements the more austere principle of clear and distinct perception, has little explicit role to play in the *Principles;* in the *Principles* the epistemology is more strictly an epistemology of clear and distinct perceptions. When sensations are discussed in the *Principles,* they are not partially validated through the no-correcting-faculty argument, as they are in Meditation VI, but through an inquiry into what it is in them that is clearly and distinctly perceived.[34]

But as important as this epistemological difference is, there is another more subtle difference, a difference that, I think, raises one of the most difficult questions with which we shall have to deal later in this book. The argument in Meditation VI starts with the observation that I have "a certain passive faculty for sensing"; what we seek is the active faculty that causes the sensations I have, and the ultimate conclusion is that that active faculty is found in bodies. But, interestingly enough, in the argument of the *Principles,* there is no appeal to an active faculty.

Indeed, the terminology Descartes uses to describe the relation between our sensation and the body that is the object of that sensation seems studiously noncausal; we all believe, he tells us, that "whatever we sense comes to us *[advenit]* from something which is distinct from our mind," that the idea of body "comes from *[advenire]* things placed outside of us." It is difficult to say for sure why the two arguments differ in this respect, and one should always be open to the explanation that, as he suggests in a number of places, metaphysical issues are taken up in the *Principles* in a somewhat abbreviated and simplified fashion, and that the *Meditations* must be regarded as the ultimate source for Descartes' considered views in that domain.[35] But it is tempting to see in this variation the shadow of an important philosophical question he was facing.

As we shall see in later chapters, when we discuss the laws of motion and their grounds in God, a serious question can be raised about the activity of bodies in Cartesian science. A certain picture of motion, in accordance with which God is the active agent directly responsible for the motion of bodies, shall turn out to be basic to the derivation of Descartes' laws of motion, as we shall see in some detail below in chapters 7–9. But, this suggests, if it is God that is responsible for moving bodies about, then bodies are strictly speaking passive, and contain no genuine causal efficacy of their own. Now, as we shall later see, while many of Descartes' followers drew this conclusion quite explicitly, it is not at all clear how he himself stood on the issue. But the change I have noted from the argument as found in Meditation VI to the parallel argument as found in Pr II 1L suggests at least some concern over the question. It is possible that he eliminated the reference to an active faculty precisely because he was no longer certain that bodies could correctly be described as active causes of our sensations. The language he substitutes is, of course, consistent with bodies being active causes of sensations, as he may well have believed; but it is also consistent with a weaker view, on which our sensations *come from* bodies, but with the help of an agent, like God, distinct from the bodies themselves, which, in the strictest sense, are inert.

The concern I have attributed to Descartes here is suggested further by a variant that arises between the Latin version of Pr II 1, which we have been discussing, and the French version published three years later in 1647. In the Latin, the crucial phrase reads as follows:

> We seem to ourselves clearly to see that its idea comes from things placed outside of us. (Pr II 1L)

In the French translation, the phrase reads:

It seems to us that the idea we have of it forms itself in us *on the occasion of* bodies from without. (Pr II 1F; emphasis added)

One must, of course be very careful drawing conclusions from variants between the Latin text and Picot's French translation; while, as we shall later see in connection with the laws of impact treated in part II, some changes are clearly Descartes', it is often unclear whether a given change is due to him or to his translator. But this change is consistent with the trend already observed between Meditation VI and Pr II 1L, and weakens the causal implications further still. Rather than asserting that the idea *comes from* the thing, the French text says only that it "forms itself in us on the occasion of bodies from without." Furthermore, while it is by no means clear how to interpret the word *occasion* in Descartes' vocabulary, the word is certainly suggestive of what is to become a technical term in later Cartesian vocabulary, that of an occasional cause, a cause whose effect is produced through the activity of God.[36] At the moment, though, I mean to draw no definite conclusion, but only to raise an issue that we shall have to face later.

OBJECTIVE REALITY AND THE NATURE OF BODY

We have seen Descartes' arguments for the real existence of bodies. But, one might well ask, why does he consider it important to offer such an argument? Part of his motivation for proving the existence of body is, no doubt, just what he says it is, to establish with certainty something that had been called into question by the skeptical arguments with which his enterprise began. In the *Principles,* just before offering his proof, he wrote: "Although there is no one who is not sufficiently persuaded that material things exist, however, since a short while ago we called this into doubt, and counted it among the prejudices of our earliest years, it is now necessary that we seek out the reasons by which it can be known for certain" (Pr II 1).

But the argument has another function as well. The bodies Descartes shuffles out of his world in Meditation I, bodies which come up from time to time in the course of the first three *Meditations,* are the bodies of common sense, bodies known to me through my senses and, like the piece of wax examined so carefully in Meditation II, endowed with scents, tastes, and tactile qualities (AT VII 30).[37] But when in Meditation VI the existence of bodies is proved, and the furniture removed by hyperbolic doubt in Meditation I is replaced, it has undergone a significant transformation. The sensual bodies we started with have been replaced by the lean, spare objects of geometry. He concludes the argument for the existence of body in Meditation VI as follows:

And thus corporeal things exist. However, they do not exist altogether as they are comprehended by sense, perhaps, since that comprehension of the senses is in many ways very obscure and confused; but indeed, everything we clearly and distinctly understand is in them, that is, everything, generally speaking, which is included in the object of pure mathematics. (AT VII 80; see also the parallel passage in Pr II 1)

In this way the argument for the existence of body plays a central role in establishing the nature of the bodies that exist outside of us.

How the argument establishes the nature of bodies in the world seems relatively straightforward. In the version of the argument given in the *Principles,* the conclusion follows from the claim Descartes makes that "we seem to ourselves clearly to see that its idea [i.e., the sensory idea of a body] comes from things placed outside of us, things to which it [i.e., the idea] is altogether similar." Since, of course, his God is not a deceiver, we must be right, and the bodies from which our sensory ideas proceed must be "altogether similar to our ideas of them." Here we must be careful to note that when he says that bodies are altogether similar to our ideas, he doesn't mean to assert that bodies resemble everything we are initially inclined to include in our ideas of bodies. As the concluding sentences of the passage make clear, what bodies resemble is that which is *clear* in our conception of body, or better, bodies contain "all of those properties which we clearly perceive to pertain to an extended thing." The idea of ours that bodies resemble is, thus, the idea we have, purged of those properties that we do not clearly perceive as belonging to body as such, and this idea is just the idea of "a certain thing extended in length, breadth, and depth," he claims (Pr II 1L); it is *this* idea that we are taken clearly to perceive proceeds to us from something it exactly resembles.

The point is made similarly in Meditation VI, using the language of objective and formal reality, first introduced in Meditation III in the course of the argument for the existence of God. As Descartes uses the terms, to exist *formally,* or to have formal reality is, simply, to exist; in this sense ideas considered as modes of mind exist formally in the mind, and rocks and trees exist formally in the world (see AT VII 40–41). But to exist objectively, or to have objective reality, is to exist in the mind as the *object* of an idea. As he explains the notion to Caterus in the *First Replies:* "'Objectively existing in the intellect' means . . . being in the intellect in the way in which its objects are normally there, so that the idea of the sun is the sun itself existing in the intellect, not formally, of course, as it does in the heavens, but objectively, that is, in the way objects normally exist in the intellect" (AT VII 102–3).[38] And lastly, to exist *eminently* in something is for the object in question to exist in something superior in perfection to it, not as an object represented, but as an object which the thing in question is capable of creating, either formally or objectively. It is in this sense that bodies exist eminently in God (AT VII 41, 79–80, 161).[39] Now, the argument proceeds, since I have a great propensity to believe that my sensations proceed from bodies, that is, from some substance "which contains formally everything which is in the ideas [of bodies] objectively" (AT VII 79), and since God has given us no faculty for correcting this propensity, no means for determining that they come from some substance, like God, that contains eminently that which is contained objectively in the ideas of bodies, bodies must exist. But, of course, the bodies whose existence the argument proves are not the bodies of common sense, but the bodies which are the formal correlates of the bodies that Descartes supposes to exist objectively in our minds, bodies, he claims, all of whose properties are broadly geometrical.

In this way the arguments for the existence of body are intended to establish that the bodies that exist in the world, outside of our minds and God's, are bodies understood in a certain very special way, things, substances, all of whose properties are broadly geometrical. But it is obvious that this argument rests on establishing that our idea of body is as Descartes says it is, that what we perceive clearly and distinctly in our idea are the geometrical properties of bodies alone, that the body that exists objectively in the mind is a geometrical object and that alone.

There are, I think, three distinct sorts of argument that Descartes uses to establish this conclusion, what I shall call the argument from elimination, the argument from objective reality, and the complete concept argument. We shall examine each of these arguments successively.

THE ARGUMENT FROM ELIMINATION

What I call the argument from elimination is suggested first in Meditation II, in connection with the wax example. Descartes begins by considering the wax, fresh from the honeycomb; it has the taste of honey and the scent of flowers, a certain color, size, and shape, it is hard and cold and makes a sound when tapped (AT VII 30). But put it by the fire, and it looses its shape, changes its size, becomes liquid and hot, changes its color and smell, and no longer makes the sound it did. But yet it is the same piece of wax. He concludes:

> What, therefore, was in that which I understood so distinctly? Certainly none of those things which I obtained through the senses, for whatever came under taste or odor or sight or touch or

hearing has now changed, while the wax remains. . . . Let us be attentive, and, having removed what does not pertain to the wax, let us see what remains: indeed nothing but something extended, flexible, changeable. (AT VII 30–31)

The nature of the wax, it appears, is arrived at by eliminating all of those properties that can be changed, without eliminating the wax altogether; it is because the wax can loose its smell, taste, color, etc., that these do not pertain to the nature of the wax, but only being extended, flexible, and changeable do.

This is obviously suggestive of the conclusion Descartes reaches about the nature of body, as numerous commentators on the *Meditations* have noted. But to what extent does it establish that conclusion, or is it intended to establish that conclusion? In the Synopsis of the *Meditations* he suggests that it does, and included Meditation II along with Meditations V and VI as passages that contribute to the formulation of a "distinct concept of corporeal nature" (AT VII 13). And there is no question but that the wax example gives the reader of the *Meditations* a foretaste of the full doctrine of body, and prepares the mind to accept the idea that bodies may really be only extended things. But it should be remembered that the primary purpose of the wax example is something quite different. Earlier in Meditation II he began his reconstruction of the world with the *Cogito* argument, establishing the existence of the self through a direct intuition of the intellect, without the help of the senses. But his common sense meditator finds this puzzling, since he is inclined to think that the first thing that we can know are bodies, and that these are known through the senses (AT VII 29). The wax example is introduced to show that since sensation leads us more directly and with more certainty to the mind than it does to the wax, the mind can be, and indeed should be known before we know that bodies exist (AT VII 33), and that bodies, strictly speaking, are known through the intellect, and not through the senses or the imagination (AT VII 31, 34).[40] Furthermore, it should be noted, the account Descartes gives of the wax is not the account he will later give of the nature of body as such; while the wax is, like all bodies, extended and changeable, it is also flexible, something that not all bodies are, it would seem. And so, he points out in reply to Hobbes: "I did not prove anything with the wax example but that color, hardness, and shape do not pertain to the formal definition *[ratio formalis]* of that wax. And there I was not concerned with the formal definition of the mind, nor, indeed, that of the body" (AT VII 175).[41]

But despite the ambiguities of its status in the *Meditations,* an argument very much like the one given in connection with the wax example comes up twice in the *Principles.* The argument comes up first in Pr II

4, where it is clearly presented as an argument for the claim "that the nature of matter, or of body regarded in general does not consist in the fact that it is a thing that is hard or heavy or colored or affected with any other mode of sense, but only in the fact that it is a thing extended in length, breadth, and depth." In the argument that follows, Descartes considers the case of hardness *(durities)*, and argues that even if we imagined bodies to recede from us when we try to touch them, so that "we never sensed hardness," things "would not on account of that lose the nature of body." He concludes by claiming that "by the same argument it can be shown that weight and color and all of the other qualities of that sort that we sense in a material body can be taken away from it, leaving it intact. From this it follows that its nature depends on none of those qualities" (Pr II 41L).[42] The argument is taken up again in Pr II 11, where it gets a slightly different treatment. There he defends the claim (which we shall take up in detail below in chapter 5) that "it is the same extension that constitutes the nature of body and the nature of space." Before making that claim, though, he reviews the claim that the nature of body is extension, and the argument he chooses to emphasize is a version of the argument by elimination. He considers as a concrete case the idea of a stone, and undertakes to "reject from it everything that we know is not required for the nature of a body." He first rejects hardness, now because a stone melted or pulverized would not be hard, and would, yet, be a body. Color is rejected because there are transparent stones. Heaviness is excluded because fire, though light, is still thought to be a body. And finally, cold, heat, and "all other qualities" are rejected since they are either not considered to be in bodies, or since they can be changed without thereby eliminating corporeal nature from a stone. Descartes concludes this portion of the argument by noting that "nothing plainly remains in its idea but the fact that it is something extended in length, breadth, and depth." This, presumably, is what "constitutes the nature of body" (Pr II 11).

This simple argument from elimination neither begins with Descartes nor ends with him.[43] But to what extent does it advance the Cartesian program? One point worth making is that from the point of view of the teachers in the schools that Descartes was attempting to address in the *Principles,* whom he was trying to convince to adopt his ideas, there is an obvious flaw in the argument. Many would certainly agree with Descartes, following Aristotle, that length, breadth, and depth are notions that pertain to all bodies as such.[44] But from the fact that extension is *inseparable* from body, it doesn't follow that it is *essential.* In addition to the essence of a thing, scholastic logicians recognized what they called inseparable properties (risibility in humans) and inseparable accidents (whiteness in a swan; see Toletus 1614, ff. 33r,

34r; Gracia 1982, pp. 253–54). While these properties and accidents are found in all individuals of the relevant sort, and, in the case of inseparable properties, necessarily so, it was denied that they are essential in the strict sense.

But while such a criticism may be telling for a scholastic adversary, it is not fair, perhaps, to criticize Descartes from the point of view of an intricate theory of predication not altogether his own.[45] But there is another more serious problem with the argument in the context of the Cartesian system: while it may establish that the essence of body is extension in *some* sense, it doesn't establish that body is extension in the sense Descartes needs for his program.

Descartes is attempting to establish that the bodies that exist objectively in the mind and thus the bodies that exist formally outside of the mind are bodies *all* of whose properties are broadly geometrical; what exists in the world of bodies is "a certain thing extended in length, breadth, and depth" (Pr II 1L),[46] *and that is all,* a geometrical object that lacks all of the purely sensory qualities, like color, taste, heat, and cold that many are inclined to impose on it. This is an extremely important claim for Descartes, one that is intended to ground his mechanistic program for the sciences, as we shall see in more detail in later chapters. But it is not a claim that the argument from elimination can support, I think. What Descartes needs to establish is that our idea of body is the idea of a thing whose *only* properties are geometrical, a thing that *excludes* all other properties. But what emerges from the argument from elimination is the idea of body as a thing at least *some* of whose properties are required to be geometrical. From the fact that there are transparent bodies it follows that there are bodies without color, and thus, it follows that having color cannot be essential to our idea of what it is to be a body. But from the fact that *some* bodies are not colored it does *not* follow that *no* body is really colored, any more than it follows from the fact that some bodies are not spherical that no body is really spherical. At best, this argument can be said to have established that bodies are essentially extended in the weaker Aristotelian sense of the term, but not in the stronger sense Descartes' own program demands. In this way the argument from elimination must be regarded as insufficient for Descartes' purposes.

THE ARGUMENT FROM OBJECTIVE REALITY

A second kind of argument to establish that the idea of body is the idea of an extended thing alone is suggested in Meditation V, whose title clearly indicates that it is supposed to deal with "the essence of material things."[47] There, having discussed mind and God in earlier meditations, Descartes finally turns his attention to an investigation of bodies. To begin such an investigation, he argues, "we ought to consider their ideas insofar as they are in my thought, and see which of them are distinct, and which confused." When we make such an examination, he claims, we find that what is distinct in our ideas of body is "the quantity that philosophers commonly call continuous, or the extension of its quantity, or, better, the extension of the thing quantized, extension in length, breadth, and depth" (AT VII 63). But on what grounds does this claim rest? What strikes Descartes as significant about the geometrical features of the idea of body is the fact that one is capable of performing *proofs* about them, and discovering properties they have that we didn't impose on them:

> I also perceive by attending to them innumerable particular [features] concerning shape, number, motion, and the like, whose truth is so evident and in harmony with my nature, that although I uncover them for the first time, I seem not to be learning them anew as much as remembering what I already knew, or, it seems that I am noticing for the first time what was already in me, although I had not previously turned my mental glance upon them. (AT VII 63–64)

He notes, for example, that one can demonstrate various properties of the triangle, "namely, that its three angles are equal to two right angles, that its greatest side is subtended by its greatest angle, and the like, which whether I want to or not I now clearly know, when I imagine the triangle, even if I may not have thought of them before in any way. And thus [these properties] were not made by me" (AT VII 64).

The geometrical features of our ideas of body thus appear themselves to have real properties that are true of them whether we notice them or not, properties that, he claims, we do not merely impose upon them. And so, Descartes argues, these features, features that are the subject matter of geometry, must in *some* sense really exist: "I find within myself innumerable ideas of certain things which, even though they may exist nowhere outside of me, cannot, however, be said to be nothing; and although in a sense they can be thought of at will, they are not formed by me, but have their own true and immutable natures" (AT VII 64).

The argument offered here is obscure. But matters are somewhat clarified when, in a slightly later passage from Meditation V, Descartes explicitly gives the premises on which this reasoning depends: "it is obvious that whatever is true is something, and I have already demonstrated at length that everything I know clearly is true" (AT VII 65). What he seems to have in mind is an argument like this. From Meditation IV we know that a clear (and distinct) perception is one in which the will is completely determined to judgment by the intellect alone, a judgment over which the will has no choice, strictly speaking (see AT VII 58–60). And so, from the fact that my mind is forced to acknowledge certain geometrical truths concerning such ideas as shape and size I find in my ideas of body, and does not make those judgments arbitrarily and impose them on my ideas, it follows that it clearly perceives those propositions. But, as Descartes established in Meditation IV, whatever is clearly (and distinctly) perceived is true. So, there then must be genuine truths about such ideas as shape and size. But, he seems to hold (and this is the crucial premise introduced only in Meditation V), "whatever is true is something," or, to put it more rigorously, and in the form that he seems to use it, whatever is true is true *of something*. That is, the truth of a proposition entails the existence, in some sense, of the objects it concerns.[48] Now, at this stage, in Meditation V, we do not know whether or not anything (except God) exists outside of the mind. So, he concludes, the objects must exist in the mind.[49] And consequently, objects with geometrical properties must exist in the mind, not, of course, as they exist outside the mind, as they might (and, by Meditation VI, do), but as objects normally exist in the mind, as objects of ideas, as objective realities. And so, Descartes takes himself to have established, the idea of body has, as its object, at very least something that really has geometrical properties, sizes, shapes, etc., which exist in the idea of body and in the mind as objective realities, genuine objects that exist in the mind in the way objects normally exist in it.

An important question is, of course, what this shows. At very least it shows that our idea of body is an idea of something with geometrical properties, that we clearly perceive that bodies have properties such as size and shape. But despite the title of Meditation V, "On the Essence of Material Things . . . ," we are still some distance from establishing the nature of these objects that have been found to exist in the mind. While we may know that they possess geometrical properties, we don't know, on the basis of the argument of Meditation V, how bodies stand with respect to an assortment of other properties. As Descartes is fully aware, there are other ideas often connected with the idea of body, ideas of color, smell, taste, and even pain. He acknowledges that, at least before careful reflection on the matter, we all tend to think that bodies really have the sensory qualities that invariably accompany their geometrical qualities, and have them in just the same sense that they have shape, size, and the like. As he puts it in the *Principles*:

There is none of us who has not from our earliest years judged that everything which he sensed is something existing outside of his mind, plainly similar to his senses, that is, similar to the perceptions he has had of them. And so, seeing, for example, color, we have thought that we see a certain thing placed outside of us, and plainly similar to that idea of color which we then experienced in us. (Pr I 66; see also Pr I 46; AT VII 38)

So regarded, the ideas of color, sound, taste, etc., seem to represent to us something existing in body, and thus appear to pertain to the object of our idea of body, the object existing in our mind in the way objects normally exist in the mind. Ultimately Descartes wants to establish that these sensations we habitually attribute to bodies don't belong to bodies at all, but to minds;[50] ultimately he wants to establish that what pertains to the idea of body, clearly and distinctly apprehended, what belongs to the bodies existing objectively in the mind are geometrical properties and geometrical properties alone. But the argument of Meditation V by itself can give us no such conclusion, it would appear.

Descartes may have thought that sensory qualities had been eliminated from bodies before the argument of Meditation V, in Meditation III, where he attempted to establish that the sensory ideas we have are, upon inspection, confused and obscure in the sense that it is not clear to us what they represent or, for that matter, whether they are representative at all. He makes this argument in the context of the main proof for the existence of God in Meditation III. The main strategy of that argument is to show that I have an idea, the idea of God, that could only have been caused in me by God himself. And in the course of that argument, he considers, class by class, the principal sorts of ideas he has, among which fall the ideas of sensation. About them Descartes makes the following remarks:

Others [i.e., other ideas], moreover, such as light and colors, sounds, smells, tastes, heat and cold, and other tactile qualities, these I think of only in a very confused and obscure way, so that I am even ignorant of whether they are true or false, that is, whether the ideas I have of them are ideas of certain things or of non-entities. . . . For example, the ideas which I have of heat and cold are so little clear and distinct that from them I cannot learn whether cold is only a lack [privatio] of heat, or heat a lack of cold, or whether both or neither are real qualities. (AT VII 43–44; see also Pr I 68)[51]

In responding to Arnauld's comments on this passage in the *Fourth Objections,* Descartes restates the point in terms of the language of objective reality:

It often happens in obscure and confused ideas (among which we must number the ideas of heat and cold) that they are attributed [*referentur*] to something other than that of which they are really ideas. And so, if cold is only a privation [*privatio*], the idea of cold is not cold itself, insofar as it exists objectively in the intellect, but something else which we mistakenly take for that privation, namely, a certain sensation which has no being outside of the intellect. (AT VII 233)

In the context of Meditation III Descartes is not, of course, entitled to say everything he says to Arnauld—that is, that the idea of cold is only a sensation existing in the mind. But the general point is still appropriate, that from an inspection of the idea of cold, heat, or any other sensations, it cannot be determined what, if anything, they represent in bodies, and whether or not they represent any genuine feature of the bodies that exist in the mind as the object of our idea of body.

In Meditation III (and, indeed, in the *Fourth Replies* and the *Principles,* where similar claims are made) the claim appears to rest on introspection. But using the style of argument Descartes introduces in Meditation V, one can advance some further grounds for the view. In Meditation V, he argues that our ideas of bodies objectively contain objects with geometrical properties because we are able to perform geometrical proofs. But outside of certain trivial propositions ("heat is not cold," "red is not yellow," etc.), no such proofs can be made concerning the ideas of sensation. And so, one might argue, this is an indication that our ideas of sensation do not represent to us objectively real properties of bodies in the way geometrical ideas do.[52]

But as important as these observations may be, they don't seem sufficient to establish the strong conclusion that Descartes wants to establish, that the bodies that exist objectively in our mind and formally outside of our mind are geometrical objects alone, things that we can be certain exclude the nongeometrical sensory qualities that the uninitiated are inclined to attribute to bodies. From the argument of Meditation V it follows that the bodies that exist objectively in the mind really have their geometrical properties, and from the argument of Meditation III it follows that from a bare consideration of our ideas we cannot establish that sensory qualities like heat and cold, color and taste belong to bodies. The confusion and indistinctness Descartes calls attention to in our sensory ideas can lead us seriously to consider the *possibility* that many of the qualities we characteristically attribute to bodies are really just states of mind that we impose on bodies, and don't really exist in the bodies themselves, considered either objectively or formally. (This, indeed, may be all that the argument of Meditation III is intended to do.) But so far as I can see, there is nothing in his observations about our sensory ideas that supports the stronger claim that our idea of body actually *excludes* such sensory qualities.

THE COMPLETE CONCEPT ARGUMENT

The two arguments we have examined so far have proved somewhat disappointing; neither the argument from elimination nor the argument from objective reality establishes with appropriate rigor the conclusion Descartes ultimately wants to establish, that the idea we have of body is the idea of a purely geometric object, and thus, through the argument for the existence of body, that the bodies that exist outside of mind are purely geometrical objects, the objects of geometry made real. But the synopsis of the *Meditations* points to three places in the *Meditations* which contribute to "a distinct concept of corporeal nature," the Second, Fifth, and Sixth Meditations (see AT VII 13). The two arguments we have examined so far are suggested in Meditations II and V. We must, then, turn to Meditation VI for the final clarification of the idea of body.

The passage in Meditation VI that Descartes no doubt has in mind in the synopsis is closely linked with the argument for the distinction between mind and body and immediately precedes the Meditation VI argument for the existence of body, the argument that issues in the conclusion that bodies are geometrical objects and geometrical objects alone. The strategy there is to show first that minds and bodies are different kinds of things, distinct substances, and then to show that sensory qualities like heat, cold, color, and taste can pertain only to mental substances, while size, shape, motion, and the like can pertain only to bodies.

In order to show that mind and body are distinct substances, Descartes must establish that they are capable of existing apart from one another; this, for him, is what it means to be a separate substance (AT VII 44, 162).[53] Descartes' strategy for establishing this is to claim that we can clearly and distinctly conceive mind, the subject of thought, as discovered in Meditation II, independently of body, and body, conceived as an object containing only geometrical properties, independently of mind (see AT VII 78). If this can be established, then he can appeal to God's omnipotence to establish that mind and body are genuinely distinct, since "everything I clearly and distinctly understand can be brought about by God in the way in which I understand it." The argument thus concludes as follows: "Since on the one hand I have a clear and distinct idea of myself insofar as I am only a thinking thing, and not extended, and on the other hand [I have a clear and] distinct idea of body, insofar as it is only an extended thing and not a thinking thing, it is certain that I am really distinct from my body, and can exist without it" (AT VII 78). From this Descartes proceeds to distinguish the two substances, mind and body, from their modes. He first notes that he has certain special faculties of thought, imagination and sensation, "without which I can clearly and distinctly understand myself as a whole, but not vice versa, [that is, I cannot understand] them without me, that is, without some intelligent substance in which they are, for they include some intellection in their formal concept, whence I perceive that they are distinguished from me as modes from a thing" (AT VII 78). Descartes then treats motion, shape, and the like in a similar way. These must be modes since extended substance can be conceived without them, but they cannot be conceived without extended substance, and it is extended, not thinking substance they belong to because "something of extension, and plainly nothing of intellection is included in their clear and distinct concept" (AT VII 79). And so, he claims, sensory qualities belong to mind alone, to a substance distinct from body, which contains only geometrical properties. And with the idea of body thus clarified, its object and its geometrical properties distinguished from the mind and its thoughts, he can proceed with the argument to show that bodies so conceived, objects that contain geometrical properties and geometrical properties alone, really exist.

The argument is, in a sense, all there, or, at least, its elements are all there. But the argument as given in the body of the *Meditations* is not altogether satisfactory. Given the crucial importance of the claim, it is disappointing for Descartes simply to declare without argument or explanation that he can clearly and distinctly conceive mind without body and body without mind, and that despite what we might previously have thought, the contents of sensation and imagination simply belong to mind and not to body.[54] But in response to dissatisfactions with the argument, Descartes expanded the argument and offered fuller explanations of his claims.

The explanations Descartes gives, mainly in the *Replies* appended to the *Meditations* and in letters from the years immediately following the publication of the *Meditations,* suggest that the following sorts of considerations lie behind the abbreviated version of the argument given in Meditation VI. When we examine the ideas we have, we notice that some of them are incomplete and depend for their full comprehension on other ideas we have. Writing to Gibieuf on 19 January 1642, Descartes noted:

> In order to know if my idea has been rendered incomplete or inadequate by some abstraction of my mind, I examine only if I

haven't drawn it . . . from some other richer or more complete idea that I have in me through an abstraction of the intellect. . . . Thus, when I consider a shape without thinking of the substance or the extension whose shape it is, I make a mental abstraction. (AT III 474–75 [K 123])

And so, Descartes writes, responding to Arnauld's objections to the *Meditations:* "For example, we easily understand shape without thinking of a circle . . . but we cannot understand any specific difference of circle without at the same time thinking about shape" (AT VII 223; see also AT VII 120–21). In this sense our ideas of particular shapes depend upon the idea we have of shape in general, Descartes would claim, insofar as we cannot have a genuine understanding of the one without the other. Following out this series of conceptual dependencies, from specific (circular) to more general (having shape), we are led ultimately to the idea of a thing that has the appropriately general property, since, Descartes holds, "no act or accident whatsoever can exist without being in a substance" (AT VII 175–76).[55]

Now, the ideas we have of incomplete things, properties, accidents, modes, sort themselves out into two distinct classes. Some depend upon the notion of extension in general and extended substance, insofar as they all presuppose the notion of extension, and others depend upon thought in general and thinking substance insofar as they all presuppose the notion of thought. Answering Hobbes Descartes wrote:

There are certain acts, which we call corporeal, such as size, shape, motion, and all others which cannot be thought of without local extension, and the substance in which they exist we call body; nor can we imagine *[fingi]* that it is one substance which is the subject of shape, and another which is the subject of local motion, etc., since all of those acts agree in the common concept *[communis ratio]* of extension. Next there are other acts, which we call acts of thought, such as to understand, to will, to imagine, to sense, etc., all of which agree in the common concept of thought or perception or consciousness *[conscientia].* And the substance they are in is a thinking thing, or a mind. . . . (AT VII 176)[56]

And so, Descartes observes, again to Hobbes, "acts of thought have no relation to corporeal acts, and thought, which is their common concept, is altogether distinct from extension, which is the common concept of the others" (AT VII 176). Thus Descartes concludes that the ideas we have of mind and of body, the ultimate subjects of the two kinds of properties, do not depend upon one another for their conception; they are ideas of distinct things that can be conceived apart from one another. In fact, he argues, they are ideas of complete things that do not depend upon any other ideas at all for their distinct conception. Responding to Caterus in the *First Replies* Descartes wrote:

> I cannot completely understand motion apart from the thing in which there is motion . . . nor, in the same way, can I understand justice apart from the [person who is] just. . . . But I completely understand what a body is by thinking of it only as being extended, shaped, mobile, etc., denying of it everything that pertains to the nature of mind, and vice versa I understand that mind is a complete thing which doubts, which understands, which wills, etc., even though I deny that there is in it any of those things contained in the idea of body. (AT VII 120–21; see also AT VII 176, 223, 444; AT III 474–75 [K 123–24])

It is important to note that what terminates the two series of incomplete concepts, those that agree in the common concept of thought and those that agree in the common concept of extension, is neither the bare concept of thought or the bare concept of extension, nor is it the concept of a bare thing, but the complete concept of a *thinking thing* or the complete concept of an *extended thing*. In separating thought or extension from the idea of a subject that is thinking or extended, we "destroy all of the knowledge we have of it," Descartes tells Arnauld (AT VII 222). Similarly, in the *Principles* he remarks:

> When [others] distinguish substance from extension or quantity, they either understand nothing by the name 'substance,' or they have only a confused idea of an incorporeal substance, which they falsely attribute to corporeal substance, and leave for extension (which, however, they call an accident) the true idea of a corporeal substance. And so they plainly express in words something other than what they understand in their minds. (Pr II 9)[57]

Since our basic complete concepts are those of a thinking thing and an extended thing, in the *Replies* of 1641 Descartes rejects out of hand the possibility of a single thing, a single substance (as opposed to a composition of two distinct substances) that is both thinking and extended (AT VII 223, 423–24, 444).[58]

And so, the careful examination of my ideas and the way they depend on one another shows that they all contain one or another of two "common concepts," thought or extension, and thus depend for their intelligibility on one or another of two basic notions, that of a thinking thing and that of an extended thing, notions that don't depend on one another or on anything else. It is this doctrine on which the argument of Meditation VI rests. Since an extended thing and a thinking thing are conceivable as complete things apart from one another, God could create them apart from one another, and thus they are distinct things, with distinct sets of properties. In particular, the things that Descartes found existing objectively in the mind in Meditation V turn out, at long last, to be things whose *only* properties are geometrical. And insofar as it is bodies so conceived that, we are inclined to believe, are the source of our sensory ideas of body, it is bodies so conceived that exist in the world, Descartes concludes. The bodies of physics are thus the objects of geometrical demonstration made real.

This, in any case, is the complete concept argument as Descartes gives it ca. 1640–42, at roughly the time of the composition of the *Meditations*. After that, though, the view undergoes an interesting change, a change apparently motivated by considerations relating to mind-body unity and the intelligibility of mind-body interaction.

Descartes' clear position throughout his mature writings, at least, is that although the mind and body are genuinely distinct, they are united in the human being, and form a genuine individual. In Meditation VI he wrote, "Nature also teaches me through the sensations of pain, hunger, thirst, etc., that I am not in my body only as a sailor is in a ship, but I am joined with it as closely as possible, and as it were, intermixed with it, so that with it I compose a single thing" (AT VII 81; see also AT VI 59, AT III 493 [K 127–28], AT III 692 [K 141]). The position, repeated with some frequency, is that in human beings, mind and body are united not *per accidens* but *per se,* and form a genuine *substantial union* (see AT III 493, 508 [K 127, 130]; AT VII 228). Indeed, Descartes suggests, the mind is the *substantial form* of the body, the only such substantial form he will recognize in nature, as we shall see in chapter 4 (see AT III 503, 505).[59] And while he seems clear in a wide variety of texts that sensations and imaginations pertain to the mind (AT VII 78; PS 17, 19, 23), he is absolutely clear that they pertain to the mind *only* because it is joined in this way to the body: "human minds separated from the body do not have sensation, strictly speaking" (AT V 402 [K 256]; see also AT III 479 [K 125–26]; Pr II 2). What precisely this unity comes to, how to understand phrases like unity *per se* and *substantial union,* and, most importantly, how the doctrine of the union relates to the doctrine of the distinction between mind and body are questions that have puzzled both his contemporaries and later generations of commentators.[60] Descartes himself seems not to have seen a contradiction here, nor should he have. The argument for the distinction between mind and body establishes that they are separate substances in the precise sense that they are capable of existing apart from one another; because we have complete concepts of mind and of body, God can create the one without the other and the other without the one. But at least during our lifetimes, this is not a capability that is actually exercised; while alive, we are genuine individuals composed of these two separable but not separated components (AT VII 228, 423–25, 444–45; Pr I 60). And, at least according to a principal strand in his thought, as expressed in the *Meditations,* even in the mind-body unity, our ideas divide into two distinct classes, those that agree in presupposing the natures of extension and thus pertain to the bodily component, and those that agree in presupposing the notion of thought and thus pertain to the mental component of the union (AT VII 78–79; 176; 423–25; 444–45). This is true even though there are some ideas, those of sensation, that owe their very existence to the unity of mind and body (AT III 479 [K 125–26]). And it is because of this division among ideas that we can know that mind and body are really distinct.

But Descartes' thought on this shifted in a subtle and interesting way. By May 1643, almost two years after the *Meditations* first appeared in print, Descartes added a third basic notion to that of mind (thought) and body (extension), that of the *union* of mind and body. Writing to Elisabeth in the celebrated letter of 21 May 1643, Descartes explained:

> First I consider that there are in us certain primitive notions which are, as it were, the originals on the model of which *[sur le patron desquels]* we form all of our other thoughts. And there are only a very few such notions; thus, after the most general, being, number, duration, etc. . . . we have for the body in particular only the notion of extension . . . , and for the soul alone we have only that of thought . . . ; and finally, for the soul and body together, we have only the notion of their union, on which depends the notion of the force the soul has to move the body, and the body has to act on the soul in causing sensations and passions. I also consider that all human knowledge consists only in carefully distinguishing these notions, and in attributing each of them only to the things to which they pertain. When we want to explain some of these notions by another [we go wrong] . . . since being primitive, each of them can only be understood by itself. (AT III 665–66 [K 138])

The view, then, seems to be that the notion of a mind-body union is a notion co-equal with that of mind, and that of body, something that, like the notions of mind and body, must be comprehended through itself, and something that, like the notions of thought and extension, can constitute a common nature for a class of ideas.

This view is not reflected, interestingly enough, in the Latin version of the *Principles,* published the following year in 1644. As in the *Meditations,* ideas are divided into two main classes:

Moreover, I do not recognize more than two highest classes [*summa genera*] of things: one is the class of intellectual things, or of thinking things, that is, the class of things which pertain to mind or thinking substance; the other is the class of material things, or things which pertain to extended substance, that is, to body. (Pr I 48L)

This twofold distinction is also consistently maintained in the passages that follow (Pr I 53, 63, 65, etc.). But the Latin version of Pr I was almost certainly drafted at the same time as the *Meditations* and Descartes' *Replies* were being finished and readied for publication, and it is no surprise that they agree in this. However, the French edition of the *Principles* that was published in 1647 is slightly different than the Latin version, and may possibly reflect the apparent change in Descartes' thought found in the letters to Elisabeth. While there is not a major revision of the passages dealing with mind and body, the sentence quoted above now reads as follows:

And the principal distinction that I note between created things is that some are intellectual . . . and the others are corporeal. (Pr I 48F)

Weakened slightly, this suggests that thinking substance and extended substance, while constituting the *principal distinction*, may not be the *only summa genera* he recognizes. More striking is a change later in the same paragraph. The Latin reads thus:

But there are also certain other things we experience in ourselves, which ought not to be referred [*referri*] to mind alone or to body alone, and which . . . arise from the close and intimate union of our mind with the body. (Pr I 48L; see also Pr II 3)

The passage in the Latin is somewhat noncommittal; the things in question (sensation and appetite) are not to be *referred* (in what sense?) to mind or body alone, but *arise from* the union. The view here *seems* to be that the feelings in question, while they may *belong* to mind, arise from neither mind nor body alone, but from their union. The French, though, is somewhat different:

But there are also certain other things we experience in ourselves which ought not to be attributed [*attribuées*] to the mind alone nor to the body alone, but to the close union that there is between them. (Pr I 48F)

Here the claim seems to be that the feelings in question are to be *attributed* to the union of mind and body. This suggests the position

Descartes expresses to Elisabeth, that the mind-body union is a primitive notion, one separate from mind and body.[61]

The doctrine of the three primitive notions and the claim that mind-body unity is a primitive notion on a par with the primitive notions of thinking and extended substance raises problems for the theory of concepts on which rests the distinction between mind and body and the characterization of body in terms of extension alone. Most notably, it is difficult to see how the notion of mind-body unity can be primitive in the sense Descartes claims it is, understood through itself, since the concept of unity of mind and body would appear to presuppose the notions of mind and of body. But be that as it may, the later elaboration of Descartes' theory of concepts, the addition of the notion of unity and the suggestion that sensations may pertain to the unity of mind and body rather than to either taken separately, does not disrupt the argument of most importance to the foundations of Descartes' physics, the claim that our idea of body, properly considered, is the idea of a thing all of whose properties are geometrical, which is capable of existing apart from mind. For whatever we say about the state of union, we can still conceive of a complete thing all of whose properties are thoughts of one sort or another (whether or not sensations and feelings are to be included among such thoughts), and a distinct complete thing all of whose properties are geometrical, and this is all we need to start the argument that ultimately allows us to say that the bodies that exist without are geometrical objects alone.

But these considerations do raise an interesting question regarding the argument for the existence of body and the conclusion that Descartes apparently wants to draw from it, that the bodies that exist outside the mind are things all of whose properties are geometrical. Whether Descartes holds the notion of union to be a primitive or not, he certainly held through his mature writings that in at least some cases, minds and bodies are united and together form genuine individuals; though minds and bodies are separable, they are not always actually separated. And so, it seems, the argument for the existence of body can at best establish that our ideas of sensation result from extended things *capable* of existing apart from mind. But nothing in the argument itself establishes that our sensations result from things that *actually* exist apart from mind. Though we have a great propensity to believe that our ideas of sensation proceed from something that has formally all of the properties that our idea of body has objectively, i.e., geometrical properties, the argument apparently leaves open the question as to whether the bodies from which our sensations arise exist apart from or united to a thinking substance. And well it should, since, as a matter of fact, some of our sensations *do* proceed from bodies united with mind insofar as some of our sensations proceed from other human beings! While the view that bodies in general may exist united to minds may sound fanciful and far-fetched, it is not; indeed, it is Descartes' interpretation of the scholastic doctrine of the body in terms of substantial form and primary matter, as we shall see below in chapter 4. And insofar as the argument for the nature of bodies in the world that is supposed to fall out of the argument for the existence of body falls short of excluding the possibility that *all* bodies in the world external to mind may be united to mind, it by itself fails to ground a purely mechanistic physics in terms of size, shape, and motion, and it fails to eliminate a principal competitor to this program.

DESCARTES AGAINST HIS TEACHERS: THE REFUTATION OF HYLOMORPHISM

IN THE PREVIOUS chapter we investigated Descartes' account of body. The account of body as extension is one of the pillars of his physics. Because body is a substance whose essence it is to be extended, and because all the properties bodies have must be understood through extension and must consequently be modes of extension, Descartes claims, physics must ultimately consist in the explanation of the properties bodies have in terms of the size, shape, and motion of bodies. Writing in the *Principles* he notes, "I openly admit that I know of no other matter in corporeal things except that which is capable of division, shape, and motion in every way, which the geometers call quantity and which they take as the object of their demonstrations. And, I admit, I consider nothing in it except those divisions, shapes, and motions" (Pr II 64; see also Pr II 23). Establishing the laws of motion will take us beyond body to God, as we shall see below in chapters 7–9. But once the laws of motion are established, explanation in Cartesian physics proceeds in terms of extended substance and its geometrical modes; Descartes' project is to explain the phenomena of this visible world in terms of the sizes, shapes, and motions of the insensible particles that make up the bodies of everyday experience. I would like further to clarify Descartes' account of body and the explanation of its properties by contrasting it with two rival seventeenth-century accounts of what there is in the physical world, accounts that Descartes meant his own to oppose. First, in this chapter we shall consider the scholastic-Aristotelian view of body as matter and form, and the related common-sense view of body from which Descartes thinks the scholastic view derives. This view, what Descartes learned in school, is the most obvious target of his account of body. Then, in chapter 5, we shall consider Descartes' opposition to a variety of mechanism different from his own, the tradition of atomism, derived from Democritus and Epicurus, and coming very much into fashion at just the time that Descartes' own philosophy was being formulated and disseminated.

MATTER, FORM, AND THE SINS OF YOUTH

Hylomorphism, the ubiquitous scholastic account of body, is, not surprisingly, a doctrine of great complexity, with a variety of different schools holding a variety of different positions on the central issues. But, by and large, Descartes was not interested in the subtleties of scholastic thought. Writing to Mersenne in November 1640 Descartes notes with great confidence, "As for the philosophy of the schools, I don't consider it at all difficult to refute it because of the diversity of its opinions, since one can easily overturn all of the foundations on which they are in agreement with one another, and once that has been done, all of their detailed disputes will appear absurd" (AT III 231–32 [K 82]).[1] What interested Descartes were the foundations of scholastic thought, and what lay at the foundations was the doctrine of body in terms of (primary) matter and (substantial) form, the doctrine of hylomorphism.

Matter is, quite generally, the subject of properties of a thing, that which remains constant as a thing changes from one sort of thing to another. Taken in this general sense, matter includes, for example, the bronze from which a statue is made or the plastic from which a pair of sunglasses is fashioned, matter that though it lacks the properties that it will have when shaped by the sculptor or manufacturer, still has the properties that make it bronze or plastic.[2] But scholastic philosophers took the doctrine a step further and recognized what they called primary matter. Primary matter is matter regarded without any properties whatsoever, matter "regarded without any form, either substantial or accidental."[3]

The notion of a substantial form is somewhat more complex. Most simply, the form is that which, added to primary matter, results in a complete substance.[4] But more substantively, substantial form is that from which the characteristic behavior of the various sorts of substances derives, and thus that in terms of which their behavior is to be explained. And so, Descartes notes, writing to Regius in January 1642, helping him to formulate his attack on the scholastic Voëtius, "they [i.e., forms] were introduced by philosophers for no other reason but to explain the proper actions of natural things, of which actions this form is to be the principle and the source" (AT III 506 [K 129]). This conception of the notion of substantial form is also expressed by the Coimbrian Fathers in arguing for the necessity of form over and above matter:

Natural things are not composed of matter alone, since if that were so, a human being, a stone and a lion, being made of the same matter, would all have the same essence and definition. Therefore, in addition to matter, they have their own forms which differentiate them from one another. . . . There are individual and particular behaviors *[functiones]* appropriate to each individual natural thing, as reasoning is to a human being, neighing to a horse, heating to fire, and so on. But these behaviors do not arise from matter which, as shown above, has no power to bring anything about *[nullam effectricem vim habet]*. Thus, they must arise from the substantial form. . . . Consequently, one cannot deny to each and every natural thing its inherent substantial form, from which it is formed, by which degrees of eminence and perfection in physical compositions are determined, on which all propagation of things depends, in which the marks *[nota]* and character of each thing are stamped . . . and finally, which distinguishes and adorns the remarkable theater of this world with its variety and wonderful beauty.[5]

And so, for example, heaviness, the tendency some bodies have to fall toward the center of the earth (universe), is taken to be a quality (what Descartes often calls a real quality) they have by virtue of having the substantial form they do.[6] Thus the Coimbrian Fathers wrote:

Since heavy and light things tend toward their natural places, though absent from that which produces them, they must necessarily have been given some means *[instrumentum]* that remains with them by virtue of which they are moved. But this can only be their substantial form and what follows from it, heaviness and lightness.[7]

And so Descartes characterized the scholastic account of heaviness:

Most take it [i.e., heaviness] to be a virtue or an internal quality in every body that one calls heavy which makes it tend toward the center of the Earth; and they think that this quality depends on the form of each body, so that the same matter which is heavy, having the form of water, loses this quality of heaviness and becomes light when it happens that it takes on the form of air. (AT II 223)

One could go on at some length on scholastic notions of matter and form. But more important than that for our immediate purposes is coming to a better understanding of how Descartes himself understood the scholastic heritage, of what he thought the common foundation of their view was and, ultimately, why he thought it ought to be rejected.

As we shall later see in more detail, Descartes often claims to find the scholastic view simply unintelligible. In *The World,* for example, he re-marks that the schoolmen did such a good job in eliminating all of the qualities from matter, extension in particular, that "nothing remains that can clearly be understood" (AT XI 33; see also AT XI 35). Similarly, Descartes tells Regius in 1642 that the explanation of the behavior of bodies through substantial forms and the real qualities, like heaviness, that they are supposed to give rise to is explaining "that which is ob-scure through that which is more obscure" (AT III 507).

But despite the unintelligibility of the scholastic view taken at face value, Descartes sometimes acknowledges that the view is not entirely empty, and offers a reconstruction of what he thinks the view comes to. In the early 1640s, while finishing the *Meditations* and beginning work on the *Principles,* Descartes seems to have turned his attention toward the scholastic view that he intended to refute. Part of the motivation was, perhaps, connected with his original intention to include in the *Principles* an explicit discussion of or commentary on Eustachius' scho-lastic manual, as we discussed above in chapter 1; also, no doubt, the growing affair with Voëtius and Regius caused Descartes to turn to an examination of matter and form, an important aspect of the growing battle.[8] But whatever the reason for his attentions, the texts of those years suggest an interesting view of his scholastic opponent.

In writing to Regius in January 1642, Descartes characterizes the scholastic view as follows:

> Just so there won't be any verbal ambiguity, we must note here that when I deny them, I have understood by the name 'substan-tial form' a certain substance joined to matter, and with it com-posing something whole that is merely corporeal, and which no less than matter, indeed more than matter, is a true substance or a thing subsisting in and of itself *[per se].* (AT III 502 [K 128])

Substantial forms, then, are conceived of as substances of a sort, joined to matter to produce bodies. And the sort of substance they are, Descartes thinks, is *mental* substance, "like little souls joined to their bodies" (AT III 648 [K 135]).[9]

The mentalistic nature of the scholastic's substantial forms is made clear in the account Descartes gives of the scholastic theory of heavi-ness in the *Sixth Replies,* in an important passage that purports to repre-sent what Descartes thought in his youth, and how he came to reject the views he was taught in school. The historical veracity of his account may be questionable; there is no direct evidence that Descartes was inter-ested in carefully characterizing or systematically refuting substantial forms much before the early 1640s. But the passage does give us a good idea of how he thought about his scholastic opponent in the years when he seems most seriously to have been interested in opposing that view. Characterizing his earlier conception of heaviness he wrote:

> Now although, for example, I conceived of heaviness like a certain real quality which is in solid bodies, since I added that it is *real*, I really thought it to be a substance, although I called it a *quality* insofar as I attributed it to a body in which it inhered; in the same way clothing, regarded in and of itself, is a substance, though it may be a quality when it is attributed to a clothed person, and even the mind, though really a substance, can be called a quality of the body to which it is joined. And although I imagined heaviness to be diffused through the heavy body, I did not attribute to it the same extension which constitutes the nature of body. For the true extension of body is such that it excludes all interpenetration of parts, while I thought that the amount of heaviness in ten feet of wood is contained in a one-foot mass of gold or other metal; in fact, I judged that all of that heaviness could be contracted into one mathematical point. Indeed, while it remained coextensive with the heavy body, I saw that it could exercise its force in any part whatsoever, since from whatever part of the body it is hung by a rope, it draws the rope down with its entire weight *[gravitas]*, in the same way as if that heaviness *[gravitas]* were only in the part touching the rope, and not diffused through the remainder [of the body]. And it is now in just this way that I understand the mind to be coextensive with the body, and the whole [mind] to be in the whole [body], and the whole [mind] to be in any part of it you like. But what especially showed that the idea I had of heaviness was derived from that of the mind was the fact that I thought that heaviness bore bodies toward the center of the earth as if it contained in itself some knowledge *[cognitio]* of it [i.e., the center of the earth]. For this could not happen without knowledge, and there cannot be any knowledge except in a mind (AT VII 441–42).[10]

Heaviness, as conceived by the scholastics, is thus mentalistic in a number of ways. It is imagined to be diffused throughout a body, yet capable of acting on a single point, just like the Cartesian soul, which is somehow thought to be diffused throughout the human body while, at the same time, it is especially connected to the pineal gland (see, e.g., PS 30–31). Like the human soul, it is extended, not as bodies are extended, but by virtue of being able to act on body, what he calls an extension of power *(extensio potentiae)* a few years later in a letter to Henry More, in contrast to the extension of substance *(extensio substantiae)* (AT V 342 [K 249]; see also AT III 694 [K 143]). And finally, the notion of heaviness is mentalistic insofar as it appears to attribute to the heavy body a kind of volition or intention, the intention to bear the body toward a particular place, the center of the earth, something that could only happen if the real quality of heaviness had some knowledge of the center of the earth.[11] This last observation, an observation that Descartes himself considers most important, cuts right to the heart of the scholastic doctrine. If substantial forms and the real qualities that are supposed to follow from them are supposed to explain the characteristic behavior of bodies of various sorts, then we must be thinking of them as intentional entities, agents of a rudimentary sort, things capable of forming intentions and exercising volition, little souls joined to matter.

Indeed, Descartes thinks, the hylomorphic body of the scholastic philosophers, form and quality joined to matter, is just the image of the Cartesian human being, immaterial soul united to extended body, projected out onto the material world. This is a theme that comes up a number of times in the 1640s. When he is pressed to explain how the human mind and its body are joined to one another and how they can interact, as many correspondents sought to understand in the years following the exposition of that doctrine in the *Meditations*, he often compares his account with the scholastic account of form, quality, and matter. Writing to Arnauld, for example, in July 1648 he explains:

> Most philosophers, who think that the heaviness of a body is a real quality, distinct from the stone, think that they understand well enough how this quality can move the stone toward the center of the earth, since they think that they have evident experience of such a thing. But I, who persuade myself that there is no such quality in nature, and thus no true idea of it in the human intellect, I think that they use the idea that they have in themselves of an incorporeal substance to represent to themselves this heaviness, so that it is no more difficult for us to understand how the mind moves the body than it is for those others to understand how such heaviness bears a stone downward. It doesn't matter that they say that this heaviness is not a substance, since they really think of it as a substance insofar as they think that it is real, and that it can exist without the body through some power, namely, through divine power. (AT V 222–23 [K 235–35]; see also AT III 424 [K 112]; 667–68 [K 139]; AT IXA 213)

Descartes sometimes also takes the comparison in the other direction and argues that the human soul is "the true substantial form of man" (AT III 505), indeed, it is "the only substantial form" he recognizes (AT III 503; see also AT IV 346; AT VII 356; and Grene 1986).

Regarded in this way, Descartes has an interesting account of how the scholastic philosophy arose and why it had exerted such a powerful influence over so many for so long. The scholastic philosophy, he argues, is the philosophical elaboration of childhood error and prejudice.[12] Childhood, he argues, is a time when we are "governed by our appetites and by our teachers" (AT VI 13), when the faculty of reason is overwhelmed by the senses, and when we acquire the storehouse of prejudice that makes it difficult to apprehend the truth later in life. As he noted in the *Principles*, "And indeed in our earliest age the mind was so immersed in the body that it knew nothing distinctly, although it perceived much clearly; and because it nevertheless formed many judgments then, it absorbed many prejudices from which the majority of us can hardly hope ever to become free" (Pr I 47; see also Pr I 71, AT IV 114 [K 148], etc.). The most basic prejudice that Descartes thinks derives from youth is a prejudice in favor of the senses, the conviction that the faculties of sensation and imagination represent to us the way the world really is. He writes in the *Principles*, "Every one of us has judged from our earliest age that everything which we sensed is a certain thing existing outside his mind, and is plainly similar to his sensations, that is, to the perceptions he has of them" (Pr I 66; see also AT VII 74f). Or, as Descartes develops the theme in a later section of the *Principles:*

> In our earliest age, our mind was so allied with the body that it applied itself to nothing but those thoughts alone by which it sensed that which affected the body, nor were these as yet referred to anything outside itself. . . . And later, when the machine of the body, which was so constituted by nature that it could of its own inherent power move in various ways, turned itself randomly this way and that and happened to pursue something pleasant or to flee from something disagreeable, the mind adhering to it began to notice that that which it sought or avoided exists outside of itself, and attributed to it not only magnitudes, figures, motions, and the like, which it perceived as things or modes of things, but also tastes, smells, and the like, the sensations of which the mind noticed were produced in it by that thing. . . . And our mind has in this way been imbued with a thousand other such prejudices from earliest infancy, which in later youth we quite forgot we had accepted without sufficient examination, admitting them as though they were of the greatest truth and certainty, and as if they had been known by sense or implanted by nature. (Pr I 71; see also Pr I 73)

In our earliest years then, aware of only what the bodily faculties tell us, but through our dealings with the world, aware that there are things outside of our immediate control and thus outside of us, we came almost spontaneously to the belief in a world of external objects similar to our sensations. These judgments became so natural to us, Descartes thinks, that we confused them with the sensations themselves, and we came to believe that it is our sensory experience itself that gives us the belief in the way the world really is. As he wrote in the *Sixth Replies:* "In these matters custom makes us reason and judge so quickly, or rather, we recall the judgments previously made about similar things so quickly that we fail to distinguish between these operations and a simple sense perception" (AT VII 438). In this way, we come, in our adulthood, to put our trust in the senses as an accurate representation of the way the world is.

The prejudice in favor of the senses, the belief that the senses represent to us the way the world of bodies really is, gives rise to a multitude of prejudices. In an obvious way it leads us to think that "seeing a color, for example, we thought that we see something which exists outside of us and which plainly resembles the idea of that color which we then experience in ourselves" (Pr I 66).[13] Similarly, when we have a painful or pleasant sensation, this epistemological prejudice leads us to believe that it is "in the hand, or in the foot, or in some other part of our body" (Pr I 67; see also Pr I 68). Such prejudices also lead us to posit vacua where we sense no objects (Pr II 18; AT V 271 [K 240]), to think that more action is required to move a body than to bring it to rest (Pr II 26), and that bodies in motion tend to come to rest (Pr II 37). Common sense also leads us to posit certain other tendencies in bodies, the tendency to fall toward the center of the earth (AT VII 441f). More generally, Descartes thinks, in trusting the senses as accurate representations of the world of bodies, we are led to confuse mind and body, and as we saw in the case of heaviness, we are led to attribute properly mental qualities to bodies, as well as corporeal qualities, like extension, to minds.[14]

This, Descartes claims, is what we get from common sense. Certainly not all of these errors are shared by the schoolmen. For example, the schoolmen rejected the vacuum every bit as much as Descartes himself did. And furthermore, the scholastic physics is not the only program to derive support from the prejudices of common sense. For example, seventeenth-century atomists like Beeckman and Gassendi accepted the vacuum, and other anti-Aristotelians, like Kepler, thought that color was really in things.[15] But what appears to interest Descartes most is the way the common-sense view underlies the doctrine of hylomorphism, and, more generally, scholastic physics. Writing in 1641 Descartes tells a correspondent:

The first judgments we made in our youth, and later also the common philosophy [i.e., the scholastic philosophy] have accustomed us to attribute to bodies many things which pertain only to the soul and to attribute to the soul many things which pertain only to the body. They ordinarily blend the two ideas of the body and the soul, and in the combining of these ideas they form real qualities and substantial forms, which I think ought entirely to be rejected. (AT III 420 [K 109])[16]

Common sense attributes to bodies the qualities and tendencies to behave in particular ways that bodies appear to have, the properties our senses tell us bodies have. The scholastic philosopher takes this one step further, and posits in bodies substantial forms and real qualities, principles of action that are intended to explain the properties that sense tells us are in bodies. Since the qualities that sense attributes to bodies are largely mental qualities, the sensations and volitions of the mind itself, projected onto the physical world as colors, tastes, and tendencies, the forms and qualities must be "tiny souls," mental substances capable of receiving the properties that common sense attributes to them. The scholastic world is, thus, nothing but the world of common sense, with sensible qualities transformed into mental substances—forms and real qualities—and embedded in the world of bodies. Put briefly, the scholastic world, as Descartes understood it, is simply a metaphysical elaboration of the world of common sense.[17]

This is Descartes' view of his scholastic opponent, at least as of the 1640s or so, the common-sense sensualist turned metaphysician who created a world in his own image, a world whose basic properties are to be explained in terms of form and matter, tiny souls attached to extended bodies. There is, of course, a serious question about the extent to which the view he represents as that of his opponents corresponds to the views that they actually held.[18] This is an extremely difficult question to address due both to the as yet largely unexplored diversity of views among late scholastic thinkers and to Descartes' explicit unwillingness to enter into the discussions and debates among his adversaries in any serious way. His characterization of scholasticism is quite clearly a rational reconstruction, and probably also something of a caricature; Descartes, it seems, could not help seeing the schoolman's ontology through his own glasses, and imposing the ontology of thinking substance and extended substance onto his opponent's conception of form and matter. But rather than trying to sort out these issues and distinguish the truth from Descartes' possible falsifications, I would like to turn to a closely related question. The characterization he gives of his scholastic opponent is the first step in an argument intended to refute him, the rational reconstruction a prelude to a rational rejection. We must now examine how this refutation is supposed to work and why it doesn't.

DESCARTES AGAINST HIS TEACHERS: THE ELIMINATION OF FORMS

Descartes' opposition to hylomorphism, the doctrine of body as form joined to matter, is not always obvious. He often adopts the language of the schools to express his own views. Writing in *The World*, for example, he describes the property of having small rapidly moving parts as the "form" of his first element, and similarly uses the term 'form' to characterize the characteristic configuration of parts that makes up flame, on his account (see AT XI 26, 27).[19] He sometimes also suggests that his conception of matter is, in important respects, connected to that of the schools. Writing to Mersenne in October 1640 he notes:

> [My opponent] ought only to conclude that both salt and all other bodies are only of the same matter, which is in accord with both the philosophy of the schools and with mine, except for the fact that in the schools they do not explain this matter well insofar as they make it pure potentiality and join to it substantial forms and real qualities, which are only chimeras. (AT III 211–12)[20]

Furthermore, Descartes often explicitly minimizes his differences with the schools. His only mention of substantial forms and real qualities in the *Discourse* and *Essays* is to say that "I don't at all want to deny what they imagine in bodies over and above what I have discussed [i.e., their geometrical properties], such as their substantial forms and their real qualities" (AT VI 239 [Ols. 268]). And, despite his original intention to write an explicit answer to the philosophy of the schools, the doctrine of hylomorphism gets almost no explicit mention at all in the *Principles*.[21] Writing in 1644 to Father Charlet, assistant to the General of the Jesuits, sending him a copy of the recently published Latin *Principles*, Descartes noted that "some have also thought that my intention was to refute the views accepted in the schools and try to make them ridiculous, but one will see that I have spoken of them [in the *Principles*] no more than if I had never known them" (AT IV 141).[22] He also tells Charlet that "one will see here that I make use of no principle which was not accepted by Aristotle and by everyone else who has ever dabbled in philosophizing," a theme he repeats in the *Principles* themselves (AT IV 141).[23]

However, these considerations should not mislead us into thinking that Descartes was not a serious opponent of hylomorphism. While he

uses the terminology of form and matter on occasion, there is no suggestion that he thinks that the world is anything but mechanist. His
matter is not the primary matter of the schools but extended substance,
and when he uses the term 'form' it is clear that it is meant only in the
weak sense of 'essence' or 'nature'. The form of a particular sort of
body or quality is to be just the particular configuration of size, shape,
and motion that results in fire, say, or in heat; it is only in the case of
the human soul where the term 'form' has anything like its supposed
scholastic meaning, a principle distinct from matter which together
with matter produces an individual.[24] And Descartes minimizes his differences with the scholastics in his published writings not because he
thinks his position genuinely compatible with theirs, but because he
thinks direct confrontation neither necessary nor prudent.[25] Writing
probably to Huygens in June 1645, concerning a request from some
unknown supporters asking Descartes to attack the schools more
openly, he noted:

> I find it very strange that they want me to refute the arguments of
> the schools, since I think that were I to undertake this project, I
> would do them a bad service. For some time the maliciousness of
> certain persons has given me an opportunity to do this, and per-
> haps they will in the end force me to do it. But since those who
> are most concerned with the issue are the Jesuit Fathers, my re-
> spect for Father Charlet . . . and for Father Dinet and for certain
> other important members of their Society, whom I believe truly to
> be my friends, has caused me to abstain from it up until now, and
> also has caused me to write my *Principles* in such a way that one can
> say that they don't contradict the common philosophy at all, but
> only enrich it with numerous things it did not have before. (AT IV
> 225)[26]

Descartes' strategy is, as much as possible, to emphasize his agreement
with the schools, and simply to pass over his disagreements in silence.
It is in this spirit that he emphasizes to Charlet that he doesn't hold
anything Aristotle wouldn't. But this is a trifle disingenuous. In saying
this he means only to say that what he holds follows from the light of
nature, something common to all human beings, Aristotle included.
While Descartes may claim that Aristotle would agree with everything
that he, Descartes believes, he *doesn't* mean to say that he accepts everything *Aristotle* believed (see AT VII 580).

Despite Descartes' attempts to camouflage his position in his published writings, there is ample evidence that Descartes set his own mechanical philosophy in opposition to the forms and qualities of the
schools. Descartes characterizes them as chimeras (AT III 212), philo-

sophical beings unknown to him (AT II 364, 367 [K 61]; AT III 648–49 [K 135]), beings from which he shrinks (AT II 74). He emphasizes to his correspondents that the hylomorphic view of body derives from confusion and from the errors of youth (AT II 212–13, AT III 420, 435 [K 120], 667–68 [K 139]), claims that his philosophy destroys that of Aristotle and the scholastics (AT I 602–3; AT III 297–98 [K 94], 470), and, indeed, boasts that it is easy to refute his opponents (AT II 384, AT III 231–32 [K 82]). This feature of his position was evident to Descartes' contemporaries, both his supporters and his opponents. Substantial forms are a central focus of the debate between Voëtius and his followers and Descartes and his in Utrecht in the early 1640s, a debate that occupied much of Descartes' attention then.[27] Substantial forms are also at issue in the *Philosophia Cartesiana,* the Voëtius camp's response to Descartes in mid-1643 and in *The Letter to Voëtius,* Descartes' answer to that attack (AT III 599; AT VIIIB 62, 120).[28] The question of substantial forms is also prominent in a similar dispute over Cartesianism in Leyden. There, on 18 July 1643, Adriaan Heereboord, a professed Cartesian, presented theses in opposition to substantial forms, which were answered by Jacob Revius.[29] Forms came up again at Leyden a few years later in 1648, when Heereboord, involved in another battle for Cartesianism, characterizes the rejection of substantial forms as one of the central tenets of Cartesianism (AT V 128). Cartesianism was so associated with the denial of substantial forms that by the end of the century these (by then) vestiges of scholastic thought plot their revenge against Descartes, the philosopher whom they thought had treated them so badly. In a satire, perhaps by one Gervais de Montpellier, *Histoire de la conjuration faite à Stokolm contre Mr Descartes,* substantial forms and real qualities are represented as gathering together in a tribunal to sentence Descartes to death for his mistreatment of them.[30]

Given the centrality of the issue to the Cartesian program, it is curious that there is so little in the way of sustained and rigorous argument against substantial forms and real qualities in the entire corpus of Descartes' writings. There is an obvious preference for mechanical explanation early on. As I noted in chapter 1, the writings that survive from Descartes' brief apprenticeship with Isaac Beeckman show that he followed Beeckman in explaining phenomena in physics in mechanical terms. This feature of his strategy in physics follows him into the 1620s, the period in which he is known to have worked out some of his most important views in optics with the help of mechanical models of light and its interaction with media. This approach in physics is clearly reflected in the *Rules.* In Rule 9 Descartes advises the methodical investigator to look to simple mechanical analogies with bodies in motion in

order to understand how a natural power can be transmitted instanta-
neously or how a cause can, at the same time, give rise to opposite
effects, rather than appeal to little understood phenomena like magnetism, the supposed influence of the stars, or occult qualities (AT X
402–3). Furthermore, in Rule 12 he claims that the only simple notions
that pertain to body as such are limited to shape, extension, motion,
and the like (AT X 419). Since, he claims in Rule 12, "we can understand nothing but those simple natures and a certain mixture or composition of them" (AT X 422), this suggests that all knowledge in physics ultimately comes down to knowledge of different sizes, shapes, and
motions, suggesting that all explanation is mechanical. But despite the
clear preference for mechanical models and mechanical explanation
in this period, there seems to be nothing like an explicit attack on his
opponents, on forms, qualities, or the hylomorphic doctrine of body.
Though he is somewhat critical of scholastic logic and the general uncertainty of the philosophy of the schools, Descartes the scientist appears simply to be going his own way, and seems relatively unconcerned
to show his teachers the error of their ways in physics.[31]

The preference for mechanical explanations is also obvious in *The
World* of the early 1630s. But as Descartes promised Mersenne in a letter
from May 1630, forms and qualities are there dealt with in a more
explicit way (AT I 154 [K 16]). For example, when in chapter VI he
introduces the matter that he hypothesizes God to have created in the
imaginary world he is there constructing, he tells his readers, "let us
explicitly suppose that it does not have the form of earth, or of fire, or
of air, nor any other more particular form, like that of wood, or of a
stone, or of a metal, no more than it has the qualities of being warm or
cold, dry or moist, light or heavy" (AT XI 33). And when in chapter II
he discusses fire, he notes:

> When it [i.e., fire] burns wood or some other such material, we
> can see with our own eyes that it removes the small parts of the
> wood and separates them from one another, thus transforming
> the more subtle parts into fire, air, and smoke, and leaving the
> grossest parts as cinders. Let others [e.g., the philosophers of the
> schools] imagine in this wood, if they like, the form of fire, the
> quality of heat, and the action which burns it as separate things.
> But for me, afraid of deceiving myself if I assume anything more
> than is needed, I am content to conceive here [only] the motion
> of parts. (AT XI 7)

And when in chapter V he explains his three elements, distinguished
from one another by shape and size alone, he remarks:

If you find it strange that I make no use of the qualities one calls heat, cold, moistness, and dryness to explain the elements, as the philosophers [of the schools] do, I tell you that these qualities appear to me to be in need of explanation, and if I am not mistaken, not only these four qualities, but also all others, and even all of the forms of inanimate bodies can be explained without having to assume anything else for this in their matter but motion, size, shape, and the arrangement of their parts. (AT XI 25–26)

Two sorts of arguments against using the scholastic machinery in physical explanation are implicit in these passages. The first is what might be called the argument from parsimony. We can, Descartes claims, explain the burning of wood in purely mechanical terms, by appealing to tiny bodies in motion, breaking up the wood into smoke, ash, and cinders. The explanation of combustion in terms of form and quality is to be rejected because such entities are not needed for explanation: "afraid of deceiving myself if I assume anything more than is needed, I am content to conceive here [only] the motion of parts" (AT XI 7). The second argument is what might be called the argument from obscurity, the argument that the scholastics' real qualities (and, Descartes would no doubt add, their forms as well) are themselves obscure and in need of explanation: "I tell you that these qualities appear to me to be in need of explanation" (AT XI 25–26). These appear to be the arguments Descartes offers against his scholastic opponent on hylomorphism. But they are only implicit, and, by Cartesian standards, rather weak. However attractive they might be, these two arguments do not seem to meet the Cartesian standards for certainty in terms of intuition and deduction as demanded in the *Rules;* they do not establish that the intuitively and deductively grounded first principles of physics (and, if they are at play at this stage of Descartes' development, the first principles of metaphysics) absolutely exclude the consideration of forms and qualities.[32]

As noted earlier in this chapter, later in the 1630s and especially in the 1640s Descartes comes to see his scholastic opponent in mentalistic terms, as imposing mind-like forms, tiny souls onto the physical world. But the same arguments used earlier persist. Though the issue is quite carefully avoided in Descartes' first publication, the *Discourse* and *Essays* of 1637, his one reference to forms and qualities suggests an appeal to the argument from parsimony. Though he says that "I don't want to deny what they imagine in bodies over and above what I have discussed," he does go on to say that "it appears to me that my arguments should be approved all the more insofar as I make them depend on fewer things" (AT VI 239 [Ols. 268]). This sort of argument appears in 1638 in a letter to Morin (AT II 200 [K 59]), in his advice to Regius on answering Voëtius in 1642 (AT III 492 [K 127]), and in a letter to Mersenne in 1643 (AT III 649 [K 136]). It may be this argument from parsimony that he has in mind when he claims that it is the very establishment of his system that defeats the Aristotelians (AT I 602–3; AT III 298 [K 94]; AT III 470). The argument from obscurity also makes a number of appearances in these years. The claim that form as the philosophers conceive it is "a philosophical being unknown to me" is made twice in a letter to Morin from 1638 (AT II 364 [K 61], 367); in 1640 it is called a chimera (AT III 212), and both to Regius and to Mersenne the argument from obscurity is characterized as one of his principal arguments against the scholastic conception of form and quality (AT III 506; AT III 649 [K 135]; see also Pr IV 201F; Pr IV 203F).

New arguments also enter in these years. The success of Descartes' program in offering explanations for phenomena is, in a number of places, contrasted with the sterility of the Aristotelian philosophy. Responding to Fromondus, who had sent a lengthy critique of his recently published *Discourse* and *Essays* in 1637, Descartes remarked:

> He adds that I can hardly hope to explain many things through position and local motion alone, which are unintelligible without some real qualities. But if he should wish to enumerate the problems I explained in the treatise on the meteors alone, and compare it with those problems in the same subject (in which he himself is well versed) that have been treated up until now by others, I am convinced that he would not find such a great opportunity for condemning my overblown *[pinguiuscula]* and mechanical philosophy. (AT I 430)[33]

This argument from sterility, as it might be called, comes up again in the dispute with Voëtius, against whom Descartes writes in the *Letter to Voëtius* of 1643 that "that common philosophy, which is taught in the schools and academies, . . . is useless, as long experience has already shown, for no one has ever made any good use of primary matter, substantial forms, occult qualities and the like" (AT VIIIB 26; see also AT III 506). And in the Letter to Picot, the introduction to the 1647 French translation of the *Principles,* Descartes remarks that "one cannot better prove the falsity of Aristotle's principles than by saying that one has not been able to make any progress by their means in the many centuries in which they have been followed" (AT IXB 18–19).

These arguments are not altogether satisfactory. The numerous critiques of Descartes' scientific program that follow the publication of the *Discourse* and *Essays* might well lead an ungenerous contemporary to question the claims about the fertility of his framework.[34] Further-more, there is good reason to think that Descartes regarded the appeal to the explanatory success as a kind of rhetorical strategy, rather than as a serious argument. As noted above in chapter 2, when he was de-fending the use of hypothetical reasoning in the *Dioptrics* and *Meteors*, he argued that the successful deduction of interesting and plausible conclusions from those hypotheses give us good reason to trust the hypotheses, and so constitute a good way of presenting his doctrine in a preliminary way. But at the same time he made it clear that this is an imperfect way of arguing, and that the hypotheses must be established by deduction from first principles in order to be fully trustworthy. One can, I think, say something similar about the argument from sterility. In order fully to establish that Descartes' framework for physics is the right one and that there are no forms, one must establish these conclusions from first principles; it is not enough, on his own terms, to show that assuming the adequacy of his mechanist framework leads to interesting results. Earlier I pointed out that the arguments from parsimony and obscurity fail in this regard insofar as they fail to establish the falsity of hylomorphism from first principles. But in the climate of the early 1640s, given the views that he came to hold at that time, they fare even worse than they did in *The World*. The doubts that Descartes came to have in these years about our ability to be certain of mechanical expla-nation further undermine the argument from parsimony. In the *Princi-ples* he admits that "even if we can, perhaps, understand in this way how all natural things could have been brought about, we ought not to conclude from that that those we see were really produced in this way" (Pr IV 204L).[35] In the Latin *Principles* he goes on to say that "it is indu-bitable that the supreme craftsman of things could have made every-thing we see in many different ways" (Pr IV 204L). In the French trans-lation of 1647 this is strengthened, no doubt with his approval, and the claim is made that not only could God have created things as we see them in many different ways but that this could have been done "with-out it being possible for the human mind to know which of these ways he had wanted to use to make them" (Pr IV 204F). What he has in mind here is, no doubt, the possibility that God could have built a different *machine* for the same end, that there may be innumerable alternative mechanical explanations for the same phenomena, and that we are incapable of establishing with certainty *which* of the possible mechani-cal explanations is the correct one. But it is not implausible to add to this the possibility that God might have constructed a system whose behavior derives from something nonmechanical as well, from a sub-stantial form or a real quality, for example. And so even though every-thing *can* be explained mechanically, from Descartes' own argument we must conclude that it doesn't follow that the mechanical explanation is the *correct* explanation. His new views in the 1640s also further undermine the argument from obscurity. His new characterization of forms and qualities as "tiny souls" suggests that he has *some* conception of them, a conception of them as something like the human soul acting on matter in something like the way the human soul acts on the human body. Descartes can complain that such things don't exist, that the projection of mentality onto the physical world is improper. But he cannot complain that he has *no* idea what his opponents are talking about.[36]

The arguments we have been examining so far are, in a sense, extrinsic to the Cartesian system; none of these arguments attempts to establish the nonexistence or the impossibility of forms from the basic principles of Descartes' philosophy. Now, one might naturally enough think that the nonexistence of forms and qualities follows directly out of the account of body as extended and extended alone, and its distinction from thinking substance, mind. If body is extension alone, then it must exclude all mentality, all thought. In particular, it must exclude the substantial forms and real qualities that the scholastic wants to impose on body. But interestingly enough, this argument is given nowhere in any explicit way in the entire Cartesian corpus. It is in a way suggested in the passage from the *Principles* quoted at the beginning of this chapter, where Descartes argues from the nature of body as extension to the claim that everything in the physical world is explicable in terms of size, shape, and motion alone (see Pr II 23, 64). It is suggested a bit more explicitly when, in the *Sixth Replies,* after discussing the mentalistic assumptions of scholastic physics, Descartes suggests that getting clear about the nature of mind and the nature of body enabled him to see that "all of the other ideas of real qualities and substantial forms that I earlier had had been put together and fashioned by myself from those ideas" (AT VII 443). It is not absolutely clear what Descartes means to be saying here, but it is not implausible to see him as suggesting that understanding the natures of mind and body and their distinction enabled him to see that the mentalistic qualities that he had previously attributed to body don't really belong there.[37] But it is *prima facie* curious that Descartes is never more explicit in giving what might at first glance appear to be the strongest and most straightforward argument he has against his opponents.

Descartes may have thought the argument so obvious that it did not need explicitly to be mentioned; he may simply have thought that anyone who really understood his thought would immediately see that it undermines the Aristotelian framework, as he suggested a number of times (AT I 602–3; AT III 298 [K 94]; AT III 470). But whatever the reason, Descartes was quite wise not to press such an argument: *for the argument from the nature of body and its distinction from mind does not succeed in eliminating forms.* The problem with the argument becomes evident when we reflect on what it entails for human beings and their bodies. If extension is the nature of body, then human *bodies* cannot think, strictly speaking, as Descartes insisted (AT VII 444). But he does not conclude from that that people don't think, or that everything in the human body is explicable in purely mechanical terms. Rather, he concludes that human beings have minds, immaterial souls distinct from their bodies, which think and, under appropriate circumstances, guide the behavior of the unthinking body. But, we might ask, why can't the scholastic argue in a parallel way to his position? Descartes' argument shows that thought is not in bodies but in the soul. This shows that a body, strictly speaking, an extended thing, cannot contain knowledge of the center of the Earth, nor can it will itself to move in that direction. But why can't we infer from that that heavy bodies must have tiny souls, souls distinct from their bodies, in order to think about the place they would rather be and will the bodies to which they are attached in the appropriate direction? And so, a scholastic might respond to Descartes' argument, the claim that the essence of body is extension no more establishes the mechanical explicability of the behavior of a falling stone than it establishes the mechanical explicability of the behavior of the human being who dropped it. To put it another way, from the point of view of the scholastic opponent, Descartes can show, perhaps, that if hylomorphism is true, it involves attributing tiny immaterial souls to extended bodies. But if the argument is to *refute* the doctrine of hylomorphism, Descartes must show why there are not or cannot be such tiny souls in nature, why human bodies are to be treated so differently from their inanimate cousins, why outside of humans there is no thought, in body or in mind.

DESCARTES AGAINST HIS TEACHERS: THE ELIMINATION OF ANIMAL SOULS

The problem we are left with is this. The consideration of the nature of mind and the nature of body leads to a dualism of mind and body, thinking substance and extended substance. But this does not necessarily undermine the scholastic view; the scholastic can simply follow Descartes and see his form as a substance attached to another substance, a soul attached to a body, a view not that distant from the one many later scholastics took of hylomorphism.[38] The real question he must confront from within his system in order to eliminate forms and qualities is quite another sort of dualism, the radical distinction between human beings, who have both bodies and souls, and the rest of the created world, where, angels perhaps excepted, we find only body. It is only by establishing this distinction that he can eliminate the abhorrent forms and qualities of his opponents.

Descartes never confronts this question in its full generality. But he does examine in detail an important special case, the problem of animal souls. Now, animal souls were, on standard scholastic doctrine, the substantial forms of their bodies, and the two problems, that of the existence and nature of animal souls and that of substantial forms and real qualities are linked in one of his earliest references to the problem of animal souls.[39] However, while forms and qualities rarely get more than passing attention in his writings, starting in 1637 with part V of the *Discourse,* the question of animal souls gets considerable attention in Descartes' writings. It is in his discussion of animal thought and animal souls that we see the grounds of the distinction Descartes draws between not only humans and animals, but more generally, between human beings and the rest of created nature, the world he sought to purge of forms and qualities, the world he sought to mechanize.

The problem of animal souls, a hotly debated question that continues to be discussed throughout the seventeenth century and into the next, is an issue of some complexity.[40] Briefly, though, Descartes' view goes something like this. We can prove the existence of a soul in each of ourselves through our immediate experience of thought. However, for animals (and, presumably, for other human beings as well) all we have to go on is external behavior. As he noted to Gassendi in the *Fifth Replies,* "the mind meditating can experience within itself that it thinks, but, however, it cannot experience whether or not brutes think. This question can only be investigated afterwards, *a posteriori* and from their behavior *[operationes]*" (AT VII 358; see also AT VII 427; AT V 276–77 [K 244]). Now, it is only in pure intellection, the exercise of reason without the aid of sensation or imagination, without the influence of the passions, that the human soul operates in a way wholly independent of the body.[41] As he remarks, again to Gassendi, "I have . . . often distinctly shown that the mind can operate independently of the brain, for, indeed, the brain is of no use for pure understanding, but it is only of use in imagining or sensing" (AT VII 358). And so, he reasons, it is only rational behavior, behavior that shows that the thing in question is capable of such abstract and nonbodily reasoning that can show us that something has an incorporeal soul like we do; as he notes in a letter to Regius in May 1642, "no actions are considered human except for the ones that derive from reason" (AT III 371 [K 102]). In the important discussion of animal souls in part V of the *Discourse,* Descartes recognizes two ways in which rationality is displayed in our behavior, and thus two ways in which we can distinguish between human beings, who have incorporeal souls, and beasts, which he claims lack such souls. The first is language:

> It is a very remarkable thing that there are no men so dull and stupid . . . that they aren't capable of arranging different words and putting together a discourse through which they can make their thoughts understood, and, on the other hand, there is no other animal, however perfect and well-endowed it might be that does anything like that. . . . This shows not only that the beasts have less reason than men, but that they have none at all. (AT VI 57–58)[42]

The second way human behavior displays our rationality and differentiates us from the beasts, who are merely machines, is our ability to respond appropriately in an infinite variety of circumstances:

> Unlike reason, a universal instrument which can be made use of in all sorts of circumstances, these organs [i.e., those of a machine or a beast] require some particular disposition for every particular action. From this it follows that it is morally impossible that there are enough different dispositions in one machine to make it act the same as our reason makes us act in all of the occurrences of life. (AT VI 57)[43]

Since animals display neither of these two signs of rationality, Descartes argues that we have no grounds on which to attribute rational souls to them, incorporeal souls capable of acting without a body. And since he assumes *only* rational behavior requires a nonmechanistic explanation, he infers that *all* of their behavior is explicable in purely mechanical terms, through the size, shape, and motion of the parts that make them up (see AT VII 230–31, 426; AT II 39–41 [K 53–54]; AT III 121; AT IV 575–76 [K 207–8]; AT V 277–78 [K 244]; AT XI 519–20; PS 50). If animals have souls, he argues, they are only corporeal souls, which he sometimes suggests identifying with their blood (see AT I 414–16 [K 36–37]; AT IV 64–65 [K 146]).

From the fact that animals appear to lack the ability to reason, and from the fact that all of their behavior is explicable mechanically, Descartes infers that animals lack incorporeal souls. But in depriving them of incorporeal souls, he deprives them of more than just reason. Insofar as they lack a soul, he claims that they lack all thought (*cogitatio, pensée*) of the sort that we have (AT VII 426; AT IV 573–76 [K 206–8]; AT V 276–79 [K 243–45]; PS 50). In particular, they don't have sensations of the sort that we do, sensations that are thoughts,[44] and while built in such a way as to seek what is beneficial for them and avoid what is harmful, they lack any consciousness *(notitia)* of these things (AT XI 519–20; PS 50). While beasts exhibit pain behavior, they lack the feeling *(sentiment)* of pain strictly speaking; indeed, they lack all feelings and passions in the sense in which we have them, strictly speaking (AT II 39 [K 53]; AT III 85).[45] And finally in lacking an incorporeal soul, they lack volition. Descartes explained his view of the will to Regius in 1641:

> When you say "will and intellect differ only as different modes of acting that concern different objects," I would prefer [that you say that they] "differ only as the action and passion of the same substances." For the intellect is properly speaking a passion of the mind, and will its action. But since we can never will anything without at the same time understanding it, and, indeed we scarcely ever understand anything without at the same time willing something, we therefore don't easily distinguish the passion in it from the action. (AT III 372 [K 102–3]; see also Pr I 32; PS 17)

Since it is the soul that wills, just as it reasons, senses, and feels, in lacking a soul animals must lack volition in the sense in which we have it.[46]

These considerations translate directly into a more general argument for distinguishing human beings from the rest of the created world, and thus an argument against substantial forms and real qualities, the tiny souls that the scholastics find scattered through all nature, on Descartes' reading of their view. If the behavior of animals displays no evidence of reason, then inanimate bodies display reason all the less; as with animals, Descartes claims to be able to explain all the behavior of inanimate bodies in terms of size, shape, motion, and the laws of motion. If the arguments he brings against animal souls establish their conclusion, then they should all the better establish the more general conclusion that there are no forms and qualities of any sort.

But it is important to note that even with animal souls, the argument does not establish the conclusion with the certainty Descartes often seems to demand, as he eventually came to see. Writing to More in February 1649 he noted, "however, although I take it as having been demonstrated that it cannot be proved that there is thought *[cogitatio]* in the brutes, I don't think that it can be demonstrated from that that they have no thought, since the human mind does not reach into their hearts" (AT V 276–77 [K 244]). He goes on to suggest that his view, the view that animals have no souls, the view that their behavior is explicable in purely mechanical terms, is simply the view that is "most probable" (AT V 277 [K 244]). Descartes' caution here is quite proper, I think. Insofar as animals appear to lack reason, the ability for abstract thought, the one ability whose manifestation *requires* us to attribute incorporeal souls to other humans, we lack one convincing argument for attributing souls to animals. But rationality is not the only manifestation of an incorporeal soul. As he noted to Gassendi, we know that we have souls *not* because we reason, but because each of us experiences thought, something that must, Descartes argues, pertain to a substance distinct from the substance that constitutes the subject of the modes of extension (AT VII 358). Though they do not reason, Descartes grants to More that it is at least conceivable that animals have thoughts in something like the way we do. Though we may never be able to "reach into their hearts" and decide one way or another, it is a possibility that cannot altogether be excluded.

It is interesting to observe that the seed of uncertainty that creeps into Descartes' reasoning here does not derive from any doubts about the mechanical explicability of animal behavior. One might question his confidence that all behavior not demonstrably rational must necessarily be explicable in mechanical terms, the premise on which he grounds his claim that all animal behavior derives from the mechanical disposition of their organs. But even in February 1649, even while admitting that his rejection of animal souls is only probable, he still insists to More that "it is certain that the bodies of animals contain . . . organs disposed in such a way that they by themselves can give rise to all of the motions we observe in brutes without any thought" (AT V 277 [K 244]). The problem arises for him even if we assume that all animal behavior is explicable mechanically. For even if we can explain all such behavior mechanically, even if all animal behavior might derive from the size, shape, and motion of their component parts, it doesn't follow that it actually does. The problem here is exactly parallel to the one we discussed earlier with respect to the argument from parsimony, the problem Descartes came to see at the end of the *Principles* in inferring the truth of a mechanical explanation from its success. Even though all of an animal's behavior can be explained mechanically, and even though there is no rational behavior that would force us to posit a soul, still we cannot absolutely exclude the possibility that God chose to create animals with souls, with thought and volition, so that the behavior that we see and can explain mechanically may really derive from an incorporeal and nonmechanical source.

Though Descartes may not have explicitly made the connection, the case must be similar for forms. Now, it does not strictly speaking follow from the fact that corporeal substance is extended and extended alone that everything in bodies is explicable in terms of size, shape, and motion, any more than this conclusion follows for us, human beings whose bodies are made up of corporeal substance, as I noted earlier in this chapter. But even if, as it turned out, everything in inanimate bodies were explicable in mechanistic terms, and even if we could be certain of that, then the elimination of forms and qualities would only be probable. For with bodies in general, as with animals, from the fact that behavior *can* be produced mechanically, it doesn't *follow* that it *is*.

And so, in the end, Descartes fails to produce an argument for the elimination of forms that meets his standards of certainty. He can show (to his satisfaction) that we are certain that forms and qualities, if they exist, must be conceived of as mental substances of a sort, distinct from corporeal substance, and through various arguments he can convince us that it is improbable that they exist. But he cannot demonstrate to the scholastics that their view is false.

So far I have emphasized Descartes' opposition to substantial forms, real qualities, and the hylomorphic conception of body common to the schools. Though his sword may have been blunter than he might have thought (or wished), this is certainly not to deny his status as an opponent of the world as pictured in the schools. Although his arguments may not have met his own stringent (and ultimately unrealistic) standards of rigor, they made an important contribution to the ultimate demise of hylomorphism in the seventeenth century. Though Descartes neither invented them nor was he unique in pressing them, arguments like ones he used, arguments from parsimony, obscurity, and sterility, as well as more strictly Cartesian arguments from the nature of body, appear repeatedly in later thinkers, who, following his lead, pressed for the rejection of the Aristotelian framework. However, it is worth pointing out that in a deeper sense Descartes never managed entirely to extricate himself from the philosophy of the schools. He rejected the illicit projection of human nature, soul, and body onto the physical world in a sense, insofar as he rejected the explanation of the particular properties bodies have in terms of form and quality, tiny souls that move bodies about as our soul moves our body. But when explaining the general properties all bodies as such have, the conservation of motion, the tendencies bodies in motion have to preserve their motion in a rectilinear path, their behavior in collision, the scholastic model of animistic explanation returns. As we shall later see in chapters 7–9, Descartes rejects the tiny souls of the schools only to replace them with one great soul, God, an incorporeal substance who, to our limited understanding, manipulates the bodies of the inanimate world as we manipulate ours (AT V 347 [K 252]). In the end, as for the scholastics, as Descartes interprets them, the ultimate explanation of the characteristic behavior of bodies takes us back to ourselves, human nature projected not downward onto the material world, but upward to God.

DESCARTES AGAINST THE ATOMISTS: INDIVISIBILITY, SPACE, AND VOID

ARISTOTELIAN IDEAS about natural philosophy dominated the schools well into the seventeenth century, long after Descartes left La Flèche;[1] it was the scholastic physics of matter and form that he would have been taught at school, as was everyone else. But there had already been numerous challenges to the authority of Aristotle in physics and cosmology by the time Descartes began to formulate his own ideas and communicate them in a series of publications that were to play a significant role in eclipsing the reputation of the Aristotelian philosophy. There were, of course, many varieties of anti-Aristotelianism in the years before Descartes began his own system. But most important with respect to Descartes, indeed, of central importance to the history of the physical sciences in the seventeenth century and beyond, was the revival of the ancient atomistic doctrines of Democritus, Epicurus, and Lucretius. Like Descartes, these ancient atomists and their later followers attempted to explain the characteristic behavior of bodies in terms of the size, shape, and motion of the small particles that make them up, and like Descartes, they argued for eliminating sensory qualities like heat and cold, color and taste from the physical world. It is no wonder, then, that he was often associated by his contemporaries with these other mechanists. But at least as important to him as the similarities were the differences between his philosophy and that of the atomists; Descartes took great care to emphasize the points on which he differed from what was probably the dominant school of corpuscularianism at the time he began to write. After a brief discussion of the revival of atomism in the sixteenth and seventeenth centuries, this is what we shall turn to, Descartes' rejection of the indivisible bodies of the atomists, and his rejection of their conception of space and void.

DESCARTES AND THE REVIVAL OF ATOMISM

The ancient atomists were by no means unknown to the medievals; at very least the views of Democritus and Leucippus could be gleaned from the hostile accounts found in Aristotle.[2] Though discussed to some extent, by and large the position was not taken very seriously.[3] Interest in ancient atomism was no doubt spurred by the rediscovery in 1417 of a copy of Lucretius' Epicurean poem, *De rerum natura,* published in 1473. The publication of Lucretius was followed shortly by the publication of the first complete Latin translation of Diogenes Laertius' *Lives of the Philosophers,* which contains the most extensive collection of Epicurus' own writings to survive, as well as brief accounts of the atomistic philosophies of Leucippus and Democritus.[4] Atomistic thought was widely discussed in the sixteenth century, and by the early seventeenth century, it had a number of visible adherents.[5] In addition to Lucretius and Diogenes Laertius the reader curious about atomism in the early decades of the seventeenth century could turn to Nicholas Hill's *Philosophia Epicurea, Democritiana, Theophrastica* (1601), to Sebastian Basso's *Philosophia Naturalis adversus Aristotelem* (1621), and to sympathetic accounts of atomism in the works of Bacon and Galileo.[6] Descartes may have become acquainted with the atomist tradition in any number of ways. But he must certainly have gotten a healthy dose of atomistic thought from his early mentor Isaac Beeckman. Beeckman seems to have been a dedicated atomist from at least 1616, and would certainly have communicated his enthusiasm for the view to his young disciple.[7] By the 1630s and 1640s ancient atomism had been more or less successfully resurrected; it could be found in the influential Daniel Sennert's *Hypomnemata Physica* (1636), in Jean Chrysostom Magnen's *Democritus Reviviscens* (1646), and more importantly in Pierre Gassendi's enormous *Animadversiones in Decimum Librum Diogenis Laertii* (1649), a massive study of Epicureanism that formed the basis for his posthumous *Syntagma Philosophicum* (1658).[8] By mid-century atomism was established as an important school of thought not only in the libraries, but also in the informal scientific circles where enthusiasts for the new natural philosophies exchanged ideas.[9]

It was against this background that many of Descartes' early readers approached his views, and quite naturally read his mechanist and anti-Aristotelian philosophy as part of the revival of ancient atomism. And, indeed, there are many important affinities between Descartes' thought and that of the ancient atomists. Basic to the atomist tradition is the idea that the sensible qualities of bodies are to be explained in terms of smaller and insensible bodies with different sizes and shapes, bodies that themselves lack most of the sensible properties, like color, taste, heat, and cold, that they are supposed to explain.[10] Furthermore, Descartes, like the ancient atomists, thought the universe to be without limit in size, and thought that it contained an indefinitely large number of worlds, suns, and planets roughly like ours;[11] indeed Descartes held as Lucretius and probably Epicurus before him did that all possible configurations of matter arise at sometime or another.[12] And finally, Descartes agreed with the ancient atomists in seeing the present state of the world, not only sun, earth, and planets, but also plants, animals, and human bodies as having evolved out of an initial chaos in a purely natural way.[13] And so Descartes, like the atomists, rejected the consideration of final causes from physics.[14]

It is not surprising, then, that Descartes was often lumped together with the revivers of atomism by his contemporaries, and assumed to be a Democritean or Epicurean.[15] Fromondus was not atypical in reacting to the *Discourse* and *Essays* with the unsympathetic comment that "he unknowingly often falls into the physics of Epicurus, crude and overblown" (AT I 402).[16] Even after the publication of the *Principles* in 1644, where he explicitly distanced himself from the atomist tradition (Pr IV 202), Descartes still had to fend off such attributions. Responding to a correspondent who had reported to him on the reaction to his recently published *Principles,* probably in June 1645, Descartes commented with some exasperation that "seeing that he says that . . . my principles in physics are drawn from Democritus, I believe that he has not read [my writings] very much" (AT IV 223; see also AT I 413 [K 36]; AT VII 381).

There can be no question but that Descartes was deeply influenced by the atomist tradition, either directly or through one or another of its later followers; the obvious correspondences between his program and that of other mechanists, ancient and modern, can be no accident. But, at the same time, we must be aware that Descartes himself did his best to disassociate himself from that tradition. Part of his motivation here is, no doubt, pride in his own discoveries. In general, Descartes did not react well to challenges to his originality; it is not implausible to suppose that he identified closely with the unnamed protagonist of the *Discourse,* who cut himself off from the past and, by himself, aided only by the method he had discovered, found a new world, a world that, but for his labors, would be utterly unknown.[17] It is this pride that I suspect lies behind comments like the one he conveyed to Mersenne in 1640: "I wonder at those who say that I have written only a patchwork of Democritus, and I would like them to tell me from what book I could have taken those patches, and if anyone has ever seen any writings where Democritus has explained salt, hexagonal snow, the rainbow, etc., as I have" (AT III 166; see also AT II 51).

But leaving the question of pride aside, there are some real issues that separate Descartes from the atomists; he quite understandably did not want his readers to think that his agreement with some aspects of the atomist program in any way entailed agreement on others. Some of the differences involved aspects of the Epicurean doctrine that were apparently at odds with Christian doctrine or tradition. According to the Epicurean doctrine, the soul is material, made up of atoms, and dissipates when the body is destroyed.[18] While some Christians, even some among Descartes' contemporaries, had held such a view,[19] this sort of materialism certainly runs counter to his thought. The Epicureans also believed that the gods are material and neither create nor regulate the world, a view at odds both with Christian thought and with the Cartesian system, a system in which an immaterial God creates the world and, as we shall see in later chapters, is the ground of order insofar as his activity determines the laws bodies in motion and impact obey.[20] In addition there are important differences in physics. The basic particles of the atomists are by their nature heavy, and tend by their nature to move in one particular direction. For Descartes, though, since bodies are only extended things, they have no innate tendencies to motion of any sort; whatever tendencies we observe in them derive from their own corpuscular substructure and the way they interact with the bodies in their vicinity.[21] But where Descartes differs most strikingly from the atomist tradition in physics is in his conception of the basic constituents of the physical world. For the atomists what there is, at root, is atoms and the void. The atoms are insensible bodies that are, by their nature, indivisible into smaller bodies, and between these indivisible bodies is void, space empty of body and anything else.[22] Descartes denies both of these doctrines. Descartes' conception of body as extended substance entails that every body, however small, is divisible into smaller parts, and that there can be no extension without body, no region of space that is in any sense empty. In the remainder of this chapter we shall explore these two anti-atomist doctrines.

DESCARTES AGAINST INDIVISIBILITY

One of the most important properties of atoms in the atomist tradition is their indivisibility, their indestructability. As Epicurus wrote:

> Of bodies some are composite, others the elements of which these composite bodies are made. These elements are indivisible and unchangeable, and necessarily so, if things are not all to be destroyed and pass into nonexistence, but are to be strong enough to endure when the composite bodies are broken up, because they possess a solid nature and are incapable of being anywhere or anyhow dissolved. It follows that the first beginnings must be indivisible, corporeal entities.[23]

Atoms are, thus, indivisible, unchangeable bodies, the ultimate parts into which bodies can be divided and from which bodies can be constructed.

It is possible that in his early years Descartes was an atomist in this sense. Beeckman most certainly was, and it is not implausible that Descartes followed his mentor in this respect. There is no record in the documents that survive of any disagreement on this point; indeed, the lack of record of any discussion at all suggests that they agreed on this basic point. Furthermore, in at least one passage from the surviving documents, a short discussion of a problem in hydrostatics that Descartes presented to Beeckman, he makes reference to "one atom of water [una aquae atomus]" assumed to travel twice as fast as "two other atoms" (JB IV 52 [AT X 68]).[24] While Descartes may not be using the term in its full technical sense here, it does, at least, suggest that he *may* have held the atomistic hypothesis in 1618.[25] Evidence is similarly scanty for the 1620s. The apparent identification of body and extension in Rule 14, discussed earlier in chapter 3, suggests the position Descartes later holds, that the indefinite divisibility of mathematical extension entails the denial of indivisible bodies. But Rule 14 is almost certainly in the very last stage of composition of the *Rules,* and Descartes' earlier concerns in the 1620s suggest a view that may be more consistent with a belief in atomism. As I noted earlier in chapters 1 and 2, the law of refraction as given in Discourse II of the *Dioptrics* probably dates from the mid-1620s. Now, Descartes' theory of light in *The World* is closely connected with the existence of vortices in a continuous, nonatomic plenum. But the model he uses in the *Dioptrics* to derive the law of refraction, tennis balls colliding with different sorts of surfaces, suggests that his original position may possibly have been an atomic view of light as a stream of atomic particles, a view that was later translated into the world of *The World,* where light is the pressure of tiny balls, Descartes' second element, in a medium, with the clumsy and highly problematical assumption that tendency obeys the same laws that bodies in motion obey.[26] Also weakly suggestive of an atomist position in the 1620s is a curious passage from Rule 12. Talking about the human mind and its relation to the brain he says, "this phantasy [phantasia] is a true part of the body and it is of such a size that different portions of it can embody many shapes distinct from one another" (AT X 414). In an atomist view, in which there are bodies of minimum size, the magnitude of the fantasy, the number of atoms it contains would set an upper bound on the number of distinct impressions it could hold at a given time. But if one were to reject indivisible atoms for indefinitely divisible matter, then this wouldn't be an issue; any body, no matter how small, could contain an indefinitely large number of distinct shapes, providing that they were small enough.

Whatever Descartes may have believed in the 1620s and before, though, by 1630 his mature view seems to have emerged. Writing to Mersenne in April 1630 concerning some details of the treatise on physics, later to be *The World,* that he was then working out, he noted:

> These tiny bodies which enter when a thing is rarified and leave when it condenses, and which pass through the hardest bodies are of the same substance as those which one sees and touches. But it is not necessary to imagine them to be like atoms nor as if they had a certain hardness. Imagine them to be like an extremely fluid and subtle substance, which fills the pores of other bodies. (AT I 139–40 [K 9–10])

Matter, the matter that makes up not only this subtle and all-pervading fluid but all matter, Descartes implies, is not made up of indivisible and unchangeable atoms. This is a position that he maintains throughout his career; it is clearly put forward in *The World* of 1633 (AT XI 12), in the *Meteors* of 1637 (AT VI 238–39 [Ols. 268]), in the *Meditations* and *Objections and Replies* of 1641 (AT VII 85–86, 106, 163), in the *Principles* of 1644 (Pr I 26; Pr II 20, 34; Pr III 51; Pr IV 202), and in a letter to More from 1649 (AT V 273–74 [K 241–42]), not to mention other discussions in numerous letters (AT I 422 [K 39]; AT III 191–92 [K 78–79]; AT III 213–14 [K 80]; AT III 477 [K 124]; AT IV 112–13 [K 147]). Rejecting the indivisible unchangeable atoms others had insisted on seems to have been a matter of some importance to him.

But it is important to understand the grounds on which Descartes rejected the atomism. His most visible argument is an argument from the nature of body. In the *Second Replies* Descartes asserts, "since . . . divisibility is contained in the nature of body, that is, in the nature of an extended thing (for we can conceive of no extended thing so small that we cannot divide it, at least in thought), it is true to say that . . . every body is divisible" (AT VII 163; see also AT III 213–14 [K 80]). And so, Descartes tells Mersenne in October 1640, "as for an atom, it can never distinctly be conceived, since the very meaning of the word implies a contradiction, namely being a body and being indivisible" (AT III 191 [K 79]; see also AT III 477 [K 124]; AT V 273 [K 241]). In more detail, the argument goes as follows. As we discussed above in chapter 3, Descartes' bodies are extended substances in the sense that the only properties they really have are geometrical extension and its modes, size, shape, etc. Now, Descartes claims, one of the obvious properties extension as such has is its divisibility; writing to Mersenne in March 1640 Descartes notes that "there is no quantity that is not divisible into an infinity of parts" (AT III 36). And so body, being essentially extended, extension made real, must *also* be divisible into an infinity of parts, it would seem; every part of matter is divisible (Pr I 26; AT III 477 [K 124]), and since body is continuous (AT VII 63; AT I 422 [K 39]), it is divisible in an infinite number of ways (AT VI 238–39 [Ols. 268]), a division that goes to infinity and never comes to end, so far as we can know (AT I 422 [K 39]; AT IV 112–13 [K 147]). And so bodies cannot be made up of indivisible parts, as the atomists claim they are. If the parts are extended, then they are not indivisible, and if they are non-extended, then they are not bodies and cannot be genuine parts of bodies (AT III 213–14 [K 80]).[27]

This argument appears to take us from the infinite, or as Descartes often prefers to put it, indefinite divisibility of geometrical extension to the infinite or indefinite divisibility of extended substance, from the fact that we cannot conceive of an extended thing that cannot be divided in thought to the actual nonexistence of an extended indivisible atom.[28] But from the point of view of the atomists, this argument is obviously inadequate.

Basic to atomists, both ancient and seventeenth century, was a distinction between two different sorts of atomism, mathematical (or conceptual) and physical.[29] There were, of course, disputes going back to antiquity about whether or not the continuous magnitudes treated by the geometers are made up of geometrical minima, whether lines are made up of points, or surfaces are made up of lines, for example. But this was a question quite carefully separated from that of the *physical* indivisibility of atoms. As Gassendi put the matter, "that minimum or indivisible that Epicurus admits is physical and of a far different nature than the mathematical indivisible, that is, what they suppose to be a point of some sort or another."[30] And so, the fact that mathematical extension is always divisible does not entail that every physical and extended body is also divisible; even if the mathematical continuum is divisible ad infinitum, it doesn't follow that there are no smallest extended bodies in nature that cannot be split by natural means.

It is quite possible that Aristotle and many later Aristotelians missed this crucial distinction, and saw the arguments directed against mathematical minima as holding equally well against physical minima, atoms.[31] It is also possible that Descartes himself missed the point for some years. But by the early 1640s, he saw the need to argue for a link between the conceptual divisibility that pertains to every body by virtue of being an extended thing, and the sort of real, physical divisibility required to refute the atomistic hypothesis. The strategy he used to establish the real divisibility of all bodies is similar to the one he used to establish the real distinction between mind and body. As I noted in chapter 3, Descartes appeals to God to link the conceptual separability of the mind from the body with the real distinction between the two.

We can clearly and distinctly conceive of mind without body and body without mind. But what makes mind and body genuinely distinct things, capable of existing apart from one another, is God, who can bring about whatever I clearly and distinctly conceive. Thus, even though mind and body exist joined together in this life, our ability to conceive them apart from one another, together with God's omnipotence, entails that they can really exist separately. The case, Descartes thinks, is similar for the supposedly indivisible atoms that some philosophers have posited. The argument first surfaces in a letter to Mersenne in October 1640, where the appeal is not to God himself but to his angels. He writes, "If something has [extension], we can divide it, at least in our imagination, which is sufficient to be sure that it is not indivisible, since if we can divide it in this way, an angel can really divide it" (AT III 214 [K 80]). However, by 1642 it is to God's power that Descartes appeals with an explicit link to the argument for the distinction between mind and body (AT III 477 [K 124]), and it is this argument that appears in the *Principles* of 1644:

> We also know that there can be no atoms, that is, parts of matter by their nature indivisible. For if there were such things, they would necessarily have to be extended, however small we imagine them to be, and hence we could in our thought divide each of them into two or more smaller ones, and thus we can know that they are divisible. For we cannot divide anything in thought without by this very fact knowing that it is divisible. And therefore, if we were to judge that it is indivisible, our judgment would be opposed to our thought. But even if we were to imagine that God wanted to have brought it about that some particles of matter not be divisible into smaller parts, even then they shouldn't properly be called indivisible. For indeed, even if he had made something that could not be divided by any creatures, he certainly could not have deprived himself of the ability to divide it, since he certainly could not diminish his own power. And therefore, that divisibility will remain, since it is divisible by its nature. (Pr II 20; see also AT V 273 [K 241])

But even though Descartes may be right in pointing out the divine divisibility of even the hardest body, this argument in an important way misses the mark. For the ancient atomists, like Epicurus and his school, the gods are themselves made up of atoms; while stronger than humans, they are not omnipotent and have no power to create the world or split the atoms it contains.[32] Descartes' claim that God could divide any portion of matter, no matter how small, does, indeed, introduce a notion of divisibility that they would deny. But the Christian atomism of Descartes'

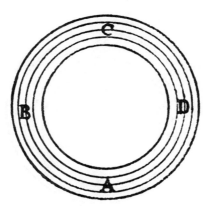

FIGURE 5.1 From *Renati Des-Cartes Principia Philosophiae* (Amsterdam, 1644), p. 51.

contemporaries is quite another matter. Gassendi, for example, goes to considerable lengths to show that while he believes in the atoms of Epicurus, his God is the Christian God.[33] In particular, Gassendi's God is omnipotent: "there is no thing that God cannot destroy, no thing he cannot produce."[34] And so Gassendi would be quite willing to admit that God could split an atom if he chose to do so. But this is entirely consistent with God creating a thing which cannot be split by any *natural* means, a thing which cannot be split by any of his creatures, a thing which cannot be split by any but him. It is this that Descartes must establish if he is to refute atomism, that there are no *naturally* indivisible bodies; *supernatural* divisibility is, in a way, beside the point.

Descartes has no arguments intended directly to show that all bodies are naturally divisible. But he does think that in certain circumstances, at least, bodies are actually divided ad infinitum, or, as he might put it, ad indefinitum.[35] To understand the grounds for this claim we must consider a circumstance he discusses in Pr II 33. There he begins with the assumption, argued earlier in Pr II 6, that "all places are filled with body and the same parts of matter always fill places of the same size." From this he claims that it follows that all motion is very roughly circular, one body displacing another, the other body displacing yet another, until the place left by the original body is occupied, at the very moment it leaves; such a circular motion seems necessary, Descartes thinks, in order either to prevent a vacuum or to prevent a single bit of matter from occupying different volumes at different times. Such motion is relatively unproblematic if we imagine the ring of moving bodies to be uniform in width, as in ring ABCD of fig. 5.1.

But Descartes thinks that there is no real problem here even if the

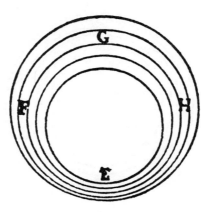

FIGURE 5.2 From *Renati Des-Cartes Principia Philosophiae* (Amsterdam, 1644), p. 51.

ring is of nonuniform width, as in ring EFGH of fig. 5.2; if the particles making up the ring move with different speeds in different parts of the ring, say faster at E where the space is narrow and slower at G where it is wider, then the matter can move in this nonuniform ring without creating a vacuum or without a given body having to expand or contract. But in the case of the ring of nonuniform width that Descartes imagines, the matter going in such a path must actually be divided into indefinitely small parts, he reasons:

> For it could not happen that the matter that now fills space G successively fills all of the spaces smaller by innumerable degrees that there are between G and E unless each of its parts accommodates its shape to the innumerable volumes of those spaces. For this to happen it is necessary that all of its imaginable particles (which are really innumerable) move the tiniest bit with respect to one another. And such motion, however small, is a true division. (Pr II 34)

The actual division of matter into indefinitely small parts, then, is required so that in moving, the incompressible and inexpandable material substance not produce a vacuum, a place void of body. This view, that matter must be indefinitely divisible and sometimes indefinitely divided in order to prevent a vacuum, goes back to Descartes' *World* and to the correspondence that precedes it.[36]

This argument is probably Descartes' best answer to the atomist, his best reason for rejecting the hard, unchangeable, and indivisible atoms from which they tried to construct their world. But the argument depends crucially on a premise that the atomist tradition explicitly rejects, the nonexistence of the void. It is to this second crucial difference between Descartes and the atomists that we must turn.

SPACE AND VOID

As important to the ancient atomists as the tiny indivisible and unchangeable atoms they posited was the space they occupied and the space, empty space, vacuum, or void that surrounded them, the emptiness *(inane)*, as Lucretius puts it, "in which they are situated and through which they are moved in different ways."[37] This emptiness can be found within the grosser bodies of everyday experience, tiny empty spaces that separate atoms from one another; this emptiness out to infinity, perhaps even beyond the region occupied by bodies.[38]

But speculation about empty space was not confined to the atomist school. Aristotle, of course, would have none of this nothingness, and quite firmly rejected the notion of space or place as independent of body, and along with it, the notion of a void, a space or place that exists without containing a body.[39] But such a view seemed to set problematic limits on God's power. For it seems, if Aristotle is taken at his word, then not even God could create an empty space. These considerations were among those that apparently disturbed Étienne Tempier, Bishop of Paris, who in 1277 issued a sweeping condemnation of the then-fashionable Aristotelianism.[40] While the condemnation held only for Paris, strictly speaking, it affected the subsequent development of scholastic Aristotelianism in important ways. As a consequence, philosophers working within an Aristotelian framework were forced to think in very non-Aristotelian ways about place, space, and the possibility of a void, both within and without the world.[41] Among the positions Tempier condemned was the claim that "God could not move the heavens [that is, the world] with rectilinear motion, and the reason is that a vacuum would remain."[42] In response to this, a number of figures consider the possibility, in some sense, of an empty space beyond the world, what came to be called "imaginary space," space that would allow God to move the finite world as a whole.[43]

The full history of ideas of place, space, and vacuum in the later middle ages is extraordinarily complex.[44] But the later medieval speculations about spatial concepts, together with the rediscovery and revival of ancient atomism, among other currents, lead to a wide variety of views on space, place, and void in the sixteenth and early seventeenth centuries. Writing to Mersenne in 1631, Descartes suggests that, unlike him, virtually everyone believes in empty space (AT I 228). This is certainly an exaggeration. But at the time Descartes was formulating his own ideas it was by no means uncommon to hold that space is in some sense independent of the bodies that occupy it, and that there can be or actually are portions of this space unoccupied by body. For a wide variety of figures, space is something in some sense distinct from body, "a certain continuous three dimensional physical quantity in which the magnitude of bodies is received," to quote Bruno.[45] For some who adopted this view, like Bruno, Telesio, and Campanella, the container space, while independent of body, is always occupied by body.[46] But for others it is not, and there are vacua both within and beyond our world. This position is strikingly held by Francesco Patrizi in the late sixteenth century, who was perhaps the first to step outside of the Aristotelian metaphysical framework and argue that space is neither substance nor accident, but *sui generis,* the container of all, God's first creation in which he placed all else, filling some places but leaving others empty.[47] In this he was followed by a number of seventeenth-century figures, most notably Descartes' contemporary and sometimes antagonist Pierre Gassendi, who easily integrated Patrizi's account of spatial concepts into his own revived Epicurean atomism.[48] Such views were even represented in some sixteenth- and seventeenth-century scholastic thinkers. Dealing with the so-called imaginary spaces that exist beyond the bounds of our finite world, an issue to which we shall later return, Pedro Fonesca, Bartholomeus Amicus, and the Coimbrian Fathers all allowed that such spaces really exist in some sense, and Suarez went so far as to characterize imaginary space as a vacuum.[49] And even though Suarez argues that the place of a body "is a certain real mode, intrinsic to a thing,"[50] he grants that God could destroy the sublunar world and leave an empty space behind.[51]

It is not entirely clear just how much of this background Descartes himself knew. At La Flèche, Father Noël, later himself to become involved with Pascal on just this issue, certainly took the young Descartes through this material from a scholastic point of view, though it is not clear how much detail Descartes would have remembered some years later when formulating his own ideas about the physical world. He certainly remembered the scholastic term "imaginary spaces"; it is in these so-called imaginary spaces that Descartes imagines God to construct his new world of *The World.* But he by no means intends us to take the scholastic conception of imaginary space at face value. He notes that "the philosophers tell us that these spaces are infinite, and they should be believed since they themselves made them" (AT XI 31–32; see also AT VI 42). And though he remembers the notion well enough to poke fun at it, it is unlikely that he remembers the complex details of the debates and positions he learned at school. Descartes' later knowledge of many of the antischolastic innovators is similarly uncertain. In 1630, writing to Isaac Beeckman, Descartes mentions the names of Telesio, Campanella, and Bruno in passing (AT I 158 [K 16]). But his point here is only to note that he can learn nothing from them, and to imply (very unfairly) that he learned nothing worth knowing from Beeckman either.

Despite Descartes' rather ungracious remarks, it probably was Beeckman who introduced him to the modern anti-Aristotelian thought on space and vacuum in 1618. Beeckman certainly believed that there are genuinely empty spaces in nature, a view he advanced in his doctoral theses in medicine, defended on 6 September 1618, just a few months before he met Descartes.[52] No doubt this was one of the subjects that he discussed with the young Descartes, whose head was at that time still filled with what he had learned from the Jesuits, as Beeckman reports (JB I 244 [AT X 52]). And while even those who believed in a plenum debated the question of motion in a vacuum, as Descartes himself was later to do, it may possibly be significant that one of the problems he and Beeckman discussed most extensively in 1618, and one of the extended pieces he wrote for Beeckman at that time, was a discussion of the motion of a falling body in a vacuum.[53]

Whatever Descartes may have believed in 1618, by the end of the next decade he seems to have opposed the void. The situation in the *Rules,* which, granted, deals with method more than substance, is suggestive of the position he will later take, though not unambiguously. In Rule 12, for example, he claims that we are mistaken "if from the fact that in this space full of air we perceive nothing by vision, by touch, or by any other sense, we conclude that it is empty, improperly joining the nature of vacuum with that of space" (AT X 424). This might be taken to suggest that Descartes opposed vacua in general. But the passage seems to imply that the notion of an empty space is intelligible, in contrast to what he will later argue, and even if we are mistaken in thinking that a jar without wine, water, or peanut butter is empty, there is nothing in this passage that rules out the possibility that there may well be empty spaces elsewhere in nature. There is also a passage in Rule 14 that is suggestive of his later views on body and vacuum. There he seems to identify body and extension and argues, "even if someone could persuade himself, for example, that if whatever is extended in nature were reduced to nothing, and that meanwhile that extension *per se* could exist alone, he would not use a corporeal idea, but the intellect alone, badly judging" (AT X 442–43). This, again, is suggestive of his later position, and the argument we shall later examine, that extension without body is inconceivable, strictly speaking. But this passage is not without its complexities. It has a role to play in the rather complex argument of Rule 14, and it is not absolutely clear that Descartes intends this to support any general point about the nature of the world.[54]

By 1631, though, Descartes has declared himself unambiguously against the void. Writing to Mersenne in October or November of that year concerning an earlier discussion they had about motion in the void, he remarks that "one cannot assume a vacuum without error" (AT I 228).[55] This is the view he takes in *The World*, of course, where God creates Descartes' hypothetical world without any empty space (AT XI 33, 49), a view that is repeated often in the letters and writings that immediately follow (AT I 301, 417; AT II 382–83; AT VI 86; AT XI 629; etc.).

When Descartes first declares himself against the void it is not absolutely clear just why he opposes what many of his contemporaries accepted. There is a chapter in *The World* entitled "On the vacuum, and why our senses do not perceive certain bodies" (AT XI 16).[56] But, unfortunately, that chapter contains no proper argument against the void. Instead Descartes offers therapeutic considerations to "free us from an error with which all of us have been preoccupied from our youth" (AT XI 17). While it is not easy to follow the argument in this section, the main aim seems to be to show that contrary to our prejudices, air is body, not empty space, and, indeed, is less likely to contain empty spaces than are hard bodies. Descartes, for example, notes that hard bodies will bend and shatter rather than allowing a vacuum, from which he concludes that air, a fluid, will all the more deform itself to prevent an empty space. Furthermore, he notes that even if, as he thinks, the invisible bodies around us filled all available space, then this would not prevent larger bodies from moving, but only require that bodies moving in such a plenum set up circular streams of moving matter (AT XI 18–20). But nothing in that chapter establishes in any rigorous way the claims he is trying to make if not plausible then at least not implausible, that there are no empty places within bodies, solid or liquid, and the chapter does not even address the question of whether beyond this world there are extracosmic empty spaces. This is something Descartes fully realizes. In the course of his discussion he notes:

> I don't want to establish in this way that there is no void in nature; I feared that my discourse would be too long if I tried to explain the whole matter, and the observations of which I spoke are not sufficient to prove it [i.e., that there is no vacuum], although they are good enough to persuade us that the spaces where we sense nothing are filled with the same matter and contain at least as much of it as the spaces which are occupied by the bodies we sense. (AT XI 20–21)

The discussion in chapter 4 of *The World* gives us reason for thinking that the world may be full, but it gives us no grounds for claiming that it is impossible that it isn't. However, in admitting this, Descartes suggests that he has stronger arguments at hand, and that it is only a matter of practical considerations that prevented him from treating the issue at greater length and giving those arguments. It is not implausible to conjecture that those stronger arguments are grounded in his conception of body. In *The World,* as in Rule 14, he claims that body and extension are intimately linked. He writes:

> If I am not mistaken, all of the difficulty they [i.e. the scholastics] have with respect to theirs [i.e., their conception of matter] comes only from the fact that they want to distinguish [matter] from its quantity and its external extension, that is to say, from the property it has for occupying space. . . . They shouldn't find it strange if I assume that the quantity that pertains to the matter I describe does not differ from its substance any more than number does from things numbered and that I conceive of its extension or the property it has to occupy space not at all as an accident but as its true form and essence. (AT XI 35–36)

If body and extension are linked in this way, then it is reasonable to suppose that there can be no region of extension that is without body, no vacuum, and that an argument something like this is what Descartes may have had in mind in the passage quoted earlier. If this is, indeed, what Descartes had in mind, then his reasons for not making it explicit may go beyond mere expository convenience. Descartes may not have been much of a scholar of scholastic philosophy in the early 1630s, but he surely would have remembered from his school days the theological problems connected with the view that a vacuum is impossible in any strong sense. And the argument suggested by the view of body in *The World,* that vacuum is impossible because space is identical with body, would seem to raise such problems, and make it impossible for even God to create a space empty of body.[57]

But if caution it was, then by the end of the decade caution is thrown to the wind. Writing to Mersenne on 15 November 1638, Descartes announced that "it is no less impossible that there be a space that is empty than that there be a mountain without a valley" (AT II 440). This view—that empty space of any sort, inside or outside the world, is impossible—is a view that he repeats often in the following years, both in his letters and in his published writings.

The argument Descartes gives comes in a number of variants. He often expresses his view by saying quite simply that the idea we have of space or extension is the same as the idea we have of body. And so, to Roberval who attempted to argue him into distinguishing body from space in a discussion they had in summer of 1648, Descartes is reported to have replied, "I clearly and distinctly see and know that body and space, which you think to be two distinct things because of some unknown blindness of intellect, are completely one and the same thing" (AT XI 689). While Descartes offers Roberval no argument, his reply is an apparent allusion to an argument given four years earlier in his *Principles*. There he considers as his example a stone, and shows how we can eliminate all of its properties except that of being extended without thereby eliminating its nature as a body *[corporis natura]*. Descartes then goes on to say that this extension "is the same thing contained in the idea of space, not only [space] full of body, but even that space which is called empty *[vacuum]*" (Pr II 11).

The argument is, of course, not entirely satisfactory. Leaving aside the problems with the way in which Descartes chose to establish the nature of body in this context discussed earlier in chapter 3, the argument is open to the obvious objection that while the extension of body may be the same as the extension of space, that is, while body and space may both be extended, it doesn't follow that they are the *same*. There were any number of opponents who were quite willing to distinguish between extension with body and the same extension without. But when developing the argument more carefully, Descartes intimates that the argument rests on an important metaphysical premise, that nothing has no properties. Writing, again in the *Principles,* he remarks:

> It is obvious that there cannot be a vacuum in the philosophical sense, that is, in which there is no substance at all from the fact that the extension of space does not differ at all from the extension of body. For since we correctly conclude that a body is a substance from the sole fact that it is extended in length, width, and depth, because it is entirely contradictory that extension pertain to nothing, we must also conclude the same concerning space assumed to be empty, namely, that since there is extension in it, there must also necessarily be substance. (Pr II 16)

Here he makes clearer the possible grounds for his identification of body and space. Since it is the nature of body to be an extended substance, whenever we have an extended something, we have a body; as Descartes told Roberval, "whatever is extended in and of itself, that I call body" (AT XI 688). And so, insofar as the idea of extension without substance is incoherent, and the idea of extension with substance is just body, our idea of space, properly, that is, coherently considered, must be the same as our idea of body, extended *something*, extended *substance*. Thus, he concludes, there can be no vacuum, no space void of body.[58]

Descartes sometimes offers a graphic illustration of his argument.

The illustration, first formulated in January 1639 in apparent response to a question Mersenne raised, received an elegant statement in the *Principles* of 1644:

> It is no less contradictory for us to conceive a mountain without a valley than it is for us to think of . . . this extension without a substance that is extended, since, as has often been said, no extension can belong to nothing. And thus, if anyone were to ask what would happen if God were to remove all body contained in a vessel and to permit nothing else to enter in the place of the body removed, we must respond that the sides of the vessel would, by virtue of this, be mutually contiguous. For, when there is nothing between two bodies, they must necessarily touch. And it is obviously contradictory that they be distant, that is, that there be a distance between them but that that distance be a nothing, since all distance is a mode of extension, and thus cannot exist without an extended substance. (Pr II 18L)[59]

If the two sides of the vessel are separated, there must be some distance between them, and if there is distance, then there must be body. On the other hand, if there is no body, there can be no distance, and if there is no distance, then the two sides must touch.[60]

The argument we have been examining suggests that Descartes' denial of the vacuum, while obviously connected with his account of body as extended substance, is independent to the extent that it appears to require an additional premise, that no properties pertain to nothing. While it is true that he appeals to this premise, and, indeed, must appeal to some such principle for his strategy to work, there is another way of looking at the argument against empty space that ties it more closely still to his account of body. Writing probably to the Marquis of Newcastle in October 1645, commenting explicitly on the vessel argument we have just examined, he remarks:

> I think that it implies a contradiction for there to be a vacuum, since we have the same idea of matter and of space, and since this idea represents to us a real thing, we would contradict ourselves and assert the contrary of what we think if we said that this space is empty, that is to say, that that which we conceive of as a real thing is not at all real. (AT IV 329 [K 184])

This, of course, takes us back to what I called the complete concept argument for the nature of body in chapter 3. The idea we have of body is the idea that terminates and grounds the incomplete ideas of modes of extension; it is the idea of a *thing* that is extended, and all notions of shape, size, and even distance depend upon that idea for their intelligibility. It is because we have such an idea that grounds all of our ideas of the modes of extension and none of our ideas of the modes of thought that we know that bodies, material substances, are distinct from minds, mental or thinking substances. And it is because we have such an idea, and because such an idea grounds our ideas of modes of extension that we know that there cannot be a mode of extension that does not pertain to an extended *thing*, a body. Regarded in this way, the metaphysical principle that extension requires an extended substance is not a metaphysical principle extraneous to his conception of body. In fact, the metaphysical principle is, in a way, built into Descartes' account of body, and is absolutely essential to the argument by which he establishes the nature of body as extended substance. It is still, perhaps, too strong to say that the denial of the vacuum follows directly from his account of body. But it is fair to say that both the doctrine on the nature of body and the denial of the vacuum flow from the same underlying position, the view that all of our ideas of modes of extension depend for their intelligibility on an idea we have of an extended *thing*. In a sense, then, the way he chose to eliminate sensory qualities from bodies and geometrize the material world, the view of conceptual dependency that leads him to view bodies as extended things that exclude all mentality also leads him to deny the very possibility of empty space.

Even though Descartes identifies space with body, he concedes that there is a way in which it makes sense to talk about space or place independently of the bodies that actually occupy them, at least by the time of the *Principles*. Basic to the discussion there is the distinction between internal and external place.[61]

Internal place Descartes equates with the space (i.e. volume) occupied by a given body. Though in reality it does not differ from body itself, we can, nevertheless, make the following sort of distinction:

> Space or internal place does not in reality differ from the corporeal substance contained in it except in the way in which we usually conceive it. For the extension in length, breadth, and depth which constitute space is plainly the same as that which constitutes body. But there is this difference. In body we consider it as an individual, and we think that it always changes whenever the body changes, but in space we attribute only a generic unity so that when the body that fills the space changes, the extension of the space is not thought to change, but is thought to remain one and the same as long as it retains the same size and shape, and keeps the same situation among certain external bodies through which we determine that space. (Pr II 10L)

Body, then, is concrete extension, the extension regarded as an individual. In this sense, when body moves, its extension moves with it. But we can also regard extension "generically," or, as he puts it in a later section, we can also "consider extension in general [*extensio in genere*]." Considered in this way, abstracted from the extended thing, we can imagine the same space occupied by a stone, "wood, water, air, or other bodies, or even a vacuum, if there is such a thing" (Pr II 12).

Descartes' account of external place is not quite so clear. "External place," he says, "can be taken as the surface that most closely surrounds the thing located" (Pr II 15).[62] This, of course, is the standard Aristotelian definition of place. Understood in this way, the place is a *mode* of a body, not a part, indeed a mode that a given body shares with the body or bodies that surround it (see Pr II 15; AT VII 433–34; AT III 387). But there is another slightly different way of understanding external place, a way that Descartes seems to prefer. In this sense the external place

> is understood to be the common surface, which is no more part of one body than of another, but is thought to remain the same when it retains the same size and shape. For even if every surrounding body changes, together with its surface, the thing is not thought to change its place on that account if it preserves the same situation among those external bodies which are regarded as immobile. (Pr II 15)

Regarded in this way, the external place is the surface taken abstractly, the surface regarded independently of what body or bodies may *actually* have that surface, the surface defined in terms of certain external bodies which are regarded as being at rest. Regarded in this way, the view of what is a given place, and what constitutes a change in place is strictly relative to the external bodies we happen to choose as our reference points, as he realizes: "In order to determine situation, we must look to certain other bodies which we regard as immobile, and as we look to different bodies we can say that the same thing at the same time both changes place and does not change place" (Pr II 13). When external place is understood in this way, the same thing can be regarded now in one place, now in another, and different things can be said to occupy the same place.

In this way Descartes is able to grant that the notions of space and place are, in a sense, independent from that of body. But his account should offer little comfort to those who posit a container space as something ontologically distinct from body, something in which bodies exist. As he summarizes his view, "the names 'place' and 'space' do not refer to anything distinct from the body which is said to be in a place, but only refer to its size, shape, and situation among other bodies" (Pr II 13). Space and place, thus, are not realities themselves, but merely abstractions from body, abstractions from that which is real. In thinking about internal place or space we abstract out size and shape, and in thinking about external place we abstract out position; indeed in dealing with external place on Descartes' preferred conception, the very specification of the place depends upon the arbitrary choice of external bodies that serve as a frame of reference. Though in a sense independent of bodies, spaces and places could not exist if there were no bodies in the world. And so, when in 1644 Samuel Sorbière on behalf of Gassendi pressed Descartes to admit that, as commonly believed, there was space before God created the matter that occupies it, he is reported to have responded that "spaces were created together with body" (AT IV 109; see also AT II 138).

This completes our basic account of Descartes on space and the void. But Descartes' position is further clarified by examining his intellectual relations with two representatives of alternative points of view, Blaise Pascal and Henry More, and his relation to the scholastic tradition from which he derived and which he appears most closely to resemble in his opposition to atoms and the void. This is what we shall do in the remainder of this chapter.

DESCARTES AND PASCAL:
EXPERIMENTS ON THE VOID

In October 1646 the Pascal family was visited in Rouen by Pierre Petit, a military engineer known to Étienne Pascal through scientific circles.[63] Petit demonstrated to the family an experiment that had recently been performed in Italy under the direction of Evangelista Torricelli, a student of Galileo.[64] Petit took a tube four feet in length, hermetically sealed at one end, and filled it with mercury. Placing his finger over the end, he lowered the open end of the tube into a bowl with mercury at the bottom and water covering the mercury. When the tube was near the bottom, he withdrew his finger, and the mercury dropped in the tube, leaving an apparently empty space at the top. The young Blaise Pascal objected that the space was not empty, but filled with air that had penetrated the tube through its pores. After some discussion of this possibility, in which Petit gave a number of arguments to show that no air could enter, Petit continued the experiment. He raised the tube little by little, and the apparently empty space grew, until lifting the open end of the tube above the level of the mercury and into the region of water, the mercury suddenly left the tube, which quickly filled with water, leaving no empty spaces. This, Petit argued, is definitive proof that the space at the top of the mercury was genuinely empty.[65]

Though by no means uncritical, the young Blaise Pascal was intrigued. He undertook a complex series of experiments to explore the phenomena, and planned a major work, a treatise on the void that he never was to complete. But the direction his work took is evident from two publications, the *Expériences nouvelles touchant le vide* of October 1647, and the *Récit de la grande expérience de l'équilibre des liqueurs* of October 1648.[66] In the first of these works Pascal reports on a series of variations he made on the original experiments Petit showed him. There, by varying the apparatus, using tubes of different widths, different configurations of tubes, different liquids, including water and wine, he attempted to exclude the hypotheses that the apparently empty space contained air of any sort, rarified mercury, water or wine, or "the more subtle air mixed with external air . . . [which is] separated and enters through the pores of the glass."[67] Pascal's somewhat cautious conclusion is that "until someone shows me the existence of some matter which fills it, my view is that it is genuinely empty, and destitute of all matter."[68] In the *Expériences* he suggests that it is a limited horror of the vacuum that supports the column of mercury in the tube. "All bodies have a repugnance toward separating from one another," but, Pascal wrote, "the force of this horror is limited," and is equal to the force with which the column of liquid of a given height in the tube "tends to fall."[69] But by the time that the *Expériences* were published, he may have suspected that it isn't a limited horror of the vacuum but the pressure of the air on the surface of the mercury that holds up the column.[70] On 15 November 1647 he wrote to his brother-in-law Florin Périer, asking him to take the apparatus to the top of a mountain, the Puy-de-Dôme, to see if the lower air pressure there would affect the height of the column.[71] The experiment was finally performed on 22 September 1648, and Pascal's suspicion was borne out; the column of mercury was shorter at the top of the mountain than it was at the bottom. He hastily published the results in the *Récit de la grande expérience,* which appeared within weeks of the experiment. He ended his account of the details of the experiment with the following conclusion:

My dear reader. The universal agreement of people and the multitude of philosophers agree in establishing the principle that nature would allow its own destruction rather than allowing the least empty space. Certain more elevated spirits have taken a more moderate view; although they believe that nature has a horror of the vacuum, nevertheless they believe that this repugnance has limits, and that it can be overcome through violence. But until now one could find no one who took this third view, that nature has no repugnance for the vacuum, that it makes no effort to avoid it, and that it admits vacuum without difficulty and without resistance. The experiments I have presented in my summary account [i.e., in the *Expériences nouvelles*] in my view destroy the first of these principles, and I don't see how the second can resist those which I present you now. So, I no longer have any difficulty in taking the third, that nature has no repugnance for the vacuum, that it makes no effort to avoid it, and that all of the effects which people have attributed to this horror proceed from the weight and pressure of the air.[72]

Descartes had worried about similar experimental phenomena as early as 1631, and had proposed atmospheric pressure as the explanation of a number of phenomena often attributed to the horror of a vacuum (AT I 205–8; AT II 399, 465).[73] It is possible that he may have heard about the Torricelli experiments that so moved Pascal as early as 1645, when Mersenne returned from Italy to Paris, and informed some of his circle of the new work.[74] But there is no record of his knowledge of the new experimental proofs of the vacuum until 23 and 24 September 1647, when Descartes visited the young Blaise Pascal in Paris. At that time the *Expériences nouvelles* were not yet published (the official permission to print is dated 8 October), but the pamphlet was certainly complete, and issues connected with that writing were no doubt very much on Pascal's mind.[75]

Descartes' response to the new experiments appears, at first glance, to be rather curious. On the one hand he clearly rejected the view that the apparently empty space at the top of the tube of mercury is a vacuum. In the brief account Blaise's sister Jacqueline gave of their meetings, an important but unfortunately none too detailed letter written on 25 September 1647 to her sister Gilberte, Madame Périer, Descartes is reported to have suggested that the space is filled with subtle matter, the fine and fast-moving tiny particles that make up his first element, which presumably enter through the pores of the glass.[76] But yet he is happy to agree, somewhat to Pascal's initial surprise, that it is atmospheric pressure, the weight of the surrounding air, that supports the column of mercury.[77] Indeed, there is reason to believe that Descartes may have suggested to Pascal in their meetings of September 1647 that he perform an experiment like the one ultimately performed at the Puy-de-Dôme at Pascal's direction in September 1648. In December 1647, before he knew of any directions Pascal might have given to Périer, Descartes told Mersenne that he had suggested this to Pascal, and later, in 1649, complained to Carcavy that he, Descartes, had originally suggested the experiment for which Pascal got full credit (AT V

99, 366, 391).[78] Now, it is quite possible that Pascal may have had such an experiment in mind in September 1647 when Descartes may have suggested it. But what is interesting about these remarks is not their bearing on the question of priority, but the fact that Descartes took this experiment that Pascal thought established that nature has no horror of vacuum, this experiment that Pascal thought proved that there could be vacua in nature, to establish quite the contrary, that there could not be a vacuum in nature. Writing to Carcavy in August 1649, apparently without knowledge of the details of the *Récit* in which Pascal reported the results of the Puy-de-Dôme experiment, Descartes noted, "I had some interest in knowing [the outcome of the experiment] since it was I who asked him to do it two years ago [i.e., September 1647], and I was sure that it would succeed since it is entirely in accord with my principles. Without this he never would have thought of it, since he held a different opinion" (AT V 391).[79]

Descartes' reaction to the Torricelli experiments and to Pascal's extension of them raises two important questions, one narrow and one broad. The first and most straightforward question is why did Descartes and Pascal draw such different conclusions from the Puy-de-Dôme experiment? How could they have thought that the experimental confirmation of the view that the column of mercury is supported by the weight of the air confirmed both that there is a vacuum at the top of the tube and that there is no vacuum there? But the affair raises a more general question as well. Pascal, Torricelli, and many others saw the experiments at issue as an experimental demonstration that vacua are not impossible. Should Descartes have taken Pascal's attack more seriously? Was he simply being dogmatic in refusing to concede Pascal's argument?

The first issue is relatively easy to deal with. Pascal tells us in the passage from the *Récit* of October 1648 quoted above how he interprets the experiment of the Puy-de-Dôme. Consider a simple tube sealed at one end, filled with mercury, standing in a pool of mercury. Under these circumstances, the mercury remains in the tube, and does not descend completely into the bowl below. Pascal imagines three possible explanations: (*a*) nature abhors a vacuum altogether and the mercury column remains in the tube to prevent such a vacuum; (*b*) nature resists a vacuum, but does not prevent it altogether, and the mercury descends only as far as it can, while it leaves a vacuum above the column, a column which is supported by the resistance nature has to a larger vacuum; and (*c*) nature freely allows a vacuum, and the column of mercury is supported by an external force, the weight of the air. Now, Pascal claims, the simple experiments reported in the *Expériences nouvelles* eliminate the first possibility, (*a*), since if the tube is sufficiently long, the mercury will drop to some extent, leaving what he takes to be a vacuum at the top. This leaves *(b)* and *(c)*. The Puy-de-Dôme experiment shows that the length of the column in the tube lowers as the air pressure decreases, establishing that it is the weight of the air on the surface of the pool of mercury and not any resistance to the vacuum that supports the column; explanation *b* is thus eliminated, leaving explanation *c*. And thus, "yielding to the force of truth, which constrains me," Pascal comes finally to the position that nature has no repugnance for the vacuum, a position he thinks is forced upon him by the results of the experiment at Puy-de-Dôme.[80]

It is somewhat more difficult to present the problem from Descartes' point of view; since he never wrote a systematic account of the experiments, we can only conjecture as to how exactly he saw the issue. But, I suspect, his view was something like this. Descartes, like Pascal, rejected the horror of the vacuum, indeed had done so a number of years earlier (see, e.g., AT II 399). And so he never considers Pascal's explanation *a* as a serious contender. On Descartes' view, I think, what the Pascal camp proposes is explanation *b*, the view that the column is supported by a restricted horror of the vacuum in nature, that it is supported by a not altogether insuperable resistance nature has to allowing a vacuum. This is the view Pascal apparently takes in the *Expériences nouvelles* that Descartes received by December 1647;[81] it is the view that Roberval likely pressed in the meeting of 23 September when he spoke for Pascal, then too ill to enter into the dispute, and was the view that Le Tenneur, another of Pascal's circle, expressed to Mersenne as late as January 1648, a view that likely made its way to Descartes.[82] It is quite possible that the view *c* that Pascal eventually came to hold after performing the Puy-de-Dôme experiment, the view that nature itself freely allows a vacuum, was never even considered by Descartes. While there is reason to think that Pascal may have considered it himself in September 1647, it is quite possible that it never came out in the discussion, dominated as it was by Roberval. Furthermore, that view is never as much as mentioned in Descartes' letters, and while it is prominent in Pascal's *Récit* of 1648, there is no reason to believe that he ever saw a copy of that work.[83] So, as far as Descartes was concerned, those who posited a vacuum did so largely in order to explain why the column of mercury stood in the tube of the Torricelli apparatus. Against this he proposed what he took to be an altogether different explanation, that the space above the column was filled with subtle matter, and that the column of mercury was supported by the weight of the air. And so an experiment like the Puy-de-Dôme experiment, where the apparatus is carried to a higher altitude, seemed like the perfect crucial experiment, I suspect Descartes thought. If the column was shorter, he thought he would be vindicated; if not, then what he took to be Pascal's view would have been. It is, I think, because he was alto-gether unaware of Pascal's alternative *c* that Descartes suggested in August 1649 that Pascal would not have considered the Puy-de-Dôme experiment without his suggestion. It is, by the way, interesting to note that the view of Pascal's thinking I attribute to Descartes was probably shared by some members of Pascal's circle. When Mersenne wrote to Le Tenneur in January 1648, not knowing of Pascal's supposed request to Périer the previous November, and asked Le Tenneur to execute a similar experiment, Le Tenneur replied, "I tell you that I think, along with Roberval, that this would be entirely useless, and that the same thing found below would be found up high" (CM XVI 56).[84]

It is, I think, understandable how both Descartes and Pascal could have thought that the experiment performed at the Puy-de-Dôme could constitute a crucial experiment, and could support their own views. But however understandable it might be, both were wrong.[85] The Puy-de-Dôme experiment suggests strongly that it is the weight of the air, greater at the bottom of the mountain and lesser at the top, that supports the column of mercury in the Torricelli apparatus. But while it may show us what supports the column, it doesn't tell us what lies above it in the tube; Descartes and Pascal can and did disagree on this question while being in perfect agreement on the other.[86]

The second question I raised earlier is somewhat less straightfor-ward: was Descartes unreasonably dismissive of Pascal's arguments for the void? Should he have taken them more seriously as an attack on his own denial of the void? Descartes was without question deeply commit-ted to his principles, and no doubt saw his identification of space and body and consequent denial of the vacuum as basic to his view, and would not give the position up lightly. And he certainly could appear dogmatic. In a public discussion held in the summer of 1648 in Paris, Roberval tried again to press Descartes with the same experiments and to argue again for the void. Baillet reports that Descartes "took no part but that of a spectator. That is why he spoke little and then only to note how these experiments are in agreement with his principles."[87]

But it was not merely the strength of his convictions that prevented Descartes from taking the attack of the Pascal camp more seriously. I think that it is fair to say that Descartes never saw in those experiments any serious challenge to his view, nor in Pascal any serious answer to his own interpretation. I think that it was simply obvious to him that the glass had pores that admitted the subtle matter pervading the universe to enter, even though they prevented the grosser particles of air to enter, a view that was explicitly pressed against Pascal by Étienne Noël, Descartes' physics teacher at La Flèche more than thirty years earlier, a view that Pascal clearly identified as Descartes' as well.[88] It is because of this view that Descartes never saw any inconsistency between his principles and the experiments Pascal and others presented him with. Pascal did answer Noël, of course, both in a letter directly to the Jesuit father and in a letter written in February 1648 to Le Pailleur.[89] But there is no reason to think that Descartes ever saw these texts. All Descartes ever got from Pascal was the promise of a refutation of his preferred explanation. At the end of the *Expériences nouvelles,* a preliminary outline of a never completed treatise on the vacuum, Pascal promised to respond to the objection "that a matter imperceptible, extraordinary, and unknown to all of the senses fills the space [above the column]," a position formulated with Descartes in mind, no doubt.[90] Descartes seems to have received the work in good humor. Writing to Huygens on 8 December, shortly after having received the *Expériences nouvelles,* he noted:

> It appears to me that the young man who wrote this booklet has the vacuum a bit too much on his mind, and is somewhat hasty. I wish the volume he promises were already available, so that one could see his reasons, which are, if I am not mistaken, insufficiently solid for what he has undertaken to prove. (AT V 653)

A few days later, on 13 December, he wrote similarly to Mersenne:

> It appears that he [Pascal] wants to oppose my subtle matter, and I am deeply grateful to him. But I beg him not to forget to give all of his best reasons on this subject, and not to take it badly if some time or place I explain everything I believe on this to defend myself. (AT V 98)

But almost two years later Descartes is still waiting for the objections. Writing to Carcavy in August 1649 Descartes says:

> And since he had sent me earlier a small pamphlet [i.e., the *Expériences nouvelles*] where he described his first experiments on the vacuum and promised to refute my subtle matter, if you see him, I would be very pleased for him to know that I am still waiting for this refutation, and that I will receive it without offence, as I always receive objections made against me without calumny. (AT V 391–92)

Carcavy wrote to Pascal on Descartes' behalf, but no answer seems to have been received before Descartes' death the following February.[91] Similarly, Baillet reports that Descartes asked Roberval to put his objections in writing, objections delivered with great passion in the public discussions of the summer of 1648, so that they could be given proper consideration, a request, Baillet reports, that Roberval denied.[92] And so, from Descartes' point of view, there never seems to have been a serious experimental attack on his belief in the plenum for him to answer.

Indeed, the partisans of the Cartesian point of view had an interesting argument for their view that the space in the tube above the mercury was not empty, an argument not easy to dismiss. The argument is set out at greatest length by Father Noël in his two letters to Pascal on the void from autumn of 1647. The objection starts from the observation that light can pass through the supposedly empty space, and, indeed, is refracted when it enters it. Noël develops the argument in a number of different ways. Sometimes he suggests that the transmission of light requires an appropriate medium,[93] sometimes that the phenomenon of refraction indicates that something in the interval is resisting the light while it passes through.[94] But the most interesting objection Noël raises is the simplest one. The passage of light freely through the glass shows that the glass has pores, pores that could also admit subtle matter as well.[95] A letter from Mersenne to Le Tenneur from January 1648 suggests that Mersenne had such an objection in mind as well, and the association of that discussion with Descartes' name suggests that Descartes may also have been aware of that consideration (CM XVI 59–61). Now, Pascal and Le Tenneur had answers to this kind of an objection. Pascal, for example, pointed out that we don't really know what light is, and so we cannot say anything about what does or does not follow from the fact that light can pass through the apparently empty space above the mercury.[96] Le Tenneur wrote Mersenne that he accepts the view common among his contemporaries that light is something corporeal. But since he believes that it is a limited horror of vacuum that holds up the column of mercury, he argues that if light or subtle matter could simply enter and fill the empty space, then the column in the tube would fall (CM XVI 60). Whether Descartes himself ever heard such answers we do not know; it is quite likely that they were discussed in his presence on a number of occasions. But it would have been no mere dogmatism to be skeptical of the supposed experimental refutations of the vacuum. It was, in fact, exactly on these grounds, the passage of light through the supposed vacuum, that Pierre Gassendi and following him, Walter Charleton, no foes of the vacuum and in Gassendi's case, among the very first supporters of Pascal's work, questioned the conclusions Pascal had drawn.[97]

Descartes in the end may have been wrong in denying the vacuum, and Pascal may have largely been correct. But it is not unsurprising that nothing in his exchanges with Pascal and his circle should have led Descartes to see the error of his ways.

DESCARTES AND MORE:
DIVINE EXTENSION AND IMPENETRABILITY

Pascal's attack on Descartes' view of space and body was experimental, or at least it was meant to be. But at roughly the same time Descartes was engaged in his discussions with the Pascal circle, there was another more metaphysical attack on Descartes' position from another young philosopher, Henry More, of Christ's College, Cambridge, a member of the group later to be known as the Cambridge Platonists. On 11 December 1648, More wrote Descartes a flattering letter that began an intense correspondence, cut short less than a year later by events surrounding Descartes' departure for Sweden, where he died in February 1650. In the few letters they exchanged, More asked a series of probing questions on a variety of subjects, questions that elicited important clarifications of Descartes' views. We have already seen some of the contents of their exchange in chapter 4 in connection with Descartes' doctrine of the animal soul, and in later chapters we shall examine their exchanges about motion, force, and relativity. But in this chapter I would like to examine another issue More raises. In the letters with Descartes, More opposes to Descartes' view of space and body an alternative point of view, an early hint of his view of space as a divine attribute that is later to prove so influential on English scientific thought.[98]

More's first question in his first letter to Descartes in December 1648 concerns extension and the nature of body: "Firstly, you frame a definition of matter or body that is much broader than is appropriate. For God seems to be an extended thing, as is an angel, and indeed, anything subsisting in itself" (AT V 238).[99] More goes on to give his reasons for thinking that God is an extended thing:

> And, indeed, I judge that the fact that God is extended in his own way follows from the fact that he is omnipresent and intimately occupies the universal machine of the world and each of its parts. For how could he have impressed motion on matter, which he did once and which you think he does even now, unless he, as it were, immediately touches the matter of the universe, or at least, did so once? This never could have happened unless he were everywhere and occupied every single place. Therefore, God is extended in his own way and spread out; and so he is an extended thing. (AT V 238–39; see also 379)

God, insofar as he is everywhere, must be extended. And insofar as he is capable of causing motion, he must be present to cause the motion and so, again, must be extended. More, by the way, doesn't seem to be worried about the problem that worried Elisabeth, that there is some metaphysical problem about mental-physical causality, that if the mind is incorporeal and unextended, then it cannot be the sort of thing to cause motion in a body. The point is simply that if God causes motion at a given place, he must *be* there, and if he is there, and indeed everywhere (since he causes motion everywhere), then he must be extended.

But if God, angels, and, indeed, human souls can be extended, then what is to differentiate them from body? More suggests the following:

> Although matter is not necessarily soft or hard, hot or cold, it is, however, most necessary that it be sensible, or, if you prefer, tangible. . . . But if it pleases you less to define body in terms of its relation to our senses, this tangibility can be taken more broadly and more widely to mean that mutual contact and power of touching between any bodies, be they animate or inanimate, and be it the immediate juxtaposition of the surfaces of two or of more bodies. And this points up another property of matter or body, which you can call impenetrability, namely, [the property of] not being able to penetrate other bodies or be penetrated by them. From this there is a quite obvious distinction between divine and corporeal nature, since divine nature can penetrate corporeal nature, but corporeal nature cannot penetrate itself. . . . These few considerations suffice to demonstrate that it would have been much safer to have defined matter as tangible substance, or as explicated above, as impenetrable substance, rather than as an extended thing. (AT V 239–40; see also 301, 378)

So, More suggests, what differentiates body from divine extension is the fact that body is impenetrable, while divine extension is not, insofar as divine extension *contains* bodies. And so, to define body as extended substance is to define it too broadly, and fail to distinguish that which pertains to body as such from that which seems, for More, to pertain to everything.[100]

One consequence of this position for More is that empty space, space devoid of body, is possible. More, unlike many who opposed Descartes' theory of space, was in complete agreement with him that one cannot have extension without a substance for that extension to inhere in: "I concede . . . that there is some substance in every space." But, More continues, that substance is "not necessarily corporeal substance, since divine extension or presence can be the subject of measurability" (AT V 302). And so, in response to the vessel example of Pr II 18, More claims that God *could* eliminate the matter in the vessel without the sides touching: "I contend that divine extension lies between [the sides of the empty vessel], and that here your assumption that only matter is extended is mistaken" (AT V 241).

Descartes' replies to More are interesting. Not unsurprisingly, he continues to defend the position articulated in the *Principles,* that there can be no space without body, that the idea of a vacuum hides a contradiction (AT V 271–73 [K 239–41]; AT V 403 [K 257]). And appealing to a distinction he first made six or so years earlier when answering Princess Elisabeth's worries about the mind's relation to the body (AT III 694 [K 142–43]), Descartes distinguishes the proper sense in which bodies are extended from the derivative sense in which God, angels, and human souls are extended:

> In God, in angels, and in our mind I understand no extension of substance but at most the extension of power, so that an angel can exercise its power now in a greater, now in a lesser part of a corporeal substance. For if there were no body, I should think that there would also be no space with which an angel or God would be coextensive. (AT V 342 [K 249])[101]

But most interesting is the discussion of the notion of impenetrability that More's questions elicit.[102] The discussion is not altogether new. Descartes does mention in passing in the *Sixth Replies* that "the true extension of body is such that all interpenetrability of parts is excluded" (AT VII 442). But the remark is just made in passing, and Descartes manages to pass through the entire *Principles* without even once confronting the issue of impenetrability directly.

As we saw earlier, More seems to conceive of impenetrability as being virtually equivalent to tangibility and sensibility; to say that body as distinct from empty space is impenetrable is to say that it is tangible, that it can be touched, discerned by the senses. Descartes, first of all, rejects this:

> If you conceive of extension through the relation of parts with respect to one another, it seems that you cannot deny that each and every one of its parts touches all neighboring parts; and this tangibility is a true property, intrinsic to the thing, not something denominated by the sense of touch. (AT V 341–42 [K 248–49])

The notion of impenetrability, Descartes claims, is a real property of body, the property by virtue of which parts touch one another without penetrating, and is not the notion of tangibility defined in terms of the senses.[103] The notion of impenetrability proper to bodies, Descartes goes on to say, is a direct consequence of the very notion of extension, or, more properly, a direct consequence of the notion of extended substance. The grounds of Descartes' claim are clearly to be found in the discussion of body in the *Principles*. There Descartes denied that a given body can come to occupy a greater space without the addition of more substance, "since any addition of extension or quantity is unintelligible without the addition of substance which has quantity and is extended" (Pr II 7). And similarly, Descartes claims, "it cannot happen that the least bit of this quantity or extension be removed without exactly that amount of substance being eliminated" (Pr II 8). And so Descartes writes to More:

> We cannot even understand one part of an extended thing penetrating another part equal to it without understanding by that very fact half of that extension eliminated or annihilated. But what is annihilated does not penetrate another thing. And thus, in my judgment, it is demonstrated that impenetrability pertains to the essence of extension. . . . Therefore, impenetrability must be admitted in every space. (AT V 342) [104]

The argument is simple and ingenious. If body is an extended thing, in the sense in which Descartes understands it, then take away extension and you take away body. But if two bodies could penetrate one another, then total volume, and thus some amount of body itself, would be eliminated. But if one or another body is in part eliminated by this supposed interpenetration, then there is no real interpenetration, since what is annihilated does not penetrate anything. Consider two spheres, A and B, approaching one another and eventually appearing to interpenetrate. If they do interpenetrate, then the total volume of A and B is less than the total volume of the two spheres before interpenetration. And so, Descartes concludes, some amount of bodily substance must be eliminated in the process of this supposed interpenetration. Descartes' claim seems to be that what happens in this case is that the area of apparent overlap belongs to one sphere alone, A, say, and that a corresponding amount of sphere B has been annihilated. And if this is the case, then in the case at hand, there is no interpenetration, only annihilation.

Descartes thought he had good grounds for holding that impenetrability followed from the notion of an extended substance. And, insofar as he thought that it was less basic than the notion of extension from which he took it to follow, he resisted More's suggestion that he define body in terms of impenetrability rather than extension:

> Let us see now whether [body] can more appropriately be called "impenetrable or tangible substance" in the sense in which you explicated it. But, on the other hand, that tangibility and impenetrability in body is only like risibility in people . . . , not a true

and essential differentia, which, I claim, consists in extension. And because of this, just as 'man' is not defined as a risible animal but as a rational animal, so 'body' is not defined through impenetrability but through extension. (AT V 269 [K 238])

Indeed, Descartes goes on to say, unlike risibility, which is always found in human beings, there are certain circumstances in which impenetrability is not found in corporeal substance, even though it is a direct consequence of a body being extended substance. Descartes continues the above quoted passage by noting that "tangibility and impenetrability are related to parts, and presuppose the concept of division and limitation." And so, if we imagine the totality of corporeal substance without any divisions, without any parts, and, presumably, without anything outside of it, it will be extended, but the notion of impenetrability will not be applicable.

It was later to become a standard criticism of Descartes that he ignored the property of impenetrability, and because of this, conflated body and space.[105] But this is not altogether fair, and shows the extent to which Descartes' letters with More, published as early as 1657, were later ignored, and, more generally, the extent to which Descartes' position was later misunderstood.

DESCARTES AND HIS TEACHERS: VOID, PLENUM, AND DIVINE OMNIPOTENCE

The enormous complexity and variety within late medieval accounts of space and vacuum, together with some uncertainty about what precisely Descartes may have been exposed to at La Flèche and later, and how much Descartes may have remembered about the numerous debates and disputes in the scholastic literature when he came to formulate his own views on the issues, make it virtually impossible to give any simple and straightforward answer to the question of his relation to the scholastic tradition on space and vacuum. In many ways it is fair to say, though, that Descartes' position fits well within the mainstream of the scholastic tradition, and that Descartes must certainly have been deeply influenced on this question by his teachers.

As in Descartes, the orthodox scholastic view is that there are no empty spaces within the world. And like Descartes, many scholastics found the idea of extension without body to be unintelligible.[106] It is interesting to note that even the thought experiment of Pr II 18 was not unlike medieval thought experiments to the same effect; instead of imagining just the water in a vessel to be annihilated, Descartes' scholastic forebears would have God annihilate everything within the sphere of the moon.[107] While some tried to figure out a way in which God would be able to preserve the extension without the body it had contained,[108] many others were in agreement with Descartes that without body, the sides of the space evacuated must necessarily touch.[109]

There appears to be a greater difference between Descartes and his teachers on the question of an extramundane vacuum. Aristotle's cosmos had been finite in size, and nothing was thought to be outside of it. But as I noted earlier in this chapter, this is a doctrine that Christian physicists found in need of alteration, and concocted the idea that beyond the cosmos are the so-called imaginary spaces, apparently empty space that give God the freedom to move the cosmos. This, of course, appears quite foreign to Descartes' thought. For Descartes, the world is not finite, but extends out indefinitely, not empty, but full all the way out (Pr II 21). But even here, Descartes' views are not altogether unconnected to those of the schools. While there were some who held that the space beyond the cosmos is in some sense genuinely real,[110] and others saw the infinite extracosmic space as filled with God, somewhat as More was later to do,[111] for many the space without body is not space in the true sense. Though all had to admit something in some sense beyond the world, for many the so-called imaginary spaces are in no way real until occupied by body.[112]

One can get a sense of the general orthodoxy of Descartes' views on space and place by comparing his views with those of Francescus Toletus, the sixteenth-century Jesuit whose commentaries Descartes is known to have studied as a student at La Flèche, a late scholastic whose views on space and place are in many ways not unlike those the erstwhile student of the Jesuits came to adopt in the *Principles*. Toletus' views on space and place are presented in his commentary on book IV of Aristotle's *Physics*. Following a series of seven *quaestiones* treating the orthodox Aristotelian position,[113] Toletus finally turns to a moderately lengthy exposition of "a probable opinion" held by "certain more recent" authors "which . . . reconciles all opinions, and resolves all difficulties, and preserves . . . all properties of place."[114] Toletus first distinguishes place into intrinsic place or space, and extrinsic place (the sort of place "Aristotle spoke of") and then divides place along a different dimension into true place and imaginary place, yielding four sorts of place, intrinsic and extrinsic true place, and intrinsic and extrinsic imaginary place.[115] Intrinsic and extrinsic true place, a distinction Toletus did not himself invent, correspond closely to what Descartes calls internal and external place in the *Principles*.[116] External true place is "that which surrounds the located body, namely, the containing body, or its ultimate surface, about which Aristotle spoke."[117] As with Descartes' external place, Toletus recognizes somewhat different senses, ranging from the immediate boundary, to the situation of a body as determined by more remote bodies.[118] Intrinsic true place, on the other hand, corresponds to Descartes' internal place, "the space itself which the thing really occupies."[119] This space, Toletus tells us "is a certain property following from its quantity, namely the extension of its quantity."[120] But just as in Descartes, this true intrinsic place, this space cannot exist apart from body, but "inheres in things themselves as a proper accident of them."[121] As Descartes will later argue, this space is inseparable from body: "Moreover, it cannot be denied that this intrinsic space is in every body . . . since they [i.e., space and body] can be mutually inferred from one another. For if there is body, then there is space, and if there is true space, then there is body in it."[122]

Toletus' account of imaginary place is also strongly suggestive of elements from Descartes. Imaginary place is "for example that imaginary space beyond the heavens which anyone can imagine to be there. And also vacuum here in the world, if it exists, would be imaginary place, indeed, imaginary space."[123] While Toletus is not very interested in the distinction between extrinsic and intrinsic imaginary place, there is another distinction here that interests him, the distinction between fictive and abstracted imaginary space. Fictive space is "a thing altogether made and constructed, which does not exist."[124] Under this description Toletus includes, as Descartes no doubt would, space beyond the world (though for Descartes, unlike Toletus, the world extends indefinitely) and so-called empty space. Abstracted imaginary space is the true (intrinsic) space, considered apart from the particular bodies that happen to possess it: "since [space] seems to remain the same while bodies are always changing, one thinks that it [i.e. space] remains numerically the same, although it is only the same in species and in its equality."[125] And so we abstract the space from the bodies occupying it; it is in this sense for Toletus, as for Descartes some years later, that space can be conceived of as the immobile container that remains unchanged as the bodies it contains change.[126] But though "this conception is not made but is true,"[127] such a space is an abstraction from what is real, and could not exist were there no bodies.

My point in discussing Toletus' views is not to suggest that Toletus is an important and original thinker; he presents his account only as a compendium of views from the *"recentiores,"*[128] and there is every reason to think that all of the views he expresses in the section we have been examining can be traced to earlier thinkers.[129] I also do not intend to suggest that Toletus is in any sense the direct source of the ideas on space and place of Descartes' to which I referred in my discussion. As I noted earlier, if he borrowed directly from any scholastic thinker it was probably from Eustachius' rather more elementary and accessible exposition of the material as given in his *Summa*, a book he was known to have been reading in 1640 and 1641 while working out the relevant sections of the *Principles*. I am appealing to Toletus only as one example of what a scholastic position on the issues might look like, and my point is only that Descartes' view on space, place, and vacuum is in many respects not so very different from what he or any other student might have learned in school.

But as similar as Descartes' position may be to that of Toletus and other scholastic thinkers, there is an extremely important difference to note. For Toletus, as for Descartes, one can have real space or extension when and only when one has body. But for Toletus, despite this fact, extension is only what he calls a *proper accident* of body.[130] A proper accident for Toletus, as for other schoolmen, is something like risibility in humans "that flows from the essential principles of the subject, since it cannot exist without the subject nor can the subject exist without it."[131] But though it flows from the essence, it is clearly distinct from the essence; risibility is in all and only human beings, but it is not part of their essence.[132] And so, in his commentary on the *Physics* Toletus claims that whether or not quantity is really distinct from substance, "it is at least essentially distinct from substance."[133] Similarly, when distinguishing physics from mathematics, he argues that the quantity treated in mathematics is only one kind of accident bodies have, and not even the one of primary interest to the physicist.[134] But, of course, matters are altogether different for Descartes. For Descartes there is *nothing* to body over and above its extension, which constitutes its very essence and nature. And Descartes fully realized his departure from scholastic orthodoxy. Responding to Regius in the *Notes against a Program*, he remarks that "I am the first . . . to have considered extension as the principle attribute of body" (AT VIIIB 348). It is, of course, from this position, the identification of extension and body, that Descartes' position on space, place, and vacuum derives. However much it may resemble the positions he was taught in school, it flows from a somewhat different source.

But even though Descartes' position is, in an important sense, radically non-Aristotelian, it raises for the Christian philosopher exactly the same sort of problem that Aristotle's philosophy raised for the Latin West in 1277. Though the actual condemnation of that year had long been forgotten by the time Descartes was writing, the issues persisted;[135] Antoine Arnauld and Henry More, both especially interested in the theological consequences of Descartes' views, saw in his strong arguments against the possibility of a vacuum a potentially objectionable restriction on divine omnipotence. Arnauld wrote Descartes in July 1648:

On the vacuum, I confess that I cannot yet digest the fact that the connection between corporeal things is such that . . . God can reduce no body to nothing unless by virtue of that he is held immediately to create another [body] equal in size, or unless, without any new creation, the space which the body annihilated occupied is understood to be a true and real body. (AT V 215)

Similarly, in December 1648, a few months later, More objected:

When you intimate that not even by divine power can it happen that there exist a vacuum, properly speaking, and if all body were eliminated from a vessel, that the sides would necessarily meet, this seems . . . false. . . . For if God imprinted motion on matter, as you have taught before, could he not impress a contrary motion, and prevent the sides of the vessel from meeting? . . . But the sides . . . will meet not by a logical necessity but by a natural necessity; only God could prevent them from meeting. (AT V 240–41).

The root problem is this. Descartes argues from the inseparability of the notions of body and extension in our mind to their inseparability in nature. But, one might ask, why should that be binding on God? Descartes answers More as follows:

You readily admit that no vacuum arises naturally. But you are concerned about divine power, which you think could eliminate everything in some vessel and at the same time prevent the sides of the vessel from meeting. Since I know that my intellect is finite and God's power infinite, I never decide anything about this, but I only consider what I could or could not perceive, and I am careful not to let any judgment of mine disagree with what I perceived. Therefore I boldly affirm that God can do anything I perceive to be possible, but, on the other hand, I don't boldly deny that he can do what contradicts my conception; I say only that it implies a contradiction. Thus, since I see that it is in contradiction with my conception that all body be eliminated from a vessel, leaving the extension behind, extension which I conceive no differently from the body previously conceived contained in it, I say that it implies a contradiction for such extension to remain there after the body is eliminated, and therefore the sides of the vessel touch. (AT V 272 [K 240–41])

Descartes' answer to Arnauld a few months earlier is virtually identical (AT V 223–24 [K 236–37]). As with the answer to More, Descartes tells Arnauld that he sets no limits on God; indeed, he says that he "dare not say that God cannot make a mountain without a valley, or make one plus two not equal three" (AT V 224 [K 236]). His only claim, he tells Arnauld, is that "God has given me such a mind" that I cannot conceive of extension without body. And so, he claims, we are forced to conclude that "wherever extension is, there is necessarily body as well" (AT V 224 [K 236–37]).

The response is a bit puzzling. Descartes says that space without body is a contradiction, so we must conclude that it cannot happen, and *necessarily* so. But, Descartes also says that he does not mean to deny that *God* could create space without body, something that seems to undermine the claim that there is no space without body. If *God* can do it, then it must be possible in a genuine sense, just as Descartes was prepared to admit the real distinction between mind and body on the grounds that God could create the two apart, or deny that there are atoms because at least God could split any particle of matter.

Matters are somewhat illuminated by an undated marginal note Descartes seems to have written in his copy of the Latin *Principles,* a note that, I suspect, may well have been penned in late 1648 or early 1649 in response to the questioning by Arnauld and More:

> About these things that involve a contradiction, it can absolutely be said that they cannot happen. However, one shouldn't deny that they can be done by God, namely, if he were to change the laws of nature. But we should never suspect that he had done this unless he himself revealed this: as, for example, concerning the infinite world, the eternal [world], atoms, the vacuum, etc. (AT XI 654)[136]

The view here seems to be close to the celebrated doctrine of the creation of eternal truths that Descartes elaborated first in letters to Mersenne in 1630, and came back to later from time to time.[137] Descartes' position seems to be that since God fixed the laws of nature (eternal truths), he can change them, and allow for the possibility of body without extension. It is in this sense that God has the power to create a vacuum, if he so chooses. But having fixed the laws, having voluntarily constrained his power, as it were, vacua are impossible. And while we should always be aware that there is a *sense* in which God can create a vacuum, only divine revelation could show us that such a thing was *really* possible.

This is the position Descartes takes in the late 1640s. But my suspicion is that worries about divine omnipotence and the vacuum may well lie behind the original articulation of the doctrine of the creation of the eternal truths in 1630. Remember, first, that in Rule 14, written most likely in the late 1620s, Descartes comes close to articulating the position that will later lead him to deny the vacuum in the *Principles* of

1644, the identification of space and body. Though the vacuum does not come up in Rule 14, he was no doubt thinking seriously about it in 1629 and 1630, as he began to formulate *The World,* where the vacuum is explicitly rejected, and the considerations advanced in Rule 14 no doubt were prominent in his mind. But if Descartes remembered anything of his education at La Flèche, he must have remembered the theological problems and worries about divine omnipotence that come with an absolute denial of the vacuum, problems and worries still very much in evidence in scholastic literature almost 350 years after the Condemnation of 1277. My suspicion is that the doctrine of the creation of the eternal truths, first formulated in 1630 and called into service in the late 1640s, intended when first formulated in 1630 to take a prominent place in the treatise on physics then in progress,[138] may originally have been formulated to reconcile divine omnipotence with the impossibility of a vacuum. The doctrine of the creation of the eternal truths does not, in the end, appear in *The World,* of course. But then, as we have seen, neither do the strong arguments against the vacuum that were to appear in 1644 and raise theological questions in the minds of Arnauld and More. My suspicion is that Descartes' caution about venturing into theological questions may have gotten the better of him, and rather than giving both the strong arguments against the vacuum and the theological view necessary to reconcile them with divine omnipotence, he chose to eliminate the arguments, in favor of weaker considerations suggesting the nonexistence of vacua rather than their absolute impossibility.

But though Descartes may have been plagued by many of the same worries about divine omnipotence and the vacuum that plagued his teachers, the solution he forged was rather interestingly different. There are, to be sure, precedents to Descartes' view on the creation of the eternal truths in scholastic thought.[139] But to Descartes, no scholar of the past, it must have seemed that considerations relating to divine omnipotence had forced his teachers and their sources into recognizing that there was *some* sense in which empty space is genuinely possible, and in that way had forced them into conceptual incoherence, trying to make some sense of an utterly unintelligible idea. There is genuine excitement in the letters of 1630 where Descartes announces to Mersenne his view, a view he *thought* to be completely new and utterly unprecedented. For, if I am right, Descartes thought that he had discovered how he could exclude the vacuum altogether without compromising divine omnipotence; in Descartes' view, what he discovered in 1630, empty space is possible for God, but *only* in the sense that it is possible that $2 + 2 = 5$.

Conclusion

There were various attempts to bridge the gap between Descartes and the atomists in the seventeenth century. Gerauld de Cordemoy, for example, a member of one of the important Cartesian schools that formed after Descartes' death, attempted to frame arguments for atoms and the void using arguments in the style of the master.[140] In Britain, Robert Boyle attempted to downplay the differences between Descartes and the atomists and forge a kind of nonsectarian mechanical philosophy.[141] But the difference between his position and that of the atomists was extremely important to Descartes, and despite attempts at reconciliation, the denial of atoms and the void continued to characterize his followers throughout the century, and separate them from others, like Gassendi and his followers, who attempted to revive the doctrines of Democritus and Epicurus.[142]

MOTION

WE HAVE EXAMINED Descartes' account of body in general, the notion of corporeal substance as geometrical extension made substantial that grounds his mechanical world view. This extended substance can support a number of different modes; body can come in different sizes, different shapes; bits of this substance can be distant or near, contiguous or separated. But the most important mode matter has for Descartes is motion, local motion, and it is in terms of this mode that all properties bodies have are to be explained. Thus he writes in the *Principles*, beginning his extended account of motion, that "all variation in matter, that is, all the diversity of its forms depends on motion . . . and all the properties we clearly perceive in it reduce to this one thing, that it is divisible and mobile with respect to its parts, and thus is capable of all of those properties *[affectiones]* which we perceive can follow from the motion of its parts" (Pr II 23). In order to understand the foundations of Descartes' physics, we must thus understand his conception of motion and the laws it follows.

My discussion of motion in Descartes will be structured around an important distinction that he draws in the *Principles*. There he carefully distinguishes between motion considered in itself as a mode of body, and the causes of motion, the "force or action which transfers [a body]," two distinct notions that "are usually not distinguished carefully enough" (Pr II 25).[1] And so, in the *Principles*, he devotes a number of sections to the nature of motion considered as a mode of body before turning to his discussion of the causes of motion, both the general cause, God, and the laws of motion, what he calls "secondary and particular causes of different motions" (Pr II 37).[2] While he himself may have been among those who confused motion with its causes in his earlier writings, as we shall see, the distinction is crucial to understanding Descartes' mature thought, and my discussion will respect that distinction.

Consequently we shall begin our discussion in this chapter with an account of the notion of motion as a mode of body. After an account of the notion of motion in Descartes' earlier writings, we shall turn to a discussion of his definition of motion in the *Principles* and later writings. The central question we shall have to take up is how Descartes thought motion to be a genuine mode of body. If all there really are in his physical world are extended bodies and their modes, then motion *must* be construed as a mode of body if it is to have the explanatory function it is intended to have in his physics. In order to show how he thought it to be a mode of body, we must discuss the so-called relativity of Descartes' account of motion, and its dependence on duration and time. I shall argue that contrary to the received wisdom, Descartes' definition of motion is carefully fashioned to allow for a nonarbitrary distinction between motion and rest, and that the duration on which the notion of motion depends is, itself, an attribute of body. The chapter will continue with a discussion of some other issues connected with the account of motion, the individuation of bodies, Copernicanism and the motion of the Earth, the notion of determination, and the rejection of scholastic conceptions of motion.

MOTION: THE DEFINITION

Motion and its laws concerned Descartes from his earliest years. The problem of the law governing the motion of a body in free-fall, a problem that brings into play many of the questions about motion that were later to concern him, is a problem he took up as early as 1618 during his brief period of apprenticeship with Beeckman.[3] But, for all his interest in the phenomenon of motion, there is little interest in any attempt to formulate a precise definition of what motion is. In Rule 12 of the *Rules*, for example, Descartes writes:

> Indeed, doesn't it seem that anyone . . . who says that *motion,* a thing well-known to all, is *the actuality of a thing in potentiality insofar as it is in potentiality* is putting forward magic words? For who understands these words? Who doesn't know what motion is? . . . . Therefore, we must say that these things should never be explained by definitions of these sorts, lest we grasp complex things in place of a simple one. Rather, each and every one of us must intuit these things attentively, distinguished from all other things, by the light of his own intelligence *[ingenium].* (AT X 426–27)

Descartes is taking the Aristotelian definition of motion to task here. And one of the main problems he sees in that definition is simply the fact that it is a definition at all. Motion, he seems to imply, cannot be defined strictly speaking; it is a simple nature, a concept that each and every one of us can grasp for ourselves in the privacy of our own minds, and we don't need a definition to tell us what it is.

Descartes displays a similar attitude toward motion in *The World* he began to work on shortly after setting the *Rules* aside. As in the *Rules,* he quotes the Aristotelian definition of motion to show its obscurity, quoting it in Latin: "[this definition] is so obscure to me that I am forced to leave it here in their language, since I cannot interpret it" (AT XI 39). But in *The World* he goes even farther than he went in the *Rules;* he argues that local motion is the only intelligible notion of motion and challenges the idea that there are some motions, recognized by the scholastics, that do not involve local motion, including motion with respect to form, motion with respect to heat, and motion with respect to quantity (AT XI 39). Later in this chapter we shall take these critiques up more carefully. But for the moment what I would like to call attention to is what Descartes substitutes for the Aristotelian definition of motion: nothing. Once again, he suggests that motion is a basic notion, indefinable and known through itself:

> The nature of motion that I intend to speak of here is so easy to understand that the geometers themselves, who are the best of all people at conceiving very distinctly things which they have considered, have judged motion to be more simple and more intelligible than their surfaces and their lines; so it appears from the fact that they have explained the line from the motion of a point, and the surface from the motion of a line. (AT XI 39)

Lines and surfaces are certainly intelligible and unproblematic, Descartes implies. But, he says, motion is simpler even than they are, since geometers use the notion of motion in order to define lines and surfaces.[4]

But despite Descartes' studied avoidance of any careful discussion of the nature of motion, it is not difficult to discern his conception of motion in these early writings, the definition of motion he might have given had he been pressed to give one. Motion, it is fair to say, was just change of place. In *The World,* for example, motion is claimed to be "that by virtue of which bodies pass from one place to another and successively occupy all of the spaces in between" (AT XI 40).[5] Similarly in *The World* he characterizes rest and its distinction from motion as follows: "For myself, I conceive that rest is also a quality which ought to be attributed to matter while it remains in one place, just as motion is a quality that is attributed to it while it changes place" (AT XI 40).

Much the same attitude toward motion persists through to the end of the 1630s. In a letter to Morin on 12 September 1638, for example, Descartes characterizes motion as "the action through which the parts of this matter change place" (AT II 364). Though the characterization of motion as an action is not without significance, as we shall see, it is also significant that as in *The World,* motion is characterized in terms of change of place. But as in *The World* and the *Rules,* it is not clear that an explicit definition is really needed. Writing to Mersenne about a year later, on 16 October 1639, Descartes notes, "Someone who walks in a room understands what motion is better than someone who says that it is *the actuality of a thing in potentiality insofar as it is in potentiality,* and so on" (AT II 597 [K 66]).

But when in the early 1640s Descartes turns to drafting his *Principles,* there is a significant change, both in his conception of what motion is and in his view on the necessity for a formal definition. It is not at all clear why he changed his attitude and adopted the new conception of motion he did; commentators, from his contemporaries down to our day, have tended to see Descartes' change in position motivated by his caution and his extreme reluctance to commit himself in print to a view on which the Earth can properly be said to move. Indeed, the claim has been made that the definition given in the *Principles* does not represent his real view. We shall examine this claim in some detail later in this chapter, where I shall argue that whatever fortuitous consequences Descartes' view of motion may have for Copernicanism, the definition of motion he offered in the *Principles* is firmly rooted in features internal to his program for grounding physics. But whatever his motivation may have been, it is clear that in the *Principles* Descartes explicitly attempts to define motion, and when he does so, the definition he offers is not simply given in terms of change of place.

Motion is, of course, one of the central concerns of part II of the *Principles,* where Descartes lays down the foundations of the natural philosophy that he will develop in the third and fourth parts. Descartes introduces his discussion of motion and its laws with two different definitions of motion. The first definition is intended to capture the notion of motion as it is ordinarily understood, *motus juxta vulgarem sensum:*

> Motion . . . as commonly understood is nothing but *the action by which some body passes [migrat] from one place into another.* (Pr II 24)[6]

In contrast to this, Descartes proposes the following definition, which is intended to capture the true sense of motion, *motus propriè sumptus,* motion considered *ex rei veritate:*

> But if we consider what we should understand by motion not so much as it is commonly used but, rather, in accordance with the truth of the matter, then in order to attribute some determinate

nature to it we can say that it is *the transference [translatio] of one part of matter or of one body from the neighborhood [vicinia] of those bodies that immediately touch it and are regarded as being at rest, and into the neighborhood of others.* (Pr II 25)

Before beginning to unpack these two definitions and to understand the differences Descartes saw between them, let me make two brief remarks about what they have in common. Both are, of course, intended to be definitions of *local* motion. In the *Principles,* as we saw earlier in *The World,* he claims that "no other [notion of motion] falls within the scope of my thought, nor [can] any other be imagined in nature" (Pr II 24; see also Pr I 69). Later in this chapter we shall investigate why he held this, and why he rejected other varieties of motion his Aristotelian contemporaries accepted. And second, both of these definitions appear to presuppose that we understand what it is that makes a region of extended substance *one* body, how it is that bodies are individuated. Immediately after giving the proper definition of motion, Descartes explains that "by *one body* or *one part of matter* I understand everything that is transferred at the same time" (Pr II 25). Bodies, thus, seem to be individuated by their motion. Just how he thought this works and the problems the account of individuation raises for the account of the nature of motion will also be discussed later in this chapter. But for the moment I would like to assume that we have a grasp of what it is that constitutes an individual body, and, given that understanding, explore the differences between the two definitions of motion that Descartes considers.

The first important difference between the two definitions concerns the notion of activity. According to the vulgar definition, motion is defined as an action, an *actio,* as Descartes himself had characterized motion only a few years earlier, in 1638.[7] But in the proper definition, he calls motion a transference, a *translatio.* There are, I think, two reasons for this change. For one, if we think of motion as an action, then we are immediately led to think of rest as the *lack* of action, as he notes in connection with the vulgar definition: "Insofar as we commonly think that there is action in every motion, we think that in rest there is a cessation of action" (Pr II 24). This, he thinks, is a mistake, one of the many prejudices we acquire in our youth (Pr II 26). On the contrary, he thinks, "no more action is required for motion than for rest" (Pr II 26). And so, he argues, the action necessary to put a body at rest into motion is no greater than the activity necessary to stop it; rest requires as much of an active cause as motion does (Pr II 26; AT V 345–46).

This observation of Descartes' will be important in later in chapters where we discuss the laws of motion. There we shall see that motion and rest, while distinct states, are, in a sense, on a par with one another insofar as both motion and rest will persist unless impeded by an external cause, and both motion and rest have associated forces, as he will call them, that oppose change. But these observations do not properly speaking belong to a characterization of the bare notion of motion. And this is the second reason why he chooses to characterize motion in terms of transference rather than action: what properly pertains to the body as a moving thing is not the *action* or *force* associated with the motion, but simply the fact that the moving body passes from one region and into another. And so he remarks immediately following his definition of motion *propriè sumptus:*

> And I say that [motion] is *transference,* not the force or action that transfers in order to show that it is always in the mobile thing, and not in what is moving it, since these two things are not usually distinguished carefully enough, and to show that [motion] is a mode of a thing, and not some subsisting thing, in just the same way as shape is a mode of a thing with shape, and rest is a mode of a thing at rest. (Pr II 25)

There are a number of features of this passage to which we will have to return later, in particular, the characterization of motion as a mode of body and the comparison Descartes draws between motion and shape. But one thing he wants to distinguish here is what it is that is in a body by virtue of being in motion, what belongs to motion itself, and what belongs to the cause of motion in the body, which, he implies, is something distinct from the body in motion.[8] This is made a bit clearer in a response he made to Henry More. In More's last letter to Descartes, 23 July 1649, he tells Descartes that in his view, "motion is that force or action by which bodies said to be moved mutually separate" (AT V 380), a view close to the vulgar definition that Descartes had rejected a few years earlier. Descartes answers:

> That transference that I call motion is a thing of no less entity than shape is, namely, it is a mode in body. However the force *[vis]* moving a [body] can be that of God . . . or also that of a created substance, like our mind, or something else to which [God] gave the power *[vis]* of moving a body. (AT V 403–4 [K 257])

In calling motion a *transference* rather than an *action,* Descartes means to distinguish the effect, motion, from its cause, the mover. We cannot, of course, *ignore* the cause; as I noted earlier, Descartes will give careful attention to the causes of motion, and will derive the laws of motion ultimately from its "universal and primary cause," God. But Descartes thinks it is important to distinguish the nature of motion from its cause, and treat the cause only after we understand the nature of the effect.

So far we have examined one important difference between the vulgar definition Descartes rejects and the proper definition he means to substitute, the move from motion as *action* to motion as simple *transference.* But there is another difference just as important, though considerably more difficult fully to understand. The vulgar definition is given in terms of the notion of place, *locus;* motion, for the vulgar, and, indeed, for Descartes himself in his earlier writings, involves passing from one *place* to another *place.* In the *proper* definition, though, motion is defined not in terms of *place,* but in terms of *neighborhoods* of contiguous bodies; motion, as he came to understand it, is the passing from one neighborhood of contiguous bodies and into another.

This is a puzzling change, one that has driven numerous commentators to look for an external cause; as we shall later discuss, it has been common to attribute this change to Descartes' desire to hide his Copernicanism. Somehow, it has been claimed, by rejecting the definition of motion in terms of place and adopting this rather queer alternative, Descartes thinks he can hold that the Earth is, properly speaking, at rest. Later in this chapter we shall investigate these claims in some detail. But for now I would like to investigate the explanation Descartes himself gives of this change and the new account of the nature of motion he offers.

An important feature of the vulgar definition that Descartes chooses to emphasize is the fact that whether or not a given body is in motion depends on the apparently arbitrary decision to consider one or another place as being unmoved. And so he comments on the vulgar definition, referring to his earlier discussion of external place in Pr II 13:

> As we showed above, the same thing can at a given time be said both to change its place and not to change its place, and so the same thing can be said to be moved and not to be moved. For example, someone sitting in a boat while it is casting off from port thinks that he is moving if he looks back at the shore and considers it as motionless, but not if he looks at the boat itself, among whose parts he always retains the same situation. (Pr II 24; see also Pr III 28)

It is this kind of arbitrariness that the proper definition of motion is intended to undermine. In explaining the proper definition of motion in terms of neighborhoods rather than places, Descartes notes:

> Furthermore, I added that the transference take place *from the neighborhood of those bodies that immediately touch it into the neighbor-*

hood of others, and not from one place into another since . . . what is taken as a given place varies *[loci acceptio varia est]* and depends upon our thought. But when we understand by motion that transference which there is from the neighborhood of contiguous bodies, since only one group of bodies can be contiguous to the mobile body at a given moment of time, we cannot attribute many motions to a given mobile body at a given time, but only one. (Pr II 28)

The move from defining motion (improperly) in terms of place, as Descartes once seems to have held, to a definition in terms of neighboring bodies is supposed to eliminate the claim that whether or not a body is in motion, and if it is in motion, what that motion is, is simply a matter of arbitrary choice, a distinction that depends on our thought. But, one might ask, *why* did Descartes want to eliminate the relativity that had accompanied the notion of local motion since its earliest discussions and *how* did he think that the rather strange definition he gave in the *Principles* accomplished that aim?[9]

It is not difficult to see why Descartes might want to eliminate the kind of arbitrariness implicit in the vulgar definition. Motion is to be a basic explanatory notion in Descartes' physics: "all variation in matter, that is, all the diversity of its forms depends on motion" (Pr II 23). Now, if all the properties bodies have are ultimately to be explained in terms of motion, then motion must *really be* in body, as a mode. It is not entirely clear just what this means. But certainly one thing that it means is that there is a real fact of the matter about whether or not a given body is in motion or at rest; whether or not a body is in motion should in no way depend upon how we happen to think about it, upon an arbitrary decision we make to consider this or that body as being at rest. But if this is the case, then motion cannot be as it is commonly conceived, as change of place, since, as Descartes noted, on that conception of motion the distinction between being in motion and being at rest is arbitrary and depends on our thought; such a conception of motion is hardly up to the task Descartes intends for it to play in his system.

It is not clear just when Descartes came to see the importance of a nonarbitrary distinction between motion and rest. Descartes' position in his earlier writings is somewhat difficult to discern. In those writings, he does appear to conceive of motion as change of place. But it is important to note that he does not offer this as a formal definition, and that he nowhere draws any of the relativistic consequences of defining motion in that way that he will draw later in the *Principles.* Furthermore there are other remarks in those writings that suggest, if only weakly, a genuine distinction between motion and rest. In the *Rules,* for example, 'rest' is listed as a simple nature, and distinguished from the simple nature of motion (AT X 420). In *The World* Descartes claims that rest is a "quality *[qualité]* which ought to be attributed to matter while it remains in one place," as much of a quality as motion is (AT XI 40). And referring to the scholastic view that things in motion tend per se to come to rest, a view that, as we shall later see, he rejects, Descartes notes that the motion of the scholastics tends toward rest and thus, "against all laws of nature, it tries to destroy itself" (AT XI 40), suggesting that motion and rest are distinct and opposing qualities of body.

Though all of these passages suggest a genuine distinction between motion and rest,[10] they might be reconciled with a relativistic view of the distinction, perhaps, and may be made consistent with the view that the distinction between rest and motion is simply a matter of our way of thinking of them. But in the *Principles* Descartes is much more explicit. Motion and rest are genuine modes of body there, and genuinely distinct: "It is obvious that this transference cannot exist outside of a moving body, and that this body has one mode when it is transferred, and another when it is not transferred, that is, when it is at rest. And so, motion and rest are nothing in it but two different modes" (Pr II 27). Indeed, he suggests that motion and rest are *states, opposite* states, that persist unless interfered with (Pr II 37, 44). And, as we shall later see, when he formulates his laws of impact, the difference between the states of motion and rest are manifested in a set of laws that clearly differentiate between the two states; the outcome of a collision may turn out to be entirely different depending on which of the bodies involved is designated as being at rest, and which is designated as being in motion.

It is important for Descartes to distinguish between motion and rest, to reject the view that whether or not a body is in motion depends upon our point of view, and make motion a genuine mode of body. But, one might well ask, how does he think it can be done? Does his preferred definition of motion allow for unambiguous attributions of motion or rest any better than the vulgar definition he rejects on those grounds?

Defining motion in terms of neighborhood is, in a clear sense, defining it in terms of place, Aristotelian place, what he himself called its external place earlier in *Principles,* part II, "the [inner] surface of the surrounding body" (Pr II 15)[11] But by defining motion in terms of a neighborhood rather than an arbitrarily designated place, Descartes seems to think that he is entitled to say that insofar as there is only one neighborhood of surrounding bodies, there can only be one *proper* motion for a given body. He, of course, recognizes that there is a sense in which a body can be said to participate in many motions:

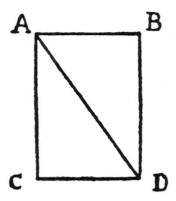

FIGURE 6.1 From *Renati Des-Cartes Principia Philosophiae* (Amsterdam, 1644), p. 50.

For example, if someone walking on a boat carries a watch in his pocket, the wheels of his watch will move with only one motion proper to them, but they will also participate in another, insofar as they are joined to the walking man and together with him compose one part of matter. They will also participate in another insofar as they are joined to the vessel on the undulating sea, and in another insofar as they are joined to the sea itself, and, finally, to another insofar as they are joined to the Earth itself, if, indeed, the Earth as a whole moves. And all of these motions will really be in these wheels. (Pr II 31)

And Descartes also recognizes that the single proper motion of a body can be decomposed into other motions that jointly can be thought of as making up the proper motion of the body:

And furthermore, that single motion of a body which is proper to it can be considered as if it were many motions. For example, in carriage wheels we can distinguish two different [motions], namely, a circular [motion] around their axle, and another straight [motion] along the length of the road on which it is traveling. . . . We can imagine any line, even a straight line, the simplest line of all, to have arisen from an infinite number of different motions. For example, if line AB [see fig. 6.1] is carried toward line CD, and at the same time point A is carried toward B, then the straight line AD which point A marks out depends no less on the two straight motions, the motion from A to B, and from AB to CD, than the curved line marked out by any point on the wheel depends on the straight and the circular motion. (Pr II 32)

But, Descartes emphasizes, the decomposition of the proper motion into other motions, is, in a sense, a fiction. Though the carriage wheel can be thought of as if it has two distinct motions,

> the fact that these motions are not really distinct follows from the fact that every point of a body that moves marks out only one particular line. . . . Although it is often useful to divide a motion into several parts in this way in order to perceive it more easily, strictly speaking, though, we must count only one motion in any body. (Pr II 32)

And similarly, in the case of the wheels in the watch, even though a body may participate in a number of different motions, "every body has only one motion proper to it, since it is understood to recede from only one [group of] contiguous and resting bodies" (Pr II 31). And so, by defining motion not in terms of place, a notion notoriously relative to an arbitrarily chosen reference point, but in terms of a neighborhood of contiguous bodies, Descartes *seems* to think that he can undermine the relativity implicit in the vulgar definition of motion, at least to the extent that he can identify at most one motion that can be said to belong to a given body, strictly speaking; though the wheel in the watch *participates* in many motions, it has only one *proper* motion, and though the motion of a point on a carriage wheel can be thought of as the composition of a straight and a circular motion, it *really* has only *one* motion, properly speaking.

But this cannot be the whole story. There are, of course, obvious difficulties in defining *precisely* what the neighborhood of a given body is, particularly when we are dealing with bodies surrounded with fluid, as Descartes takes the Earth to be. But even this problem aside, it seems obvious (and has seemed obvious to most of his commentators) that however much the definition restricts the attribution of motion to a body, however many possible motions Descartes' curious definition might eliminate, at root, his definition still fails to ground a genuine distinction between motion and rest, it appears. Motion, he says, is transference. But Descartes also tells us in the *Principles* that transference is reciprocal:

> Finally, I added that the transference take place from the neighborhood not of any contiguous bodies, but only from the neighborhood *of those regarded as being at rest.* For that transference is reciprocal, and we cannot understand body AB transferred from

the neighborhood of body CD unless at the same time body CD is also transferred from the neighborhood of body AB. . . . Everything that is real and positive in moving bodies, that on account of which they are said to move is also found in the other bodies contiguous to them, which, however, are only regarded as being at rest. (Pr II 29, 31)

Now, if motion is transference, and transference is reciprocal in this sense, then it seems obvious that motion must be reciprocal in this sense. And so, even if there may be only one motion proper to a given body, it still seems to be an arbitrary decision whether to say that a body is moving and its neighborhood at rest, or to say that the body is at rest and its neighborhood in motion.

The reciprocity of transference that Descartes clearly acknowledges has convinced many commentators that his conception of motion does not allow for a genuinely objective distinction between motion and rest. But I think that this is a misunderstanding. Although transference is reciprocal, he still thinks that there is a genuine distinction between motion and rest. But it is not immediately obvious how he means to draw the distinction.[12]

The question of the distinction between motion and rest, and its relation to the doctrine of the reciprocity of transference is broached directly in the notes on the *Principles* that Descartes seems to have written sometime after the publication of the Latin *Principles* in 1644. The relevant portion of the text reads as follows:

Nothing is absolute in motion except the mutual separation of two moving bodies. Moreover, that one of the bodies is said to move, and the other to be at rest is relative, and depends on our conception, as is the case with respect to the motion called local. Thus when I walk on the Earth, whatever is absolute or real and positive in that motion consists in the separation of the surface of my foot from the surface of the Earth, which is no less in the Earth than in me. It was in this sense that I said that there is nothing real and positive in motion which is not in rest [see Pr II 29–30]. When, however, I said that motion and rest are contrary, I understood this with respect to a single body, which is in contrary modes when its surface is separated from another body and when it is not. . . . Motion and rest differ truly and modally *[modaliter]* if by motion is understood the mutual separation of two bodies and by rest the lack *[negatio]* of this separation. However, when one of two bodies which are separating mutually is said to move, and the other to be at rest, in this sense motion and rest differ only in reason *[ratione]*. (AT XI 656–57)

Descartes is reasonably straightforward about motion and rest in this passage. There is, indeed, a sense in which the distinction between motion and rest is a distinction of reason, depending on our conception. In this sense, when I lift my foot it is correct to say that my foot is in motion and the Earth at rest, and equally correct to say that the Earth is in motion and my foot is at rest. In this sense, the distinction between motion and rest is only a distinction of reason, a distinction that depends on our conception, the way we think about things.[13] But, he emphasizes in this passage, it is not the *only* way of thinking about motion and rest. One can also think of motion as the mutual separation of a body and its contiguous neighborhood. Understood in this sense, there is a genuine distinction between motion and rest that goes beyond our manner of conception. For, if a body is in transference with respect to its contiguous neighborhood, no mere act of thought can correctly construe it otherwise, and if it is *not* in transference, no mere act of thought can correctly construe it as if it were. Because of the doctrine of the reciprocity of transference, whenever a body is in motion, we must say that its neighborhood is as well, properly speaking; a body A cannot separate from its neighborhood B without, at the same time, B separating from A. And so he notes in the *Principles*:

> If we want to attribute to motion its altogether proper and nonrelative nature *[omnino propriam, & non ad aliud relatam, naturam]* we should say that when two contiguous bodies are transferred, one in one direction, and the other in another direction, and thus mutually separate, there is as much motion in the one as there is in the other. (Pr II 29)

This, indeed, is the main thrust of the doctrine of the reciprocity of transference, not to introduce relativity and undermine the distinction between motion and rest, but to emphasize that a motion properly speaking belongs equally to a body and its contiguous neighborhood. But this in no way undermines the kind of distinction between motion and rest that Descartes wants to draw. If motion is understood as the mutual separation of a body and its neighborhood, then it is impossible for a body to be both in motion and at rest at the same time insofar as it is impossible for that body both to be in transference and not in transference with respect to the same contiguous neighborhood. Understood in this way, motion and rest are different and distinct modes of body, and the distinction between motion and rest not merely a distinction of reason, but a *modal* distinction in Descartes' technical terminology, a distinction between two different modes of the same substance (Pr I 61).

As clear as the text I quoted earlier is, one must admit that it is not always easy to see the reading of the distinction between motion and rest in the passages discussing motion in the *Principles* itself. One of the difficulties is clearly due to Descartes. Though his ostensible purpose is to distinguish the proper notion of motion from the vulgar conception, he finds it difficult to set aside the common conceptions of motion and rest, and there is at least a hint of vulgarity in his own use of the terms, even after introducing the proper notion of motion. It is with this in mind that we should approach the curious phrase he uses in his proper definition of motion, where he comments that the contiguous neighborhood must be "regarded as being at rest" (Pr II 25). A few sections later, when explaining this phrase, he asserts that, properly speaking, motion pertains as much to the neighborhood as it does to the body in question; though *considered* at rest in one sense, the neighborhood is *really* in motion. This suggests that 'rest' must, in this context, be understood as resting in the sense of not changing *place*, that is, rest *in the vulgar sense.*[14] What Descartes is doing, in essence, is *using* aspects of the vulgar and improper (though not unintelligible) sense of motion and rest, what we might call v-motion and v-rest, to define the *proper* sense of motion and rest, what we might call p-motion and p-rest, making use of notions we understand to define a new and less familiar notion. If we think of his enterprise in this way, as I think we must, then his definition of motion in the proper sense, p-motion, can be reconstrued as follows:

A is in p-motion if whenever the neighborhood of contiguous bodies B is considered at v-rest, then A will be in v-motion.

Of course, any body, like B, that we choose to regard as being at (v-)rest can also be regarded as being in (v-)motion. But this does not undermine the distinction between motion and rest considered in the proper sense as the mutual separation of a body and its neighborhood. Even though it is arbitrary whether or not B is at v-rest, once we *consider* B at v-rest, it is *not* a matter of arbitrary choice whether or not to consider A as being in (v-)motion; if A is *really* in motion in his proper sense of the term, if it is *really* separating from its neighborhood B, then no mere change of perspective will allow us to set A at rest in the *proper sense.* As confusing as it might be to make use of vulgar notions in explaining the proper sense of motion, the arbitrariness that attaches to the vulgar notion does not undermine the distinction between motion and rest properly understood.

Part of the difficulty of understanding Descartes derives from the apparent carelessness of the text, and the confusing mixture of the two different conceptions of motion and rest. But some of the blame must rest with us, and the inevitable tendency to impose later ideas of absolute and relative motion and rest back onto his text. When we consider the question of a genuine distinction between motion and rest, it is quite natural for us to think of the issue as Newton posed it later in the seventeenth century. In his celebrated scholium to the definitions of book I of the *Principia,* Newton distinguishes absolute from relative time, space, and motion. Newton writes:

> Absolute space, in its own nature, without relation to anything external, remains always similar and immovable. Relative space is a measure or some movable dimension of this absolute space. Our senses define [relative space] by its position with respect to bodies. . . . Absolute motion is the transference *[translatio]* of a body from one absolute place into another; and relative motion, the transference from one relative place into another.[15]

As commonly interpreted, Newton seems to be claiming that there is one uniquely privileged point of view with respect to which places are to be picked out, and it is this point of view that defines absolute motion and absolute rest. On Newton's view, it seems, there is a real distinction between motion and rest, considered absolutely; no mere change in point of view can transform a body at rest with respect to the privileged frame of reference into a body in motion with respect to that frame.

Descartes' way of breaking the arbitrariness found in the common conception of local motion is altogether different from Newton's, and for that reason, I think, may have been altogether missed by his post-Newtonian commentators.[16] Newton tried to make sense of a nonarbitrary notion of motion by positing a nonarbitrary reference point for the determination of place and change of place, a global rest frame in relation to which real motion and real rest can be defined; he tried to make sense of motion conceived of as change of place in a way that will allow a nonarbitrary distinction between motion and rest. But this is not Descartes' strategy, and if we read his text with Newtonian glasses, looking for a Newtonian rest frame, we will certainly be disappointed. Descartes breaks the arbitrariness implicit in the vulgar definition of motion as change of place, but he does so by changing the subject, as it were, replacing the notion of motion as change of place with a radically different conception of what motion is; rather than trying to make sense of motion as change of place, he focuses on the separation or nonseparation of a body with respect to its contiguous neighborhood, and calls *that* motion.

By doing this Descartes does get a distinction between motion and rest that is not simply dependent upon our conception. But the distinction comes at some cost. Had he given the proper definition of motion the careful attention that he should have given so basic a building block of his system, he would have discovered, I think, that even if his new definition made it easier to distinguish between motion and rest, motion so conceived is not altogether appropriate for the physics that he is attempting to build upon it. On the vulgar definition of motion as change of place, notions like speed and direction are well-defined, given the choice of a rest frame; once we specify such a rest frame, either by (arbitrarily) designating some bodies to be at rest or by abstractly designating points to be at rest, the speed and direction of motion of a body at any given time are reasonably straightforward to understand. For Descartes, there is, in a sense, a privileged frame for determining the motion and rest of a given body, just as there is for Newton; it is its contiguous neighborhood. But as a body moves in the plenum, its contiguous neighborhood will change from moment to moment. And without a common frame of reference from one moment to the next, it is very difficult to see what sense can be made of the speed or direction of a given body. But even if the speed and direction of an *individual* body could be determined on Descartes' conception, there would still be problems. While each individual body may have a natural rest frame with respect to which its motion is to be determined, the rest frames are *local;* each body has its *own* rest frame, and there is no reason to think that these different rest frames are at rest with respect to one another. But without a common framework in which to conceive of the relative motions of more than one body, it is difficult to see how we could give an adequate treatment of the phenomenon of impact, say. Consider two bodies, A and B, colliding at some angle with one another, and then going off in different directions. There the outcome would seem to be determined not by the relation each body has to its own contiguous neighborhood, but by the relation that A and B have to one another; however A and B may be related to their neighborhoods, if those neighborhoods are in (v-)motion with respect to one another, one would rightly expect an altogether different outcome than if they were at (v-)rest.

How, then, are we to regard the definition of motion in the *Principles?* My suggestion is this. Descartes for many years before the drafting of the *Principles* held that there is a genuine distinction between motion and rest, while at the same time tacitly holding to a very traditional conception of local motion as change of place. Come the early 1640s and the project of writing a full and orderly account of his philosophy for presentation to the world, he finally faced up to the problem of squaring his definition of motion with the distinction between motion and rest that had become so central to his thought. The result is the definition of motion that he frames in Pr II 25. Imperfect as it is, it does allow for the distinction he feels he needs. But, I think, Descartes never got around to working out the problematic consequences of such a definition of motion as the one he was proposing; indeed, I think that one can find clear indications that he is still thinking of motion in terms of change of place, even elsewhere in the *Principles*, most notably in the treatment of impact, where he assumes that the speed and direction of a pair of colliding bodies is measured with respect to the same frame of reference (Pr II 46–52).[17] But even though it is not altogether successful, and not carefully worked out, the new definition of motion seems clearly to be an attempt to respond to serious problems from within his natural philosophy.

MOTION AS MODE OF BODY

In the earlier part of this chapter I argued that since motion is to be a primary explanatory concept in Descartes' physics, motion must be a mode of body, something that really pertains to body. It is because of this, I suggested, that he must reject the relativism implicit in the vulgar definition of motion, and replace it with a conception of motion on which there is a genuine distinction between motion and rest. We have seen how the conception of motion Descartes is trying to capture in his definition of motion admits such a distinction, how he can say on the proper definition of motion that it is not merely a matter of our conception that determines whether a given body is in motion or not. But there are other features of the notion of motion that make it somewhat odd to consider motion to be a mode of body, features that differentiate motion from other modes of body like, say, shape. Unlike shape, motion seems to be inherently relational; though there *may* be a genuine distinction between motion and rest, motion seems to be a property that pertains not to an individual, but, in a strange way, to both an individual body and its surrounding neighborhood. And second, unlike shape, motion involves a physical magnitude that would seem to lie outside of body, properly speaking, insofar as it involves time. These, it seems, would undermine any claim that motion is a genuine mode of body.

Descartes often compares motion to other modes like shape in arguing that it is a mode of body. When introducing the proper definition of motion, he notes that "is a mode of a thing, and not some subsisting thing, just as shape is a mode of a thing with shape" (Pr II 25). Similarly, when writing to More, he emphasizes that "that transference that I call motion is a thing of no less entity than shape is, namely, it is a mode in body" (AT V 403 [K 257]). Other passages also suggest that he wants the reader to think of motion as a simple mode of an individual moving thing. For example, in the *Principles,* when attempting to get clear on the "altogether proper and nonrelative nature" we ought to attribute to motion, he notes that where two bodies are mutually separating, "we should say that . . . there is as much motion *in the one* as there is *in the other*" (Pr II 29, emphasis added; see also AT XI 656). But I think that we should not read Descartes too literally here, and demand that motion be construed as a mode in the narrowest sense, an intrinsic and strictly nonrelational property of an individual body. His point in comparing motion to shape in these passages is to differentiate motion, the effect, from its cause, the force that either puts a body in motion or sustains it in motion, as we discussed earlier in this chapter, and as we will discuss in the chapters to come. The point is that while the *cause* of motion may be "some subsisting thing," *motion itself* is not; it is in this sense that motion is taken to be like shape. And his point in emphasizing that there is as much motion in the contiguous neighborhood as there is in the body usually taken to be moving is simply to emphasize that motion (at least as he wants to construe it) is the *mutual* separation of a body and its neighborhood. Indeed, the doctrine of the reciprocity of transference would seem to lead quite directly to the view that if motion is to be construed as a mode, a real property of anything, it must be construed as a mode not of the individual moving body but of the system composed of the body taken together with its contiguous neighborhood. Nor is there any real problem with this for Descartes' program. What is important for him is that motion and rest be real features of the physical world, not arbitrary, not dependent upon the way in which we think of things. And conceived of in this way, as a mode of the system composed of a body together with its contiguous neighborhood, motion would seem to have that status.

Motion as mode differs from shape, then, insofar as it involves at least two bodies. But it also differs insofar as it seems to involve time, a feature of the physical world that would seem to lie outside of extended substance taken narrowly; though motion, like shape, presupposes extension,[18] it seems to presuppose something more. One way of dealing with the apparent anomaly is to try to take the time out of motion by taking seriously what Descartes wrote in *The World* about his notion of motion. There he claims that his notion of motion is just the motion of the geometers, the motion by which "they have explained the line through the motion of a point, and the surface through the motion of a line" (AT XI 39; see also p. 40). Considered in this way, as an abstract geometrical operation, motion does not require time strictly speaking; its presence is manifested not so much by the change observed in time as by the geometrical objects it results in. This is the view that Alexandre Koyré takes of motion in Descartes, at least as it is treated in *The World.* Koyré writes:

Cartesian motion, on Descartes' own account of it, has only an indirect relation with time. . . . The 'motion' of a point which makes a line, the 'motion' of a line which makes a plane, 'motions' such as these have no speed. Having no speed they do not take place in time. Now it is on the model of these non-temporal 'motions' that Descartes fashions his idea of motion. . . . Thus what is left of motion when its temporal character has been suppressed is precisely whatever in it is immobile: position, direction, trajectory, functional relations. The thorough-going geometrization Descartes yields to undoes the work of time.[19]

But as suggestive as this reading is, it isn't Descartes. Descartes' appeal to the motion of the geometers is, indeed, quite deliberate; it is meant at very least to differentiate his own conception of motion, local motion, the clear notion in terms of which everything is to be explained, from the complex Aristotelian notion of motion against which he argues in these passages of *The World.* But whatever the significance of the appeal to the geometrical conception of motion in *The World,* it is unlikely that he meant to separate motion from time. Already in Rule 12 Descartes noted that there is a necessary connection between motion and duration, just as there is between shape and extension, "since one cannot conceive of a shape that lacks all extension, or a motion that lacks all duration" (AT X 421). And in Meditation V, when discussing the modes that pertain to his idea of body he notes that "I assign certain durations to the motions" that belong to bodies (AT VII 63). Motion and duration are similarly linked in the *Principles,* where, Descartes says, we use "the durations of the motions of the greatest and most regular things" as a standard by which to measure the duration of other events (Pr I 57). As we shall see in later chapters, reasoning about the instantaneous states of bodies in motion is crucial to his strategy for deriving the laws of motion. But, Descartes is quite explicit in holding, from *The World* to the *Principles,* that "no motion takes place in an instant" (Pr II 39; see also AT XI 45; AT II 215).

Motion is, thus, inextricably linked with duration in Descartes' mind. But this in no way undermines the claim that motion is a mode, a genuine property of body, something that belongs to body as such. To understand how, though, we must recall our earlier discussion of the notion of duration. As noted in chapter 3, Descartes holds that duration, along with other notions like existence and unity, pertain to substances, both mental and material, even though they are not comprehended through the principal attributes of the two substances. In particular, he claims, "since any substance that ceases to endure ceases to be, there is only a distinction of reason between it [i.e., a substance] and duration" (Pr I 62). And so, I argued, Descartes' claim that all accidents of body must be "referred to" the principal attribute, extension, are, in the strictest sense, false. The Cartesian world is a world of geometrical objects made real. But as he construes them, the objects of geometry, even as they exist objectively in the mind, are taken to be enduring things and thus are at least capable of objective motion in objective time. And so, the world of bodies, the objects of geometry existing formally outside of our conception, can have real duration and motion as well. Though not strictly a mode of *extension,* motion as Descartes construes it would thus seem to be a proper mode of *extended substance,* a real, thought-independent feature of the world of bodies.

Motion and Individuation

Everything in Descartes' world must be explained in terms of bodies in motion. But motion itself is defined in terms of individual bodies; motion is the transference "of one part of matter or of one body" with respect to its contiguous neighborhood (Pr II 25). And so, it appears, the explanatory adequacy of Descartes' physics depends upon making sense of the notion of an individual body.

Before discussing Descartes' notion of an individual body, it is important to distinguish the notion of individuality from a closely related question, that of the substantiality of finite bits of material substance. Descartes is usually quite happy to treat the individual bodies of common sense as separate substances. In various contexts he refers to a stone, clothing, and a hand or an arm as separate substances.[20] But finite material substances are not limited to what we (or he) would call individual bodies. Writing about body in the context of an exposition of the real distinction, he notes:

> From the mere fact that we already have an idea of extended substance or body (though we don't yet know for certain that any such substance really exists) we are, however, certain that it can exist. And if it exists, we are certain that every part of it marked out by us in thought is really distinct from the other parts of that substance. (Pr I 60)

And so, he implies, every imaginable portion of material substance is really distinct from every other, that is, it is distinct in the way "two or more substances" are distinct from one another (Pr I 60). Later in the *Principles* he refers to each portion of a hard body as a distinct substance (Pr II 55). Similarly, in a letter to Gibieuf on 19 January 1642, he notes that we consider "the two halves of a portion of matter, however small it may be, as two complete substances," concluding from that that God can separate them, in opposition to the atomist thesis that some small bodies cannot be split (AT III 477 [K 124–25]).

Now, Descartes' apparent adherence to this view of each and every imaginable portion of matter as a separate substance is not unproblematic, either as a position to adopt or as an interpretation of Descartes' own considered view. As we noted earlier in chapter 5, he holds that a body cannot simply be annihilated, leaving an empty space in its place. And so, he argued, if God were to annihilate all body within a vessel, the sides must necessarily touch. But if this is the case, then it would appear that the portions of matter are not really independent, as later commentators observed.[21] Furthermore, some commentators have found reason to believe that his real view was that only body *as a whole*, infinite or indefinite extension, constitutes a genuine substance, and not its individual parts.[22]

But whatever complexities may surround the doctrine of a multiplicity of finite extended substances and its attribution to Descartes, they are beside the point for our purposes, insofar as we are concerned with the definition of motion. Material substances, should they exist, are *not* the same as individual bodies. While individual bodies may count as substances, *any* imaginable portion of material substance may count as a substance, be it an individual body or just a portion of one, delimited by thought alone; being a substance (if individual bodies are indeed substances) doesn't differentiate actual bodies from bodies that are only individuals in thought. And, Descartes is clear, it is only *actual* bodies that are relevant to physics. Writing in the *Principles*, again, Descartes notes that "a partition that exists only in thought changes nothing" (Pr II 23). For something to be a substance is for it to be capable of independent existence. But what is relevant to physics is not independence of this sort but individuality. And what is relevant for individuality is motion. Descartes writes in the *Principles:* "By *one body* or one *part of matter* I understand every thing that is transferred at the same time, even if the thing in question can be made up of many parts which, in themselves, have other motions" (Pr II 25; see also AT XI 15). The idea is, on its surface, a fairly simple and plausible idea; the thought is that from the point of view of physics, an individual is a portion of matter that moves together.

Before unpacking this account of individuality, it is important to note, first of all, that this definition should be understood as limited to a special kind of individuality, that which pertains to body as such, what we might call physical individuality, to distinguish it from a broader notion of individuality. As Descartes fully recognizes, a complex body can alter its constituents without thereby becoming a different body: "We can say that the Loire is the same river it was ten years ago, even though it no longer has the same water, and even though perhaps not a single part of the same earth that surrounds that water remains" (AT IV 165 [K 156]). He continues by giving a number of biological examples, animals and humans who change the constituent parts of their bodies as they absorb nutrition and grow, while at the same time maintaining their identity as the same individual bodies (AT IV 165–68 [K 156–58]). It is obvious on the face of it that the definition he gives of an individual in the *Principles* cannot accommodate this conception of an individual. The notion of an individual body he is concerned to define there is concerned with the notion of a physical individual, the sort of thing that can enter into the basic laws of nature. From other perspectives (morality, property law, medicine, animal husbandry, agriculture, etc.) we may be interested in individuals that persist as their constituent parts alter significantly. But, one might argue, from the point of view of physics, an alteration of constituent parts is a change from one individual to another; the snake that swallows a lamb may be, in one sense, the same snake, but it may well behave differently in collision with other bodies than it did before. And so, we should distinguish between the strict, physical sense of individuality that Descartes is trying to capture in connection with the definition of motion in the *Principles,* and other notions of individuality relevant to other concerns.

But even when we restrict our attention to the notion of a physical individual, one does not have to think too hard to realize that the surface simplicity of the definition hides a tangle of complexities. One set of complexities arises from the fact that individual bodies may be made up of smaller parts, parts which have their own motions. Descartes is fully aware that not everything within the surface of a complex body, one made up of smaller parts, necessarily belongs to that body. Characteristically between the parts that belong to a given body are pores. Because there are no empty spaces in his world, these pores cannot be empty, strictly speaking, but must be "filled with other bodies," the subtle matter or ether that pervades the entire universe (Pr II 6). And so, he makes the following claim about the Host, a claim that should hold for any body made up of parts: "Since the bread remains always the same, even if air or other matter contained in its pores changes, it follows that that these things do not belong to its substance" (AT VII 250). However, it is not clear that his definition of a single body can make sense of such bodies as individuals.

Descartes holds that a collection of parts in motion can constitute a single body as long as all of the moving parts are "transferred at the same time." But what can this possibly mean? It is, first of all, very difficult to define what it might mean for Descartes for a system of particles to have a common motion, to be "transferred at the same time." A natural suggestion is that what Descartes has in mind by common motion is the motion of a center of gravity of the particles that make up the body in question. Even if we could specify the relevant parts that make up a body, the center of gravity for a system of bodies in motion may well lie outside of any of the constituent parts that make up the body. But if so, then it is difficult to see what sense could be made of the notion of a system of bodies being in motion, on Descartes' proper definition. For motion, Descartes says, is the transference of one *body* with respect to contiguous bodies, and if no body corresponds to the center of gravity of a system of bodies, then, it would seem, a system of bodies can have no motion of its own, properly speaking.[23] And even if we could make sense of the common motion of the particles that make up a body, any arbitrary collection of particles can be constituted into a system, with its own center of gravity and its own common motion. As a consequence, every arbitrary collection of particles would seem to be a Cartesian body.

So far I have emphasized considerations that derive from Descartes' definition of an individual as it applies to complex bodies, bodies made up of smaller parts in motion with respect to one another. But the problems are, if anything, more serious if we consider what might be called simple bodies, bodies that while divisible are not actually divided into smaller parts.

The most obvious problem derives from the evident circularity of the definition of body. Body is defined in terms of motion; an individual body is, by definition, something that is transferred, that is, a quantity of material substance that moves all together. But the definition of motion *presupposes* the notion of an individual insofar as motion is defined as the transference of *one body* from one neighborhood into another (Pr II 25). Not all circles of this sort are objectionable, of course. But the particular way in which the definitions of motion and body are related gives rise to a rather unwelcome consequence. Since bodies are individuated through motion, where there is no motion, there can be no distinct individuals. That is, two bodies that we are inclined to say are at rest with respect to one another must, for Descartes, really be two parts of a *single* body. And so, it would seem, Descartes is committed to the rather paradoxical view that all individual bodies must be in motion. This difficulty was well known to later seventeenth-century thinkers. Gerauld de Cordemoy notes in his *Six Discours,* a Cartesian tract on physics and metaphysics first published in 1666:

> Another difficulty that I note in the opinion of those who say that matter itself is an extended substance is that they cannot conceive a body as separate without assuming a motion. So that, in accor-

dance with their doctrine, one cannot conceive a body at rest with respect to other bodies, since assuming that it touches them, this doctrine teaches that together with them it makes up only one body. However, it appears to me that we have a sufficiently clear and natural idea of a body completely at rest with respect to other bodies, none of which is in motion.[24]

For this reason (among others), Cordemoy was led to adopt a kind of atomism, and despite many other Cartesian commitments, posit as the basic building blocks of the physical world bodies that cannot be divided, and cannot change their shapes.[25] In short, Cordemoy argues that we must appeal to something outside of motion to individuate bodies. This is similar to a view Martial Gueroult takes. When responding to a problem similar to the one that worried Cordemoy, Gueroult suggests that in order to individuate bodies, Descartes must step outside of motion, considered purely as a mode of body, and appeal to cohesion, a force that arises from God's activity on body.[26]

This consequence, that a single body can never properly be said to be at rest, would seem to cause some difficulty for Descartes' strategy for deriving the laws of motion. As we shall see in later chapters, Descartes considers the distinction between bodies in motion and bodies at rest crucial for deriving the laws of impact. But if there are no bodies at rest, then this distinction can play no role in his derivation of the laws of impact. However, among the crucial cases Descartes discusses in connection with his law of impact are those in which a moving body collides with a body at rest. It is especially curious, then, that the one place where Descartes appears to acknowledge this consequence we have been discussing is in the context of a long letter on his laws of impact, and it is especially curious that he finds the consequence unproblematic. Writing to Clerselier on 17 February 1645 in an important letter to which we will later return, Descartes notes:

> In these rules [i.e., the rules governing specific cases of bodies in impact] by a body which is without motion I understand a body which is not at all in the act of separating its surface from those of the other bodies which surround it, and which, as a consequence, makes up a part of another larger hard body. (AT IV 186–87)[27]

But there is another apparent problem with his account of the individuation of simple and uncompounded bodies, a problem that Descartes never seems to have been aware of, a problem that, I think, is far more damaging than any that we have considered up until now. The last problem we considered concerned the difficulties distinguishing two contiguous bodies at rest. But, one can argue, given Descartes' conception of body and motion, even bodies in motion are not really individuated. The argument is due to Leibniz, one of the sharpest critics of Descartes' physics, and appeared first in 1698 in his essay, "De Ipsa Natura." Now, on the Cartesian view, all body is extension and extension alone, and there are no empty spaces. So, Leibniz notes, whatever distinctions may possibly arise through motion, at any given instant, there can be no distinction between one body and another, one portion of matter and another: "in the present moment (and, furthermore, in any moment whatsoever) a body A in motion would differ not at all from a resting body B."[28] At any instant, it would seem, the Cartesian world would have to be a homogeneous, undifferentiated extension; while we would be able to distinguish one region of space from another three feet away, there would not be any intrinsic difference between the one and the other, if we only have extension and motion at our disposal to make the distinction. But if there are no intrinsic differences between portions of the extended world at any instant, then, Leibniz argues, there can be no differences over time:

> If no portion of matter whatsoever were to differ from equal and congruent portions of matter . . . and, furthermore, if one momentary state were to differ from another in virtue of the transposition of equal and interchangeable portions of matter alone, portions of matter in every way identical, then on account of this perpetual substitution of indistinguishables, it obviously follows that in the corporeal world there can be no way of distinguishing different momentary states from one another.

And so, if there are no intrinsic differences between bodies at any given moment, motion cannot be called upon to provide any. Leibniz is clearly correct in holding that any given moment of the world will be indistinguishable from any other, as far as *we* are concerned; all will equally well be undifferentiated and homogeneous extension. But his point goes deeper than that: "Since everything substituted for something prior would be perfectly equivalent, no observer, not even an omniscient one, would detect even the slightest indication of change." Leibniz's worry seems to concern the reidentification of bits of Cartesian matter at different times; when comparing a present state with a future one, "under the assumption of perfect uniformity in matter itself, one cannot in any way distinguish one place from another, or one bit of matter from another bit of matter in the same place." Since at any given time there are no intrinsic properties to differentiate one portion of the homogeneous whole from any other, there are no possible grounds for saying that *this* hunk of extension is now in one place with respect to *that* hunk, and at a later time *this same hunk* is in a different place; if all we have to work with is extension, there seems to be no sense to the claim that the *same* bit of matter changes position with respect to others. And so, it would seem, motion can be used to individuate bodies *only* if there is some way of reidentifying bits of material substance across time, and this can only happen, Leibniz argues, if there is something in body over and above extension. Indeed, without such properties, without some means by which the same bit of matter can be reidentified, the very notion of motion appears to be incoherent. And so, if the world is as Descartes says it is, a plenum of bodies whose only property is extension, one cannot appeal to motion to individuate bodies.

I shall continue to talk as if Descartes is dealing with a world of individual bodies, colliding with one another, at motion and at rest with respect to one another. But, in the end, I suspect that this is something that he is not entitled to, and this is something that, if true, would seriously undermine his whole program.

MOTION AND COPERNICANISM

In the Preface General to his *Collection of Several Philosophical Writings*, published in 1662, Henry More, one of Descartes' acutest correspondents and one of his most careful readers made the following remark:

> *I cannot but observe the inconvenience this external force and fear does to the commonwealth of Learning, and how many innocent and well-deserving young Wits have been put upon the Rack, as well as* Galilaeo *in prison. For his Imprisonment frighted* Des-Cartes *into such a distorted description of Motion, that no mans Reason could make good sense of it, nor Modesty permit him to phansy any thing Non-sense in so excellent an Authour.*[29]

The charge stuck. It is now a commonplace of Cartesian scholarship that the definition of motion Descartes offered in the *Principles* was not a true representation of his position, but rather a device for enabling him to avoid Church condemnation by allowing him to say that there is a genuine sense in which the Earth is at rest; the claim is that the definition we have been carefully examining in the earlier part of this chapter is merely prudential, and that his real view of motion is quite different than the way he represents it. Now, there is no question but that Descartes was deeply affected by the condemnation of Galileo in 1633. He withdrew his own *World*, deeply Copernican, from the press, and delayed the publication of his system of physics for more than ten years. And there is also no question but that in his earlier writings

Descartes came out unambiguously in favor of the claim that the Earth *moves*: "if [the motion of the Earth] is false, all of the foundations of my physics are also," Descartes wrote to Mersenne in 1633 (AT I 271; see also AT I 285 [K 25–26]); AT III 258). But to what extent did his worries about Copernicanism motivate him to frame the new definition of motion in the *Principles* the way in which he did?

Let us begin examining this question by reviewing Descartes' discussion in the *Principles* of the main competing cosmologies, the Ptolemaic system, the Copernican system, and the Tychonic system. These discussions constitute a kind of preface to part III of the *Principles,* where, starting in §46, Descartes begins a careful exposition of his own view. He begins by quickly dismissing the Ptolemaic system, the system in which all planets and the sun travel around a motionless Earth:

> The first [such system] is that of Ptolemy. Since this is in conflict with many phenomena (to take the most important example, [it conflicts with] the waxing and waning of the light observed in Venus, just as in the Moon), it is now commonly rejected by all philosophers, and therefore I shall set it aside here. (Pr III 16)[30]

The serious alternatives, then, are to be Copernicus and Tycho, as well as Descartes' own cosmological theory, which he understands as a variant of Copernicus' system.[31] As pure hypotheses, that is, considering only the relative motions they attribute to the sun, stars, and planets, Copernicus and Tycho both "satisfy the phenomena in the same way, and do not differ greatly from one another, except for the fact that Copernicus' hypothesis is somewhat simpler and clearer" (Pr III 17). Indeed, the same can be said for Descartes' own system; all three can be represented in terms of the planets and Earth moving in roughly circular paths around a central sun.[32] Where the three differ is on the question of what is really said to move:

> Tycho had no reason *[occasio]* to alter [Copernicus], except for the fact that he was trying to explain things not only hypothetically but also in accordance with the very truth of the matter. . . . Although Copernicus had not hesitated to attribute motion to the Earth, Tycho, finding this completely absurd as [a claim] in physics, and far from people's common sense, wanted to alter this. (Pr III 17–18)

And so Tycho rejected Copernicus' moving Earth, and instead claimed that all planets move around the sun except for the Earth, which remains immobile as the sun turns around it, with both a diurnal motion producing the cycle of night and day, and an annual motion, producing the cycle of the seasons. It is on this point, the question of what is really moving and what is really at rest, that Descartes will differ with both Copernicus and Tycho:

> Since he had not sufficiently considered the nature of motion, [Tycho] asserted that the Earth is at rest merely in words, but in reality *[re ipsa]* he yielded more motion to it than the other [i.e., Copernicus] did. . . . Therefore, differing from both of them only insofar as I shall remove all motion from the Earth more truely than Tycho and more carefully than Copernicus, I shall propose here this hypothesis which seems to be the simplest of all, and the one most suitable both for understanding the phenomenon and discovering their natural causes. (Pr III 18–19; see also AT V 550)

Descartes doesn't discuss Copernicus' own version of his theory any further; Copernicus holds quite explicitly that the Earth has a number of motions, and there is nothing more to be said, as far as he is concerned. But he does confront the Tychonic system, and argue that contrary to what its inventor claimed, in that system the Earth must be said to move. First he discusses the diurnal motion. On Tycho's view the sun and planets revolve daily around a resting Earth, resulting in day and night. Descartes first notes that "since this transference is reciprocal as noted above [e.g., Pr II 29] and plainly the same force or action is required for it in the Earth and in the heavens, there is no reason why that motion should be attributed to the heavens rather than to the Earth" (Pr III 38). But there is, indeed, some reason why it may be preferable to attribute the motion to the Earth rather than to the heavens on Tycho's view. Appealing to what Gueroult called the criterion of displacement as a whole, Descartes argues since the transference occurs "on the whole surface [of the Earth], and not in the same way on the whole surface of the heavens, but only on a concave part contiguous to the Earth, which is tiny in comparison with the convex [surface]," the motion ought to be attributed to the Earth rather than to the heavens, if we are going to attribute it to the one rather than to the other. He treats the annual revolution in the Tychonic system similarly. In the Tychonic view, all of the planets revolve around the sun, while the sun revolves around a supposedly stationary Earth. But in revolving around the Earth, the sun must carry the fluid filling the spaces between the planets with it. And so, as with the case of the diurnal motion, the criterion of displacement as a whole would lead us to attribute the annual motion to the Earth rather than to the sun or the heavens, if one or the other must be said to be moved (Pr III 39).

When discussing the Tychonic system, the claim is that if we take Tycho's view seriously, motion should be attributed not to the heavens, or to the sun, as Tycho actually does, but to the Earth. The claim, though, is a weak one. As Descartes realizes explicitly in the context of his discussion of diurnal motion, strictly speaking, it is as correct to attribute motion in the Earth as it is to the heavens, on Tycho's view; when he claims that Tycho's Earth must move, he is appealing not to the strict conception of motion but to the sorts of rules of thumb that he introduced earlier in Pr II 30 to explain why we are inclined to overlook the reciprocity of transference, and assert that when we walk on the surface of the Earth is it us and not the Earth that moves.

But on Descartes' own version of the Copernican cosmological hypothesis, he asserts that in the *strictest* sense, the Earth is at rest. Descartes' view is that the heavens are fluid, filled with ether, and swirling around the sun in a great vortex; they carry the planets with them.[33] This explains the annual revolution of the Earth around the sun (Pr III 24–27). But on this view, he claims, in its annual revolution, the Earth is really at rest, in the proper sense of the term. After reminding his readers of the proper definition of motion in Pr II 25, and the distinction between motion properly understood and the common conception of motion, Descartes goes on to remark:

> Whence it follows that no motion, properly speaking, is found in the Earth or even in the other planets, since they are not transferred from the vicinity of the parts of the heavens which are immediately contingent to them, insofar as those parts of the heavens are regarded as being without motion. (Pr III 28).

Since the Earth and other planets are at rest in the heavens and carried along by them, they are not in motion, according to his proper definition (Pr III 26–27; see also AT V 550 and Pr II 62). Furthermore, he goes on to say, "even if motion is improperly understood in accordance with common usage, no motion is to be attributed to the Earth." Now in the Cartesian version of the Copernican cosmology, as in the strict Copernican cosmology, the Earth does move with respect to the stars. But, he claims, "the common people determine the places of the stars with respect to the parts of the Earth, regarded as being at rest, and regard them as moving to the extent that they recede from places determined in this way." And so, the common sort will regard the Earth as being at rest and the *stars* as moving, a view corresponding to the "philosophical sense" of the notion of place, in accordance with which place is determined by what is nearby (Pr III 29).[34] Descartes ends the French version of this argument by saying:

> If nevertheless in order to accommodate ourselves to common usage, we later appear to attribute some motion to the Earth, it must be realized that we are speaking improperly and in the same

sense as one can sometimes say of those who sleep and lie in a boat, that nevertheless they pass from Calais to Dover because the boat carries them there. (Pr III 29F)

The final conclusion of the argument is nicely summarized in a marginal comment on the *Principles* Descartes left:

> The Earth does not move according to Copernicus but rather according to Tycho. And thus are the [Holy] Scriptures exonerated: for if it is said that they spoke in accordance with common usage, then there is nothing in opposition to Copernicus, or if it is said that they spoke from knowledge of the truth then unknown by the common people, then they stand in favor of Copernicus. (AT XI 657)[35]

The argument with respect to the diurnal motion would be similar.[36] In addition to the great vortex that carries the Earth and the other planets around the sun, each planet (at least each planet with a moon) has its own vortex, of which it is the center. So, the Earth is at the center of a vortex that at the center turns once every twenty-four hours, while at the distance of the moon turns roughly once per month (Pr III 33). Presumably the Earth can be considered as being at rest in the center of that vortex in just the way that the Earth (or, perhaps, the Earth-moon system) can be considered at rest in the great vortex that carries it around the sun along with all of the other planets.

This, then, is Descartes' account of the motion of the Earth on the various planetary systems. There is no question but that Descartes makes explicit use of his formal definition of motion and the discussions that surround it in order to argue that on his system the Earth is at rest and that on the other competing systems, it isn't. Furthermore, it is highly unlikely that Descartes would have worried about such an issue had it not been for the condemnation of Galileo, and it is quite possible that Descartes would have continued to assert that the Earth moves. But the central question is this: was the definition of motion in the *Principles* formulated *specifically* for the purpose of allowing Descartes to deny that the Earth moves? Is it fair to say that the concept of motion he defines there is not his own, but one he adopted merely out of prudence?

There are several arguments found in Descartes' commentators. The first more widespread but more obviously unsatisfactory argument goes as follows. As we shall see in more detail in later chapters, Descartes' laws of motion in the *Principles* clearly presuppose a genuine distinction between motion and rest. For example if body A were one pound and body B three, then, it turns out on Descartes' laws of impact, we would get incompatible outcomes if we set B at rest and imagine A to collide with it with speed v, as opposed to if we set A at rest and supposed B to collide with it at speed v; in the first case, A would rebound from B, while in the second, the two bodies would travel off together at the same speed, after the collision (Pr II 49–50). So, as far as the laws of motion are concerned, we are not free simply to decide to set one or the other body at rest; there must be a physical fact of the matter. But, the claim is made, Descartes' preferred definition of motion in the *Principles* is *relativistic,* and does *not* admit a genuine distinction between motion and rest. And so, it is claimed, while Descartes himself recognized a genuine distinction between motion and rest, as manifested in the laws of impact, the relativistic definition of motion in the *Principles* is set out only for political reasons. Alexandre Koyré wrote in his *Galileo Studies*:

> The Copernicanism which had been so openly on display in *The World* had disappeared from the *Principles,* or rather had been hidden behind an odd and peculiar theory of motion. . . . The ultra-relativism of his idea of motion was not original with Descartes. It is our opinion that he only adopted it so as to be able to reconcile Copernican astronomy, or more simply the mobility of the earth, which was manifestly implied by his physics, with the official doctrine of the Church. It was an initiative which succeeded only in making Cartesian mechanics self-contradictory and obscure.[37]

Similarly Richard J. Blackwell has claimed:

> If we are to give Descartes all the benefit of the doubt here, we must conclude that the relativity of motion does not play a systematic role in Cartesian physics. Rather, his appeal to the relativity of motion seems to have a polemical purpose enabling him to reconcile his heliocentric theory of astronomy with the prevailing attacks against Copernicanism.[38]

Blackwell goes on to suggest that the doctrine of relativity functions in the answer to the Church by allowing Descartes to hold *both* that the Earth is at rest in its vortex, *and* that it is moving around the sun, claims both true depending upon which frame of reference one chooses to work with.[39]

It should be clear that this reading is simply and evidently false; if Descartes fashioned the definition of motion in the *Principles* in order to allow himself to say that the Earth is at rest, it *cannot* be in this way. First of all, as I argued in detail earlier in this chapter, Descartes did not think that the preferred definition of motion he gave in the *Principles* was relativistic in the sense these commentators suppose it is; indeed, it was carefully crafted to admit a genuine distinction between motion and rest, and on this score, explicitly contrasted with the common definition that he rejected.[40] And so, I think, there is no inconsistency between the definition of motion and its laws to be explained away by any appeal to politics or prudence; the distinction between motion and rest Descartes appeals to in framing his laws is just a reflection of the distinction he built into the definition, a distinction important for the grounding of motion as a real mode of body. And secondly, the relativity of motion and rest plays little role in Descartes' discussions of the motion of the Earth. As we saw, he does appeal to the doctrine of the reciprocity of transference in order to argue that at very least Tycho must hold that there is as much motion in the Earth as there is in the heavens that, on his view, turn around it. But when discussing the motion of the Earth on his own view, the main point is that on his strict definition of motion, the Earth *must* be regarded as being at rest, and that one has *no choice* in the matter. When we say that the Earth moves around the sun, as we will from time to time, Descartes urges us to remember that at best, we are only speaking with the vulgar, and that in the *proper* sense, the Earth is at rest.

This eliminates one set of reasons for suspecting Descartes of dissimulation with respect to his definition of motion. But even when his definition of motion is properly understood and his discussion of the motion of the Earth more carefully considered, there is room for some worry on the issue. One might argue, for example, that the dissimulation goes the other way, that Descartes was really a relativist in motion, and that the apparent distinction between motion and rest that his proper definition allows him is not genuinely in accord with his own views, that it is there only to allow him to deny a moving Earth.[41] This in a way accords better with the texts, and does better justice both to his definition of motion and to the use to which he puts it in the discussion of Copernicanism. But it does have one odd feature. On that view, though Descartes *appears* to be consistent insofar as the definition of motion allows a kind of distinction between motion and rest, as needed for the laws of impact, in *reality* he is inconsistent; if beneath the mask of the proper definition of motion he is really a radical relativist, then his preferred account of the nature of motion is inconsistent with the account he gives of the laws of nature. The earlier reading we discussed and rejected makes Descartes coherent at the cost of making him dishonest; this one would have him come out *both* dishonest *and* dumb. Any principle of charity in historical interpretation should make us suspicious of that.

This, of course, does not altogether eliminate the possibility that both the definition of motion in the *Principles* and the distinction between rest and motion embodied in the laws of motion are part of a carefully contrived strategem to enable Descartes to deny the motion of the Earth. It is indeed possible that the entire theory of motion and its laws is an elaborate mask. But such a claim is hardly plausible. For one, it is not clear what it is supposed to explain. The first argument we considered purports to explain the supposed inconsistency between Descartes' supposedly relativistic definition of motion and his nonrelativistic laws of impact, and the second is supposed to explain the contradiction between the apparently nonrelativistic definition of motion and Descartes' supposed commitment to relativism. At best, this last argument might be taken to explain why Descartes built into his physics, both the definition of motion and its laws, a genuine distinction between motion and rest. But there is another, better explanation for that, I think, the one I emphasized earlier in this chapter. If motion is to be a genuine *mode* of body, and is to be the basic explanatory principle in Descartes' physics, then there must be a genuine difference between being in motion and being at rest. And secondly, if the account of motion and its laws is supposed to be a mask, then Descartes has given us little hint of the face supposedly lurking underneath; if there is *another* Cartesian position the crafty and world-wise reader is supposed to see in the *Principles,* Descartes has hidden it so well that it may as well not be there at all. If mask it be, then it is an interesting mask, historically more important and more worthy of careful study than the face that supposedly lies beneath.

Descartes was no doubt pleased that he could argue that on his definition of motion, the Earth is at rest within his cosmological system, and, no doubt, he was looking over his shoulder at the Church when making this argument.[42] But, I suspect, he would have tried to make this case no matter *what* definition of motion he adopted. The denial of motion to the Earth may well be political and prudential, without thereby undermining our faith in the sincerity of the account of motion on which it is based.

DETERMINATION

So far we have been talking about Descartes' notion of motion. Before leaving the subject, we must examine one of the most important (and difficult) notions that is connected with Descartes' conception of motion, that of determination. The notion of determination, together with that of speed, will prove to be crucial to understanding Descartes' laws of motion, as we shall attempt to do in the following chapters.

The notion of determination, the Latin noun *'determinatio'* and the corresponding verb *'determinare'* have a long and complex history in scholastic texts; surely Descartes was well aware that he was not the first to use these terms, and knew something of their usage in scholastic parlance, even if he may not have been a careful scholar in the years after he left La Flèche.[43] But the difficulties various of his contemporaries had with Descartes' terminology here, the many times Descartes finds he has to explain himself to them, suggest that there is something idiosyncratic about the way Descartes understands that notion. And so we should begin by turning directly to Descartes' characterization of determination.

The notion of determination is intimately linked with the directionality of a body in motion, and is most often presented in contrast with other related notions that don't involve directionality. For example, in the *Dioptrics,* published in 1637 and, perhaps, the chronologically earliest use of the notion, Descartes writes, "The motion [of a body] differs entirely from its determination to move in one direction rather than another" (AT VI 97 [Ols. 77–78]; see also AT IX 9; Pr II 41; AT III 75, 289; AT IV 185). Elsewhere Descartes emphasizes the "difference between the determination to move this way and that, and the speed" (AT II 17).[44] Elsewhere still the contrast is between determination and force, and, at one point, between "the determination and the speed or force" a moving body has (AT II 18).[45] The determination to move in a particular direction is, for Descartes, a mode of a mode. Descartes writes to Clerselier on 17 February 1645 that "we must consider two different modes in motion *[movement]*: the one is motion *[motion]* alone or speed, and the other is the determination of this motion *[motion]* in a certain direction" (AT IV 185).[46] Similarly, writing to Mersenne on 26 April 1643 he noted that in motion, a mode, "there are only two differentia *[varietez]* to consider in it, the one, that it can be more or less fast, and the other that it can be determined in different directions" (AT III 650).[47] Determination is thus linked to motion, just as shape is linked to quantity or extension linked to body; without motion there can be no determination.[48]

Determination is closely linked with direction. But it is wrong simply to *identify* determination with direction, as many readers have done.[49] Descartes himself warned against the misunderstanding. In a letter to Mersenne from June or July 1648 Descartes comments on something Mersenne had sent him, an essay by Roberval that dealt with Descartes' account of the composition of motion. One of the first things he takes Roberval to task for is borrowing his ideas but altering the terminology: "When he calls 'impression' what I call speed and 'direction' what I call the determination to move in a particular direction, this serves only to confuse himself" (AT V 203).[50]

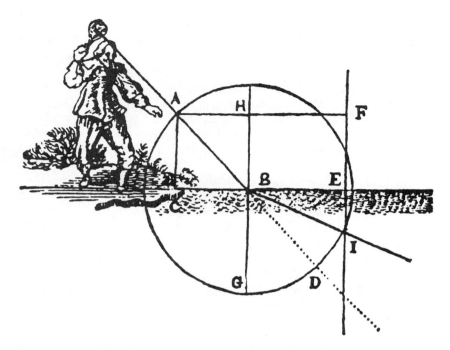

FIGURE 6.2 From [René Descartes,]*Discours de la méthode...plus la dioptrique, les méteores,et la géométrie...* (Leyden, 1637), *Dioptrique,* p. 17.

Descartes' notion of determination can further be clarified by examining the way in which he uses this notion in the derivation of the law of refraction in the *Dioptrics,* perhaps the original motivation for the distinction. In deriving his law of refraction, Descartes considers three different analogies. A tennis ball is hit obliquely to a horizontal surface CBE, from A to B (see fig. 6.2).

In the first case the surface CBE is imagined to be a cloth that hinders the motion of the ball while allowing it to pass through; in the second, it is the surface of a body of water; and in the third, it is a racket that adds motion to the ball by striking downward (AT VI 96–100 [Ols. 77–80]). The precise application of these analogies to the refraction of light is complex, as is the derivation of a sine law of refraction from them. But Descartes' treatment of the behavior of the ball itself is clear enough, and clearly involves an appeal to the notion of determination.

Since Descartes' reasoning in these three cases is parallel, I shall consider only the first case, where the discussion is most explicit. The ball is impelled from A to B (see fig. 6.2). At B, though, it encounters a cloth which permits the ball to pass through, but which slows it down, by half, for example. To figure out how the ball would travel after hitting the cloth, Descartes appeals to the distinction between motion (speed) and determination:

> To know the path it ought to follow, let us take into consideration again the fact that its motion completely differs from its determination to move in one direction rather than another, from which it follows that their quantity should be examined separately. (AT VI 97 [Ols. 77–78])

And so, he continues:

> And thus let us consider that, of the two parts of which one can imagine that this determination is composed, it is only the part which makes the ball tend downwards which can be changed in some way by the collision with the cloth, and that the part which makes it tend to the right ought always to remain the same as it had been, since this cloth does not in any way oppose it in this direction. (AT VI 97 [Ols. 78])

Descartes begins determining the postcollision path of the ball by describing a circle with center B and radius AB, and drawing three perpendiculars to the impacted surface CBE, AC, HB, and FE, specifying that the distance from HB to FE be twice the distance from AC to HB. Now, since the ball moves only half as fast after it passes through the cloth at B, it will take twice as long for it to intersect the circle after passing through B as it did to travel from A to B.[51] But, Descartes claims:

> Since it lost nothing at all of the determination it had to go to the right, in twice the time that it took to pass from the line AC to HB, it ought to go twice the distance in same direction, and consequently arrive at some point on the straight line FE at the same instant that it also arrives at the circumference of circle AFD. (AT VI 98 [Ols. 78])[52]

And so, he concludes, the ball must intersect the circle at point I, where the extension of FE intersects the circle below the cloth, since by assumption, the ball passed through cloth CBE. Using a similar strategy Descartes treats the other two cases connected with his analysis of refraction.

These arguments, which turn on the distinction between speed (motion, force) and determination, reveal a great deal about Descartes' conception of this key notion. For one, it is clear that a body does not have just one determination at a given time; the tennis ball under discussion in the *Dioptrics* has both a determination to move from left to right, and a determination to move downwards. This Descartes emphasizes in a letter he wrote to Mersenne on 21 April 1641, part of an exchange with Hobbes over these questions earlier in that year. He writes: "What he then says, namely, that a motion has only one determination, is just as if I were to say that an extended thing has only one shape, which doesn't hinder the fact that this shape can be divided into many parts, just as determination can be" (AT III 356). Just as a body has one shape, a motion has one determination. But just as a shape can be divided into other shapes which constitute its parts, just as a cube can be divided into two pyramids, say, so can a given determination be divided into other determinations, determinations that constitute its parts. Indeed, for any motion we can arbitrarily choose a direction, and consider the determination of the motion in that direction.[53]

Another feature of determination that comes out of these arguments is that it is a *quantity*. For example, in the case of refraction, the determination from left to right is said to remain the *same*. And while Descartes talks about a numerical decrease in the *speed* of the ball as it passes through CBE, and never deals directly with the *magnitude* of the change in the downward determination,[54] it is clear that the downward determination has a magnitude, and that it is decreased or increased by passing through CBE in the different cases Descartes considers. And so, discussing the laws of impact with Clerselier in the letter of 17 February 1645, Descartes can talk about one body passing to another "more than half" of its speed and "more than half" of its determination to go from right to left (AT IV 186).[55]

But even though determination, like speed, is a quantitative notion, it plays a radically different role in Descartes' laws of nature. The total arithmetical sum of speed times size remains constant in Descartes' world. Now, the notion of determination certainly enters into his accounts of the lawlike behavior of bodies; Descartes asserts that a change in determination requires a cause and cannot spontaneously occur, and the notion of determination enters into what I shall call his laws of persistence of motion (see chapter 7 below). But even though determination is a quantitative notion, Descartes proposes no general *quantitative* law to govern the changes that can occur in determination. It is significant that in the arguments concerning refraction, Descartes certainly implies that the vertical determination changes in magnitude. But the actual arguments make use of *only* assumptions concerning the persistence of the same horizontal determination and change in speed; Descartes has no general machinery for dealing directly with changes in the magnitude of a determination in a particular direction or calculating it directly.[56]

What, then, is determination for Descartes? My suspicion is that the notion of determination emerged sometime in the 1620s as he was struggling with particular problems in optics, like the analysis of reflection and refraction; and though it has some role to play in the laws of nature presented both in *The World* and in the later *Principles*, only in the face of specific criticisms of the arguments in his *Dioptrics* did Descartes attempt to clarify it, and then in a relatively ad hoc and unsystematic fashion. Unlike the notions of body or space or the laws of motion, Descartes never seems to have turned his full attention toward making clear just what *exactly* the notion of determination means for him. In attempting to give an account of the notion, we should thus be sensitive to the fact that we are almost certainly dealing not with a clear and distinct conception but with a notion in the process of emerging. But for all the complexities, it is fair to see in Descartes' notion of determination a serious attempt to deal in a quantitative way with the directionality of motion, and something akin to the modern notion of velocity (the vector quantity corresponding to the scalar speed) or momentum. Gabbey calls it "the directional mode of motive force," and Sabra claims that the distinction between the determination of a moving body and its speed correspond to "the distinction between vector and scalar quantities."[57] These interpretations are both illuminating, and correlate reasonably well with the closest thing to a definition Descartes himself gave. Writing to Mersenne on 3 December 1640, in connection with Father Bourdin's objections to the *Dioptrics*, he noted that "in speaking of determination to the right, I understand the entire part of the motion which is determined to the right" (AT III 251).

However problematic the notion of determination has proved to be for Descartes and his readers, his conceptions of directionality, determination, and their relations to the notions of speed, motion, and force will turn out to be crucial to the derivation of the laws of motion in the chapters to come.

DESCARTES AGAINST HIS TEACHERS: MOTION AND CHANGE

Motion is central to Descartes' physics, he tells us; it is the basic explanatory notion, that in terms of which all of the different properties bodies have are to be accounted for. But, he also tells us, the motion he has in mind in connection with this claim is not what his contemporaries might think it is. We have already examined one contrast he intends to draw, the contrast between the proper, philosophical definition of motion and the vulgar conception of motion, a contrast he comes to see as central in the early 1640s while drafting the *Principles*. But there is another contrast important to Descartes, a contrast he emphasized from his earliest writings. In *The World*, discussing the philosophers of the schools, he remarked that "the motion that they discuss is so very different from the motion I conceive that it can easily happen that what is true of the one is not true of the other" (AT XI 38–39). In this section we shall explore this important difference.

As with the other questions in late scholastic thought that we have touched upon, the doctrine of motion in the schools was a matter of great complexity.[58] Briefly, though, motion for the scholastics meant something significantly broader than it does for us, and, indeed, something significantly broader than it did for Descartes. Motion for the schools signified change in the broadest sense, the passage from one attribute, accident, or form (the *terminus a quo*) to another (the *terminus ad quem*). And so, for example, Eustachius characterizes motion as follows: "Motion, properly speaking is in its essence *[formaliter]* the very acquisition of a form, or the flow, path, or tendency *[fluxus, via, seu tendentia]* toward a form."[59] This is the sense of the common Aristotelian definition of motion Descartes often dismissively quotes: *motus est actus entis in potentia prout in potentia est,* motion is the actuality of a thing in potentiality insofar as it is in potentiality (AT XI 39).[60] A body in passage from form X to form Y is in potentiality insofar as it is no longer X but not yet Y, and the condition of being in motion is precisely this condition of being in the process of becoming Y. Since there are any number of ways a body can change, there are any number of kinds of motion. But, according to the schools, all motion falls into three, possibly four categories, depending on the categories of that which is changing: there is motion with respect to quantity, quality, place, and, on some accounts, motion with respect to substance. Motion with respect to quantity is intended to include increase and decrease in size. Motion with respect to quality contains a variety of changes, from hot to cold, from black to white, etc. Motion with respect to place, local motion, is change of place. And finally, motion with respect to substance, should it be considered a genuine variety of motion, is generation or corruption, the departure of one substantial form, that which makes a body the sort of thing it is, and the acquisition of a new substantial form, resulting in a new sort of substance.[61] All of these changes (with the possible exception of change of substance) are properly called motions for the schoolmen.

Descartes' objections often center on the general definition of motion commonly given in the textbooks. Writing in *The World,* Descartes gives his most common complaint:

> They themselves admit that the nature of their [notion of motion] is very poorly understood, and to render it in some way intelligible, they cannot explain it any more clearly than in these

terms: *Motus est actus entis in potentia, prout in potentia est,* which is
so obscure for me that I am forced to leave it here in their language, since I cannot interpret it. (And, in fact, their words [in
translation] are no more clear for being in French). (AT XI 39;
see also AT X 426, AT II 597 [K 66])

In addition to this general complaint about the unintelligibility of the
Aristotelian definition, in a letter probably from the mid-1640s
Descartes expresses a rather more specific criticism:

> As for the definition of motion, is is obvious that a thing said to
> be in potentiality cannot be understood to be in actuality, so that
> when anyone says that "motion is the actuality of a thing in poten-
> tiality insofar as it is in potentiality," motion is understood to be
> the actuality of a thing which is not in actuality, insofar as it is not
> in actuality. This includes either an apparent contradiction, or, at
> least, it includes a great deal of obscurity. (AT IV 697–98)

But Descartes' objections to the Aristotelian conception of motion
in the schools goes deeper than that. In a way, Descartes objects to the
very conception of motion that gives rise to the general Aristotelian
definition of motion, the idea that motion is a notion that embraces all
sorts of change, of which local motion is but one. For Descartes, of
course, all change is explicable in terms of local motion alone; the
change from small to large, from hot to cold, from black to white, from
wheat to bread is explained in terms of the local motion of the small
corpuscles that make up the visible world. And so, Descartes often emphasizes that by motion he means *only* local motion, "for no other sort
falls within my understanding, nor, therefore, do I think that we should
imagine any other sort in the natural world" (Pr II 24; see also Pr I 69;
AT X 39–40, etc.). The rejection of all varieties of motion but local
motion is one of Descartes' basic commitments; it is, indeed, closely
linked to his entire mechanist program for physics.

But even with respect to local motion, Descartes thinks the schoolmen have gotten it wrong. Now, motion is the passage from one terminus to another; indeed, without termini, the notion of motion is unintelligible, as far as the schoolmen were concerned.[62] And so, a portion
of elemental earth in free fall is in passage from its initial position
toward the center of the universe, and elemental fire from its initial
position to the sphere of the moon. For Descartes, as we shall see when
discussing his laws of motion, a body in motion will have a definite
direction toward which it tends at any given instant. But a body in
motion has no determinate goal toward which it tends, no *terminus ad
quem*. Motion is thus not the passage from one determinate state to

another, but is itself a state of body, a mode of body, "a thing of no less reality than shape" (AT V 403 [K 257]).[63] This will lead Descartes to the radical view that a body in motion remains in motion unless an appropriate cause alters that state, a view we shall examine in detail in later chapters, where we shall see further ways in which Descartes came to differ from the conception of motion he learned in school.

MOTION AND ITS LAWS:
PART 1, PRELIMINARIES AND THE
LAWS OF PERSISTENCE

IN THE PREVIOUS chapter we treated motion as a mode of body, motion considered simply as the transference of a body from one neighborhood of contiguous bodies and into another. In this chapter we turn from motion, the effect, to the cause of motion in the physical world, God, and to the laws that bodies in motion obey as a result of the way God causes them to move. We shall begin with an overview of Descartes' treatment of the laws of motion, mainly in *The World* and the *Principles,* followed by more detailed discussions of the laws in this and the next chapter. While we shall have to deal with the way Descartes grounds his laws in God in these discussions, the question of the derivation of the laws of motion from the immutability of God will not be taken up in full detail until chapter 9.

THE LAWS OF MOTION: AN OVERVIEW

The characteristic behavior of body, the laws governing bodies in motion were of concern to Descartes at least since his discussions with Beeckman in autumn of 1618. In his earliest attempts to give a mathematical account of a body in free fall, Beeckman reports that Descartes "proceeded in accordance with my basic principles, that is, that in a vacuum, something once moved always moves" (JB I 263 [AT X 60]).[1] No doubt, this principle, known to Beeckman as early as 1613 (JB I 24), was a subject of their discussions; as we shall see, it will later become one of the cornerstones of Descartes' own theory of motion. Other likely topics of conversation include the conservation of motion, and the tendency of a body in circular motion to proceed in a straight path, a question that appears in the *Cogitationes Privatae,* and was probably a question Beeckman suggested to Descartes in December 1618.[2]

But even though the laws governing motion were important to Descartes' earliest efforts in the new physics, it will be more than ten years later before he will attempt to set them out in a coherent and orderly way, and formulate arguments in their defense. Writing to Mersenne on 18 December 1629, in what must have been the earliest stages of the composition of *The World,* he notes that "I will try to demonstrate in my treatise" the principle that "in a vacuum what is once moved, always moves," the principle he had learned from Beeckman eleven years earlier (AT I 90, note a).[3]

In *The World* Descartes made good his promise to Mersenne and provided a lengthy discussion of the laws governing bodies in motion. The laws of motion are discussed in chapter VII of *The World.* In chapter VI he had asked the reader to imagine God to have made a new world in the imaginary spaces of the philosophers, a world without vacuum, filled with matter without the forms of the schoolmen, matter conceived of as "a true body, perfectly solid" (AT XI 33). Now, when God created the world of bodies, he created it together with motion: "From the first instant in which they were created, some began to move in one direction, others in another, some faster, others more slowly (or, indeed, if you please, not at all), and they continue afterwards in accordance with the ordinary laws of nature" (AT XI 34). He continues:

> For God has established these laws so wonderfully well that even if we suppose that he created nothing more than what I have spoken of, and even if he didn't put into it any order or proportion, but only made of it a chaos . . . , these laws are sufficient to make the parts of this chaos sort themselves out and dispose themselves in such good order that they will have the form of an exceedingly perfect world in which one can see not only light, but also everything else, both general and particular, which would appear in the true world. (AT XI 34–35)

So, he claims, by the laws of nature alone the initial chaos will sort itself out, and evolve into the world as we know it, the sun, stars, planets, and everything we see around us on the Earth.

This is what Descartes tries to do in the rest of *The World,* derive order from chaos by appeal to the laws of nature, that is, the laws that govern change in a world whose only changes are local motions. But before he can do this, Descartes must give the reader some idea of what these laws are. For this he turns directly to God, who is the first and sustaining cause of motion, just as he is the first and sustaining cause of everything in the world. "It is easy to believe," he says, "that God . . . is immutable, and always acts in the same way" (AT XI 38). From this he derives three laws in the following order:

[Law A:] Each part of matter, taken by itself, always continues to be in the same state until collision *[rencontre]* with others forces it to change. . . . [And so,] once it has begun to move, it will continue always with the same force, until others stop it or slow it down. (AT XI 38)

[Law B:] When a body pushes another, it cannot give it any motion without at the same time losing as much of its own, nor can it take any of the other's away except if its motion is increased by just as much. (AT XI 41)

[Law C:] When a body moves, even if its motion is most often on a curved path . . . , nevertheless, each of its parts, taken individually, always tends to continue its motion in a straight line. (AT XI 43–44)

These are the three laws Descartes gives in *The World*. In addition, though, hidden in the informal argument he gives for the first two laws is another law:

Now, these two rules follow in an obvious way from this alone, that God is immutable, and acting always in the same way, he always produces the same effect. Thus, assuming that he had placed a certain quantity of motions in the totality of matter from the first instant that he had created it, we must admit that he always conserves in it just as much, or we would not believe that he always acts in the same way. (AT XI 43; see also p. 11)

This, of course, is the principle of the conservation of quantity of motion, a principle that will play an explicit and important role in the later development of his laws of nature.

The laws Descartes formulated in *The World* and the basic strategy he used for proving them by appeal to God remained very much the same throughout his career. Though the details changed, and, most visibly, the collision law, law B, was developed more carefully in later writings, the fundamental inspiration remained of a piece. But when in the early 1640s Descartes taught *The World* to speak Latin, the laws took on a new and somewhat more coherent shape.

The problem situation in the *Principles* is much the same as it was in the earlier *World*. In the *Principles,* as in *The World*, Descartes is attempting to derive the present state of the world from an initial chaos, and as in *The World,* he must appeal to laws governing the behavior of bodies in order to do so. But in the *Principles,* instead of being sandwiched between the postulation of chaos and the derivation of the effects, the discussion of the laws governing bodies precedes the postulation of chaos, and is part of a general discussion of the notion of body and motion. Descartes begins part II of the *Principles* with a careful discussion of body and space, before focusing on the notion of motion, the basic explanatory principle in his physics. The discussion of motion is, in turn, divided into two parts. He begins with a discussion of motion considered as a mode of body, taken independently of its cause, and regarded simply as the transference of a body from one neighborhood into another. Only then does he investigate its cause, and it is in this context, starting in Pr II 36, that the laws governing bodies in motion are investigated.

Prominent in the account of the laws Descartes gives in the *Principles* is a distinction not found in the earlier *World*. He begins:

> Having taken note of the nature of motion, it is necessary to consider its cause, which is twofold: namely, first, the universal and primary cause, which is the general cause of all the motions there are in the world, and then the particular cause, from which it happens that individual parts of matter acquire motion that they did not previously have. (Pr II 36)

He characterizes the "universal and primary cause" as follows:

> And as far as the general cause is concerned, it seems obvious to me that it is nothing but God himself, who created motion and rest in the beginning, and now, through his ordinary concourse alone preserves as much motion and rest in the whole as he place there then. (Pr II 36)

Though it is not explicitly identified as a law, Descartes goes immediately on to state a version of the same conservation principle introduced earlier in *The World:*

> Whence it follows that is most in agreement with reason for us to think that from this fact alone, that God moved the parts of matter in different ways when he first created them, and now conserves the whole of that matter in the same way and with the same laws *[eademque ratione]* with which he created them earlier, he also always conserves it with the same amount of motion. (Pr II 36)[4]

After discussing the universal cause of motion, Descartes turns to the particular causes:

> And from this same immutability of God, certain rules or laws of nature can be known, which are secondary and particular causes of the different motions we notice in individual bodies. (Pr II 37)

He then introduces three laws of motion, the recognizable successors of the laws he presented earlier in *The World,* though presented in a different order. The first law corresponds closely to law A of *The World:*

[Law 1:] Each and every thing, insofar as it is simple and undivided, always remains, insofar as it can *[quantum in se est]*, in the same state, nor is it ever changed except by external causes. . . . And therefore we must conclude that whatever moves, always moves insofar as it can. (Pr II 37L)[5]

The second law concerns rectilinear motion, and corresponds to law C of *The World:*

[Law 2:] Each and every part of matter, regarded by itself, never tends to continue moving in any curved lines, but only along straight lines. (Pr II 39)

The third law pertains to collision, and it is a further development of law B of *The World:*

[Law 3:] When a moving body comes upon another, if it has less force for proceeding in a straight line than the other has to resist it, then it is deflected in another direction, and retaining its motion, changes only its determination. But if it has more, then it moves the other body with it, and gives the other as much of its motion as it itself loses. (Pr II 40)

Law 3 is then followed by a series of seven rules in which Descartes works out the specific outcomes of various possible cases of direct collision (Pr II 46–52).

Later we shall examine these laws in some considerable detail, to see what precisely they say and how Descartes argues for them. But for the moment I would like to clarify how the discussion in the *Principles* is structured.

The first distinction to note is between the "universal and primary cause" of motion, which Descartes identifies as God, who created motion and rest in the beginning and now sustains the same quantity of it, and the "secondary and particular causes of the different motions we notice in individual bodies" (Pr II 36, 37).[6] There are, admittedly, some puzzling aspects of the distinction; the general cause is an *agent*, God, while the secondary causes are *laws*, which, we are told, derive from that same agent.[7] But leaving that puzzlement aside, there are some clear differences between the conservation principle and the three laws that follow that are probably what he has in mind here. The conservation principle is universal and general in the sense that it is supposed to apply to all of the physical world, and it does so unconditionally.[8] However, the conservation principle tells us nothing at all about the behavior of any *particular* body in the world; as long as the total quantity of motion remains constant, it does not matter how motion is distributed among particular bodies, as far as the conservation principle is concerned. The three laws that follow, however, do determine the behavior of specific bodies; they tell us how the motion God conserves is to be distributed among the individual bodies in the world under various circumstances.[9] The first law tells us that motion persists not in general, but in the specific body that has it; the second tells us that every particular body tends to move in a straight line; the third tells us that in a given collision, bodies A and B will go off in such-and-such directions with such-and-such specific speeds. Furthermore, while the three laws are intended to determine the specific behavior of particular bodies, they seem not to be general and universal laws. Rather, they are conditional in the sense that they only hold in certain specifiable conditions. The first and second laws talk about the persistence of motion and directionality not categorically, but insofar as a body can, *quantum in se est*.[10] And the third law only comes into play when two (or more) bodies come into contact with one another.

It might be tempting to read the "secondary and particular" causes as causal agents distinct from the "universal and primary cause," and to see here a distinction between what God causes and what finite things cause.[11] Later, in chapter 9, we will have to discuss in some detail the question of whether anything other than God, in particular, finite minds or extended bodies, have any genuine causal efficacy for Descartes; while God sustains the world and its motion, it is yet to be determined whether he is a *direct* cause of motion, or whether he sustains motion by sustaining things that are, themselves, genuine causes of motion. But important as this question may be, I don't think that it is at issue in connection with this distinction. These laws that he introduces as his particular causes do not appear to require the introduction of any new causal agents into his world. Descartes makes absolutely clear at the very beginning of his exposition of the three laws that they, too, are derived from the immutability of God, just as the conservation principle is (Pr II 37). And God and his sustenance come up explicitly in the arguments he offers for two of the three laws, suggesting that these laws as particular causes rest on God in just the same way as does the general conservation principle (Pr II 39, 42). But though the particular causes do not necessarily introduce any causal agents into the picture beside God, they do not exclude them either. The three laws of motion, as we shall see, are noncommital about whether God determines the motion of particular bodies directly or through the intermediary of subsidiary causes.

Descartes structures his discussion of the laws of motion in the *Principles* around the distinction between the general and particular causes of motion. But there is also a distinction worth noting among the particular causes themselves, the laws of nature. The first two laws of the *Principles* are importantly similar to the general conservation principle whose statement they follow. Like the conservation principle, they concern the persistence of some feature of the physical world; while the conservation principle mandates the persistence of total quantity of motion, the first two laws mandate the persistence of motion, rest, and directionality. These principles of persistence are laws of a different sort than law 3. Law 3 is not a principle of persistence, but what might be called a principle of reconciliation. Often it will happen, indeed, in the Cartesian plenum it will *always* happen that the conditional principles of persistence will come into conflict. For example, if body A and body B are both in motion in opposite directions along the same line, law 1 tells us that each will continue to move, "insofar as it can," and law 2 tells us that this motion will tend in a straight line. But if A in its motion encounters B in its motion, then something has got to give; the impenetrability of the two bodies prevents rectilinear motion from persisting in both at the same time. This is where law 3 enters, and tells us what is to happen next, how the two incompatible conditionally persisting motions are to be reconciled with one another.

I should add a note about the terminology I have chosen to adopt, and, perhaps more importantly, the terminology I have chosen to avoid. What I have called law 1 and law 2, the first two laws from the *Principles,* are most often grouped together and called his law of inertia, in recognition of the similarities between Descartes' laws and the Newtonian principle of inertia. But matters here are extremely complicated; when we look below the surface formulations at the grounds of the principles and the roles they play in the systems, there are substantial differences between Descartes and Newton, and between Descartes and other thinkers, like Beeckman, Gassendi, and Huygens, who formulated similar principles. Furthermore, as we shall discuss later in chapter 8, through Kepler the term 'inertia' had a specific meaning for natural philosophers working in the early seventeenth century, a meaning different than it was to have later in the century with Newton, and Descartes has a number of significant remarks to make about inertia as understood in that way. For these reasons I have chosen to break the tradition and *not* use the term 'inertia' in connection with Descartes' first two laws of motion. They are, more properly, principles of persistence, principles that tell us what features of bodies God sustains in the world.

I have tried to set out in a very preliminary way Descartes' views on the laws of motion. Now we must examine them more carefully. In the remainder of this chapter we shall discuss the conservation principle and the first two laws of nature. In chapter 8 we shall examine his account of impact. Then in chapter 9 we shall examine the common strategy that he seems to use in arguing for those laws, the appeal to God's immutability.

THE CONSERVATION PRINCIPLE

As we noted earlier, the conservation principle, the principle that mandates the constancy of the total quantity of motion in the world, has a prominent place in the *Principles;* it is an expression of God as the general cause of motion in the world, and it governs the more specific laws he goes on to give in senses that we shall later make more precise. Indeed, in later years the conservation principle Descartes announces in the *Principles* will become almost definitive of the Cartesian school of physics, and the center of a controversy between those who identified themselves as the followers of Descartes, and others who were formulating the doctrines of what was to become classical mechanics.[12]

But it is interesting to note that the conservation principle did not always hold such a position in Descartes' thought. Earlier I remarked that a version of the conservation principle can be found as early as *The World.* While it is true that one can find something like the conservation principle stated there, it is also important to note that it is not isolated *as* a principle, and that it is not really distinguished from other statements that are quite distinct from the conservation principle that was later to be formulated.

In *The World* the conservation principle is enunciated in chapter VII as part of the justification he offers for his laws A and B. Law A states that "each part of matter, taken by itself, always continues to be in the same state . . . [and so] once it has begun to move, it [i.e., a body] will continue always with the same force," and law B that in a collision, one body cannot transmit motion to another body "without at the same time losing as much of its own, nor can it take any of the other's away except if its motion is increased by just as much" (AT XI 38; AT XI 41). Descartes' claim is that both rules follow from the fact that God "always conserves in it [i.e., the totality of matter] just as much [motion]" (AT XI 43). Now, it is clear that these laws do *not* follow from this consideration alone; God may preserve the same amount of motion in the world (however that is construed) without conserving motion in any *individual* body. For example, God may conserve the same amount of motion in the totality by accelerating some bodies while slowing others. Similarly, he may conserve motion in the totality of the world without conserving it in any *particular* collision. But the infirmities of Descartes' argument aside, what is interesting here is the *role* that the conservation principle plays in *The World.* The conservation principle appears in *The*

World as simply an intermediary link between divine immutability and laws A and B. In this respect it seems no more important to Descartes than the somewhat different links he uses to take him from divine immutability to the law of the persistence of rectilinear motion, law C (AT XI 44–45). In all three cases what is of ultimate importance is the establishment of the laws on the basis of divine immutability and constancy of action; what we focus on as an early statement of the conservation principle would seem to have no more importance within the context of the chapter than do many other details of the arguments that take Descartes from divine immutability to the three specific laws that interest him in that chapter. The statement of the conservation principle in chapter VII of *The World* is important not because of the position it occupies there, but because of the central position it will come to occupy in his system.

In the later 1630s, and into the period in which the *Principles* was being drafted, the conservation principle maintains its connections with the more particular laws that govern motion, particularly the collision law. For example, writing to Debeaune on 30 April 1639 Descartes calls attention to the fact that "there is a certain quantity of motion in the whole of created matter which never increases or decreases" as a prelude to a discussion of how motion is transferred in collision (AT II 543 [K 64]).[13] Similarly, in writing to Mersenne on 17 November 1641, while the *Principles* were being drafted, he defends a claim he had earlier made about a particular case of impact by noting that "you will easily understand this if you consider motion or the force to move itself as a quantity which never increases or decreases, but which only transfers itself from one body into another" (AT III 451).

In the *Principles,* the conservation principle emerges from the shadows, from a subsidiary role within the context of arguments for other principles, and takes on its role as an independent principle. But despite the prominence the principle comes to have, it is never *entirely* clear just what the principle says, or how exactly Descartes intends the appeal to God and divine immutability to establish the law. In *The World,* he offers no numerical measure of what it is that God conserves in the world, characterizing it merely as a "certain quantity of motions" (AT XI 43). The phrase he uses, "quantité de mouvemens," curiously enough in the plural, may be a typographical error, but, as Pierre Costabel has suggested, it *may* indicate that what his God is preserving is, quite literally, a certain number of motions, perhaps the fact that such-and-such a number of bodies is moving.[14] But it is also quite possible that he was simply unclear about what precisely it was that God was conserving at this point. As I pointed out, Descartes' apparent interest in the conservation principle in this context is to support laws A and B.

However the conservation principle might support these laws (not very well, as I have suggested), he does not feel called upon to give a numerical measure of what it is that God conserves in this context.

But later, as the law becomes more independent, Descartes attempts to specify a numerical measure for what it is that God conserves. No such measure appears in the *Essays* of 1637. As our discussion of the law of refraction in chapter 6 suggests, in the *Dioptrics* Descartes was more interested in persistence and change in determination and speed so as to determine, say, the path of a ray of light under various circumstances; what happens to any motion lost is of no real concern to him. And in the *Meteors,* while he is interested in the behavior of bodies of various sizes, shapes, and motions, there is no serious attempt to apply quantitative laws to the phenomena in question, with the exception of the question of the rainbow, where the laws in question are borrowed from the *Dioptrics* (AT VI 325ff [Ols. 332ff]). But as early as 1639, it is evident that Descartes has quantification in mind. Writing to Debeaune on 30 April 1639, he explains that "I hold that there is a certain quantity of motion in the whole of created matter which never increases or decreases" (AT II 543 [K 64]); see also AT V 135 [K 228]). He goes on in this letter to indicate at least two of the parameters involved in calculating this quantity of motion:

> When a rock falls onto earth from a high place, if it does not rebound but stops, I conceive that . . . it transfers to it its motion. But if the earth it moves contains a thousand times as much matter as it does, in transferring to it all of its motion, it gives it only a speed one one-thousandth as much as it had at first. . . . If two unequal bodies each receive the same amount of motion, this same quantity of motion does not give as much speed to the larger as it does to the smaller. (AT II 543 [K 64])[15]

This implies that quantity of motion, that which is conserved, is proportional to size and to speed and is measured by size times speed. This is, indeed, the measure that Descartes explicitly adopts in the *Principles*:

> Although motion is nothing in moving matter but its mode, yet it has a certain and determinate quantity, which we can easily understand to be able to remain always the same in the whole universe of things, though it changes in its individual parts. And so, indeed, we might, for example, think that when one part of matter moves twice as fast as another, and the other is twice as large as the first, there is the same amount of motion in the smaller as in the larger. (Pr II 36)

Similarly, Descartes wrote to Mersenne about quantity of motion on 23

February 1643, most likely after this passage was drafted. Although, he claimed, "impression, motion, and speed, considered in one body, are the same thing, in two different bodies motion or impression are different from speed." He went on to explain: "if these two bodies cover as much ground in the same time, one says that they have the same speed. But that which contains more matter . . . needs more impression and motion to go as fast as the other" (AT III 636). Here, too, Descartes has the same measure of quantity of motion in mind: "if ball A is quadruple ball B, and if they move together, one has the same speed as the other, but the quadruple body has four times as much motion" (AT III 635).

It is not accidental, I think, that the mathematical expression of the notion of quantity of motion arises at the very time that Descartes is attempting to work out the details of his collision laws, which, as we shall see, he is doing in this period. While the conservation principle is global and governs the universe as a whole, Descartes very quickly comes to interpret it as holding for any closed system, any system free from the causal interference of external bodies, though given the Cartesian plenum, it is not clear that anything short of the universe as a whole counts as such a closed system. From his earliest statements of the collision law, Descartes assumes that in impact between two bodies, one body can transfer motion to the other only by losing some of its own. Thus he wrote in law B of *The World* that "when a body pushes another, it cannot give it any motion without at the same time losing as much of its own, nor can it take any of the other's away except if its motion is increased by just as much" (AT XI 41). But, of course, if this law is to be precise, and enable him to say something specific about the postcollision state of bodies, as he wanted to in the later 1630s and early 1640s, then some content must be given to the notion of quantity of motion. And whatever content was given the notion in the context of collision would, of course, hold for the context of the more general conservation principle.[16]

In the most visible formulations of the conservation principle, it is quantity of motion that God is supposed to conserve in the world. But Descartes also gives a somewhat different account of what remains constant in the world, not motion, quantity of motion, size times speed, but *force* or *impression*. For example, in chapter III of *The World* Descartes writes that "the ability *[vertu]* or power *[puissance]* to move itself which is found in a body . . . cannot entirely cease to be in the world" (AT XI 11). Here he seems to be talking about the ability a body has to persist in its motion, force in the sense he will later use it in the collision rules of the *Principles*, where he will attribute to each body a "vis ad pergendum," a force of continuing (Pr II 40). This seems to be what he has in mind when he writes to Mersenne on 17 November 1641, indicating that one should consider "motion or the force *[force]* to move itself" as a quantity conserved in the world (AT III 451). But elsewhere the force conserved is identified with the presumably external force necessary to cause motion and cause it to continue. This seems to be Descartes' view of conservation in the important letter to More from August 1649. There he is explicitly concerned with the force *[vis]* that impels bodies and keeps them in motion. In the case of inanimate bodies, bodies unconnected with souls or angels, "the force moving [things] . . . can be that of God, conserving as much transference in matter as he placed there at the first moment of creation" (AT V 403–4 [K 257]). But, he goes on to say later in the same letter, "when . . . I said that the same amount of motion always remains in matter, I understood this as concerning the force *[vis]* impelling its parts" (AT V 405 [K 258]).

There is not necessarily a contradiction between these different ways of characterizing what it is that God conserves in the world, quantity of motion (size times speed), the force a body exerts in collision (force of going on), and the force necessary to put a body into motion and keep it there. As we shall see below in chapter 8, the force of going on, the force relevant to determining the outcome of a collision, has the same measure as its quantity of motion; if one is conserved, then so is the other. Furthermore, however Descartes might have conceived of an impelling force, and however he may have thought it could be measured, it is plausible to suppose that in his view a constant application of this force will result in a constant quantity of motion (size times speed); again, if one is conserved, then so is the other.[17]

It is important to note in this context that though the conservation principle involves the speed and size of moving bodies, the law does not govern *directionality* at all. As I noted in chapter 6, Descartes draws a radical distinction between speed and the determination of a body in a particular direction. Determinations have different magnitudes, of course, and can be combined geometrically. But it is only speeds and quantities of motion, size times speed, that can be summed arithmetically, and about which one can intelligibly speak of a certain quantity in the totality of the world, Descartes thinks. While he does not emphasize this point in the context of his explicit discussion of the conservation principle, it comes out clearly in the application of that principle to the problem of impact, as we shall later see in more detail. There he will assume that the conservation principle holds for any closed system, like a pair of bodies colliding in a nonresisting medium. In treating impact Descartes makes it quite clear that the conservation principle determines *only* the postcollision *speeds* of bodies; questions about what directions they move in are handled quite independently.

Later in chapter 9 we shall examine Descartes' arguments from divine immutability for the conservation principle. But for the moment this completes my account of Descartes' views on the conservation of motion. However, before going on to examine his other laws, I will comment briefly on some of the later history of the conservation principle.

Though it may have taken Descartes himself a while to come to his own statement of his conservation principle, once stated in the *Principles,* it became central to the later Cartesian approach to physics. It was one of the great blows to the Cartesian program in physics when it was demonstrated that the conservation principle is false. Building on the work of others, Christiaan Huygens in particular, Leibniz formulated a series of arguments intended to show the insufficiency of the Cartesian conservation principle.

Leibniz's basic argument is rather straightforward, though he formulated a number of different versions of the argument in response to objections from those who still held onto the Cartesian orthodoxy.[18] In one version, Leibniz's argument goes roughly as follows. Let us call what God conserves force. Whatever quantity God conserves, Leibniz reasons, should be such that its conservation results in conservation of the ability of the bodies in the system to accomplish an effect. That is, any two systems that contain the same quantity of force should be able to produce the same effects. But if force is identified with Cartesian quantity of motion and it is size times speed that is conserved, then this will not hold. Consider a body A of four size-units and one speed-unit, and let us suppose that it has the ability to raise itself a vertical height of one foot on the surface of the earth (imagine it climbing an inclined plane, say). Now, Leibniz reasons, lifting four size-units one foot is equivalent to raising one size-unit four feet.[19] Consider, now, a body B of one size-unit, which has the same Cartesian quantity of motion as A does. So, B must have four speed-units. But what effect can B produce? Leibniz assumes that a heavy body falling acquires a speed sufficient to raise itself back up to the height from which it fell, resistance aside.[20] And so, to calculate how high B could raise itself, we need only calculate what distance B would have to fall through to acquire four units of speed. By Galileo's law of free-fall, the speed acquired in free-fall is proportional to the square root of the distance fallen. And so, the height to which a body can raise itself is proportional to the square of its speed. So, if one unit of speed can raise a body one foot, then four units of speed can raise a body *sixteen* feet. And consequently, it appears, body B, with the *same* quantity of motion A has, has the ability to produce an effect *four* times as great as A can. Were quantity of motion conserved and were A to be able to transfer its total quantity of motion to a smaller body like B, then we could increase the ability to produce effects, and, in that way, build a perpetual motion machine.[21]

Leibniz goes on to argue for the conservation of other quantities in the physical world, mv^2 and mv, size (mass) times velocity, a vector notion that involves both speed and directionality. But the very idea of a quantity in the world that is conserved was a very important idea, despite Descartes' mistake about what exactly it was that is conserved. Problematic as it may have been, Descartes' formulation served to get the idea into circulation, and, in that way, laid the grounds for the classical mechanics that was to follow.

Laws 1 and 2: Some General Remarks

Earlier in this chapter I noted that the conservation principle, considered as an explicitly stated quantitative law, was a fairly late development in Descartes' thought. Not so for the first two laws of motion. Indeed, they are evident in Descartes' first serious work in physics. Descartes seems to have learned from Beeckman the principle that "in a vacuum, what is once moved, always moves" as early as fall 1618 (JB I 263 [AT X 60]). This principle, later to be generalized as law A in *The World* and law 1 in the *Principles*, plays a central role in his early accounts of the problem of free-fall, from 1618 onward, enabling Descartes to think of gravity as adding increments of speed to a falling body, which retains the speeds previously imposed on it.[22] The same principle also appears in connection with the discussion of the rebound of a ball off of a surface without losing any of its speed, the model Descartes uses in deriving the law of reflection for light (AT VI 93f [Ols. 75f]). Writing on 25 February 1630, he tells Mersenne that the speed of the ball remains constant, before and after reflection, "because of the fact that once a thing has begun to move, it will continue to move as long as it can *[quamdiu potest]*" (AT I 117; also p. 107).[23] It is very likely that he made use of this same principle when, in the mid-1620s, he first successfully attacked the problems of reflection and refraction. While the texts are somewhat less clear on this, my suspicion is that law 2 of the *Principles* may also date from the late 1620s, and may also have arisen in the context of a specific problem he was concerned with, in this case the problem of the nature and properties of light. Light (sunlight) for Descartes, is the centrifugal pressure that the bodies of celestial matter exert as they turn around the center of the vortex that constitutes a planetary system like ours.[24] Now, when accounting for this centrifugal tendency, both in *The World* and later in the *Principles*, Descartes turns to the law first stated in chapter VII of *The World*, the law that states that rectilinear motion persists unless interfered with, that a body in motion will tend to move in a straight path (AT XI 84f; Pr III 55–60). It is quite possible that this account of light, together with its crucial appeal to the tendency bodies have to travel in straight lines, may have been part of Descartes' conception of the world well in advance of the actual beginnings of *The World,* perhaps even as early as his work on the laws of reflection and refraction.

Descartes' original interest in these laws of nature may have derived from the particular problems that interested him in physics. But starting with *The World,* they have a life of their own; he attempts careful statements of them, and attempts to argue for them from more basic considerations.

LAW 1

Let us begin with the first law. In its earlier statements, it is a law that concerns motion alone. In a note on free-fall that Descartes wrote for Beeckman in 1618, the claim is made that "in a vacuum, what is once moved always moves" (JB IV 51 [AT X 78]). Eleven years later, the principle has changed little. Writing to Mersenne on 13 November 1629 about free-fall, again, he sets forth his assumptions:

> First of all I assume that motion which is once imprinted on some body remains there forever, unless it is removed by some other cause. That is, I assume that in a vacuum, what once began to move, always moves with the same speed. (AT I 71–72)

A month later, on 18 December 1629 Descartes tells Mersenne that "I shall try to demonstrate [this principle] in my treatise," presumably *The World* (AT I 90 note a). But when the principle is stated in *The World,* it is considerably more general:

> Each part of matter, by itself *[en particulier],* always continues to be in the same state, until the collision *[rencontre]* with others forces it to change. That is to say, if it has some size, it will never become smaller unless other things divide it; if it is round or square, it will never change this shape, unless others force it to; if it is stopped in some place, it will never leave, except if others drive it off; and if at one time it has began to move, it will continue always with the same force, until others stop it or slow it down. (AT XI 38)

The law is given a similar statement much later in the *Principles:*

> Each and every thing, insofar as it is simple and undivided, always remains, insofar as it can *[quantum in se est],* in the same state, nor is it ever changed except through external causes [French version: collision with others]. And so, if some part of matter is square, we can easily persuade ourselves that it will always remain

square unless something else comes along and changes its shape. If it is at rest, we do not believe that it will ever begin to move, unless it is impelled to motion by another cause. Nor is there any more reason if it is moving, why we should think that it would ever interrupt its motion by itself *[sua sponte]*, and with no other obstacle. And therefore, we must conclude that whatever moves, always moves insofar as it can *[quantum in se est]*. (Pr II 37)

The law, as stated both in *The World* and in the *Principles*, is a *very* general law, a law that concerns *all* states of bodies, at least all states of bodies considered individually *[en particulier]* or considered as simple and undivided.[25] The persistence of motion in a body, the principle that, no doubt, motivated this law, arises as a special case of a principle that mandates the persistence of at least shape and rest, in addition to motion.

The law Descartes presents has a number of features worth commenting on. Puzzling at first glance is a restriction that Descartes introduces into the law. In *The World*, the law is taken to govern bodies taken by themselves *[en particulier]*, and in the *Principles*, the law is restricted to bodies that are simple and undivided *[simplex et indivisa]*. Descartes is by no means clear about this. But I suspect that this feature of the law is intended to restrict the scope of the law to genuine states of the genuine individuals in Descartes' physical world; that is, the states that persist are to be modes of extended substance and those alone. These are the states that pertain to the genuine particulars, to bodies insofar as they are simple and undivided. And these are precisely the examples Descartes gives in stating the law. In *The World* the examples of states that persist include size, shape, rest, and motion, and in the *Principles* it is shape, rest, and motion, all modes of extended substance. What Descartes means to exclude here are complex states like being hot or cold, which, it may have seemed to Descartes, can change without any *obvious* external cause. Insofar as heat and cold involve the greater or lesser motion of parts, for Descartes, those states fall outside the scope of this law; they are not states that one can talk about in connection with "simple and undivided" bodies. But while excluded from this law, they are by no means excluded from Descartes' physics. Writing to Mersenne on 26 April 1643, and discussing the apparent exclusion of certain physical states from this law, Descartes remarks: "and heat, sounds, or other such qualities give me no trouble, since they are only motions which are found in the air, where they find different obstacles that hinder them" (AT III 649–50 [K 136]). While such states are not governed directly by the laws of persistence, they are treated in

Descartes' physics insofar as they are analyzable into genuine states of body which are governed by the basic laws.[26]

Motion, of course, is one of the basic modes that the law is certainly intended to cover. But Descartes' statement of the law appears to leave open precisely what features of motion are meant to be included. At *very least* it is obvious that Descartes means to say that a body moving will continue to move with *some* speed or other and with *some* determination or other. But what of the speed and determinations, the two modes, varieties, or properties that, Descartes holds, characterize a given motion? (AT III 650 [K 136]; AT IV 185). Does Descartes intend to say that these persist as well?

Every indication is that the law is meant to hold that a body in motion will move with a constant speed, unless acted upon by an external cause. While this proviso is missing from the version of the law Descartes gives in Pr II 37, it seems to be what he has in mind when in *The World* he mandates that a body in motion will remain in motion "avec une egale force," "with the same force." Furthermore, when Descartes appeals to the principle of the persistence of motion in a number of contexts outside of the formal presentation he gives in *The World* and the *Principles,* it is clear that constancy of speed is included. Writing to Mersenne on 13 November 1629 about the problem of free-fall, Descartes states his principle as follows: "whatever once begins to move in a vacuum, always moves with the same speed *[aequali celeritate]*" (AT I 71–72). It is not unreasonable to suppose that Descartes had this proviso in mind in all of the passages we cited earlier where he appeals to the persistence of motion in connection with the problem of free-fall. His strategy in dealing with that problem is to imagine the falling body as receiving equal increments of speed at equal intervals, and to derive the law of free-fall by summing those equal increments. This, of course, presupposes that a body continues, at any instant, to move with the speeds previously imprinted on it. This is where Descartes appeals to the principle of the persistence of motion, and this principle would do him little good unless what persists is not only motion but also a particular speed. This assumption can also be found in later texts, texts written at roughly the time the *Principles* was being drafted. For example, writing to Huygens on 18 or 19 February 1643, Descartes notes that a motion, once begun, "will always continue with the same speed," unless interfered with (AT III 619).

The question of determination is somewhat more difficult. This question is not unconnected with the meaning of law 2 of the *Principles* and corresponding law C of *The World,* where Descartes argues that motion is, in its nature, rectilinear, and so a body tends to move in a straight line tangent to the circle and thus recede from a curvilinear path. Later in this chapter we shall discuss this law in detail, its meaning, and its relation to the law of persistence under discussion now. But setting these difficult questions aside, for the moment, I think that it is plausible to say that the law of the persistence of motion is intended to cover the determination of a body to move in some particular direction in the principal sense which we discussed above in chapter 6, the component of a body's speed in some particular direction. Descartes is, if anything, less explicit about this than he is about the persistence of speed. The law of the persistence of motion is what explicitly stands behind the claim that a ball bouncing off of an immobile surface will retain its speed. Now, important to the derivation of the law of reflection from this model is the further assumption that in colliding with the surface, there is no change in the horizontal determination of the ball, the determination the ball has to move from left to right (AT VI 94–95 [Ols. 76]). While Descartes does not make any appeal to any general principle to justify this assumption, it is not implausible that he thought it followed in an unproblematic way from the principle of persistence of determination; if motion persists unless interfered with, then so should the determination of that motion in a given direction, a genuine *part* that composes that motion.

There is one last remark to make about the law itself. The law, of course, is conditional; it asserts that a state of body (mode of extension) like size, shape, motion, or rest will persist until something causes it to change. However, the law gives no *precise* account of the conditions under which change will occur. This is a condition that Descartes never really specifies in *The World*, as I shall later argue. In the *Principles*, though, it is dealt with in law 3, which, Descartes tells us, "contains all of the particular causes of change that can happen to bodies, . . . at least, those that are corporeal" (Pr II 40). This connection between law 3 and law 1 is signaled in the French version of law 1, where Descartes indicates that everything remains in the same state until changed not merely by "external causes," as the Latin version puts it, but "through collision with others" (Pr II 37F; cf. Pr II 37L).

Writing to Mersenne on 18 December 1629, Descartes brought up his principle of the persistence of motion, and told him that "in my treatise I shall try to demonstrate it" (AT I 90 note a). The treatise in question was, almost certainly, *The World*, and in *The World* Descartes does attempt to argue for the persistence of motion, now subsumed into a more general law concerning the persistence of all states of body. The argument comes after his statement of a law of impact, law B, the second of the three laws he presents in *The World*. Descartes begins with a general consideration that is supposed to underlie both the law of persistence of states, law A, and the collision law B: "now, these two rules follow in an obvious way from this alone, that God is immutable, and acting always in the same way, he always produces the same effect." He then goes on to give the special grounds for his first law: "thus, assuming that he had placed a certain quantity of motions in the totality of matter from the first instant that he had created it, we must admit that he always conserves in it just as much, or we would not believe that he always acts in the same way" (AT XI 43).

It is, then, what will later become Descartes' conservation principle in the *Principles* that is supposed to support the first law of *The World*. It is fairly obvious that this will not do. First of all, whatever this argument might tell us about the persistence of motion, it does not tell us anything about any of the other states of body that, Descartes holds, also persist; though it may support the persistence of the special case of motion, it does not address the more general persistence principle that Descartes frames in *The World*. But, perhaps more importantly, it does not even support the special case of the persistence of motion. As I pointed out earlier, the conservation principle to which Descartes appeals here is a very general principle that governs the world as a whole. But it says nothing at all about how motion is to be distributed among individuals in the world, whether it is to persist in individual bodies, or whether it is to redistribute itself promiscuously and arbitrarily from body to body. The two principles are linked in an obvious way. The conservation principle is a consequence of the fact that once God causes motion in the world, by his immutability he is committed to continue the motion he created. The law of the persistence of motion simply adds that the motion created persists in the body that has it, unless something causes it to change. But though linked, they are not linked deductively; one cannot derive the persistence of motion from the conservation principle, as Descartes suggests in *The World*.

However, the discussion of persistence in *The World* suggests another kind of argument, an argument suggested by the very generality with which the law is stated there. Descartes' real interest is in the special case of motion, in the persistence of that particular state of body. Now, there are many states of body, like size, shape, and rest, that *everyone* would admit persist unless caused to change. Descartes, I think, was attempting to argue for the law of persistence of motion, a law to which he had appealed many times in his earlier writings, by making that controversial law a trivial consequence of a broader but widely accepted principle, and, in *The World*, mocking those who would except motion from the general law of the persistence of states. Immediately after stating law A, he contrasts the treatment of motion he is developing for the new world he is creating with the way motion is treated in the "old world," the world of the schools:

> No one does not believe that this same rule isn't observed in the old world with regard to size, shape, rest, and a thousand other similar things. But the philosophers have excepted motion. . . . But don't think, on account of that that I desire to contradict them: the motion of which they speak is very different than that which I conceive. (AT XI 38–39)

Descartes goes on to poke fun at the scholastic conception of motion, their definition of motion, the varieties of motion that they recognize, and contrast it with his own geometrical conception of local motion, as we examined in chapter 6. But one comment he makes about the schools is especially relevant to the issue at hand:

> And finally, the motion they speak of has such a strange nature, that unlike other things, which have their perfection as a goal, and try only to conserve themselves, [their motion] has no goal but rest, and contrary to all of the laws of nature, it tries to destroy itself. But on the other hand, [the motion] I assume follows the same laws of nature that all of the dispositions and qualities found in matter in general follow. (AT XI 40)

All other properties of body tend to persist. Why, Descartes suggests, should we make an exception of motion? Why should motion alone tend toward its own destruction and reduction to its opposite, rest?

What is rhetoric in *The World* becomes the official argument in the period of the *Principles*. As in *The World* the persistence of motion is presented as a consequence of a more general principle, the persistence of all states in bodies (at least insofar as they are "simple and undivided"). And similar to his treatment in *The World* Descartes speaks against those who would except motion from this law, attributing the mistake now to the errors of the senses and the rash judgments of youth. As in *The World* Descartes points out that those who except motion from the general principle of the persistence of states hold that "[motions] cease of their own nature, or tend toward rest. But this is, indeed, greatly opposed to the laws of nature. For rest is contrary to motion, and nothing can, from its own nature, proceed toward its own contrary, or toward it own destruction" (Pr II 37). But when Descartes appeals to the immutability of God to prove the law, it is the *law as a whole*, the general principle of the persistence of all appropriate states, that he means to prove, and not just the special case of motion. This is not entirely clear in the sections of the *Principles* where law 1 is presented and defended. But it comes out quite explicitly in a letter to Mersenne from 26 April 1643. Descartes gives Mersenne the law in its most general form, "everything that is or exists always remains in the same state that it is, unless something external changes it." He then continues:

> I prove this through metaphysics. For since God, who is the creator of all things, is entirely perfect and immutable, it seems repugnant to me that any simple thing that exists and, consequently, of which God was the creator, has in itself the principle of its own destruction. (AT III 649 [K 136])

The consideration Descartes explicitly put forward with respect to motion both in *The World* and in the *Principles,* that a state should not tend toward its own opposite, toward its own destruction, is presented here as the general consideration on which the general law rests. Put positively, Descartes holds that God's immutability and constancy of operation entails that if an individual body is in a particular state, then God will sustain that body in that state, unless there is a reason external to that body for altering the state. Later, in chapter 9, we shall examine in more detail the way in which divine immutability supports this and the other laws Descartes posits.

For Descartes, in the end, the persistence of the state of motion would seem to depend on its *opposition* to a state of rest. This is closely connected with a point I argued in chapter 6. There I emphasized the importance to Descartes of making motion a genuine mode of body, a mode of extended substance. If everything is to be explicable in terms of motion alone, then motion *must* be a genuine mode of extended substance. But if motion is a genuine mode, then it cannot be an arbitrary matter of point of view whether a body is in motion or not; there must be a genuine distinction between motion and rest. Earlier I emphasized the foundational aspects of the doctrine of motion, the extent to which it is required in order to ground Descartes' mature conception of the physical world. But here we see another motivation. In construing motion as a genuine mode of body, Descartes can then include it among the states of body that an immutable God sustains and, in that way, prove a central principle of his physics, a principle he first learned from Beeckman in autumn 1618. To be sure, it did not all come together at once. Descartes began with the special case, the persistence of motion, and it took him more than ten years to link the persistence of motion to the more general principle of the persistence of states of body, as he did in the early 1630s with the composition of *The World.* Though motion is not carefully defined there, it is clearly designated as a state, a state whose opposite is rest. But with the composition of the *Principles,* Descartes makes clearer the sense in which motion and rest are modes of body and distinct modes of body, the sense in which motion and rest are states and opposite states. With this the position is complete, and the proof of the persistence of motion in individual bodies rests solidly on its Cartesian foundations.

LAW 2

The first law of persistence, law A in *The World* and law 1 in the *Principles,* asserts that motion persists along with the other states of a body. To this Descartes adds a second law of persistence, both in *The World* (law C) and in the *Principles* (law 2), a law that asserts that motion persists in a rectilinear path unless interfered with by an external cause. This law is not found in any surviving Cartesian texts before *The World.* But once stated there, it appears in an almost unaltered form in the *Principles;* though there are some differences of detail, the exposition of law 2 in the *Principles* is one of the few places where Descartes has virtually translated an earlier discussion into Latin. In both works, the development of the law takes place in two different contexts. The law is first announced and proved together with the other laws of nature, in chapter VII of *The World* and part II of the *Principles* (AT XI 43–47; Pr II 39). But Descartes makes some important later clarifications of the law in both works, when the law gets its principal application in grounding the theory of light, for light turns out to be the centrifugal pressure of celestial matter turning in vortices. This further discussion is found in chapter XIII of *The World* and part III of the *Principles* (AT XI 84–86; Pr III 55–60).[27]

As with law 1, this law is taken to be grounded in divine immutability; this we shall discuss at greater length below in chapter 9. But our first problem in dealing with this law is determining what precisely it is meant to assert. The basic law is stated as follows in *The World*:

> When a body moves, even if its motion is most often on a curved path, . . . nevertheless, each of its parts, taken individually, always tends to continue its motion in a straight line. (AT XI 43–44)

The law is given in a similar form in the *Principles*:

> Each and every part of matter, regarded by itself, never tends to continue moving in any curved lines, but only along straight lines. (Pr II 39)

In the postil to the Latin edition of this section, the law is summarized by saying that "all motion, in and of itself, is straight."

But closely linked to the tendency of a body in motion to move in a straight path are some assertions about the behavior of bodies in circular motion, which are regarded sometimes as consequences of the law (Pr II 39 postil) and sometimes as part of the law itself (Pr III 55). Consider a body revolving around a center; for example, a stone in a sling. If we consider all of the causes that determine its motion, then the stone "tends" *[tendere, tendre]* circularly (AT XI 85; Pr III 57). But if we consider only "the force of motion it has in it" (Pr III 57), then, Descartes claims, it "is in action to move," or "is inclined" to go, or "is determined to move" or "tends" to move in a straight line, indeed, along the tangent to the circle at any given point (AT XI 45–46; Pr II 39; Pr III 57; AT XI 85). And, Descartes concludes, "from this it follows that everything which is moved circularly tends to recede from the center of the circle that it describes" (Pr II 39, postil).[28]

In order to understand what the law asserts, we must understand what Descartes means when he asserts that a body has a tendency or an inclination. Descartes explains the notion as follows in a passage from *The World*:

> When I say that a body tends in some direction, I don't want anyone to imagine on account of that that it has in itself a thought or a volition that pushes it there, but only that it is disposed to move in that direction, whether it really moves or whether some other body prevents it from moving. (AT XI 84)

This same view is also found in the *Principles*, where Descartes also attempts to dementalize the term "tendency," latinized there as "conatus" (Pr III 56).[29] But in the *Principles*, he is a bit clearer about what this tendency is. A body with a tendency or a conatus to move in a particular direction "is situated and incited to motion in such a way that it would really go [in that direction] if it were not hindered by some other cause" (Pr III 56). So, a body tends in a particular direction (a straight line, a path along the tangent, a motion away from the center of a rotation) insofar as it would actually go there unless otherwise impeded. And so, this law asserts the conditional persistence of a certain sort of motion in exactly the same way that the first law asserts the conditional persistence of the states of a body in general: the motion in question persists unless an external cause prevents it from persisting. Indeed, in stating one consequence of this law, Descartes uses exactly the same formula he uses in connection with the first law: "all bodies which move in circles recede from the center of their motion insofar as they can *[quantum in se est]*" (Pr III 55).

We should be very careful to distinguish the notion of tendency from the notion of determination, with which it might be confused. Determination is an aspect of the motion a body has. In its primary sense, it is a directional component of the motion a body has, a measure of the extent to which a body moving with a particular speed and with a particular direction advances in a direction that may or may not be identical with the direction of the body's motion. But Descartes does not use the word "tendency" in this way. For him a tendency is not a motion or an aspect of motion, but a property a body has by virtue of which it would move if it were unimpeded. It is not motion, but, as he suggests in one context, motion in potentiality, and in another, a "first preparation for motion [prima praeparatio ad motum]" (AT I 450–51 [K 42]; Pr III 63). Like motion, tendency has determination, a potential determination, as it were. A body in motion that tends in a particular direction is said to be "determined [determinatum] to continue its motion in a certain direction, following a straight line" (Pr II 39).[30] And, as we shall see, Descartes will decompose the potential determinations to go in different directions that a body that tends to go in a particular direction has; as with bodies in motion, a body that tends in some direction as a whole will have tendencies in other directions as well, though Descartes has no clear technical terminology to express them.

It is obvious why Descartes chose to express this law in terms of tendencies rather than more straightforwardly in terms of a state of body that persists conditional on a lack of interference. In his plenum, this condition of noninterference can *never* be met. Since all is full, a body in motion must push another out of its way in order to move, and the space it leaves must be filled by another body that comes and takes its place. This, he argues, entails that all motion must be circular (Pr II 33).[31] Descartes clearly has this in mind when framing this law (AT VIIIA 63, ll. 23–26). Though a body would go straight if it were not interfered with, other bodies are always interfering with it.

As noted earlier in this section, the rectilinear tendency of a body in motion is closely connected to two features of bodies in circular motion. First of all, Descartes asserts, a body in circular motion tends away from that circle along the tangent. It is interesting to note that this is something that does not follow directly from the claim that bodies tend to move rectilinearly. Though a body moving in a circular path may, at any moment, tend to move in a rectilinear path, from that fact alone nothing follows about which rectilinear path it tends to move in at any given moment. Descartes takes his claim to be "confirmed by experience, since if [the stone] is then ejected from the sling, it will not continue to move toward B [on the circle] but toward C [on the tangent]" (Pr II 39). But he never offers any serious argument for the claim that the body tends to move along the tangent, and he seems to have regarded it as simply obvious.[32]

But wherever the conclusion may come from, it leads Descartes to

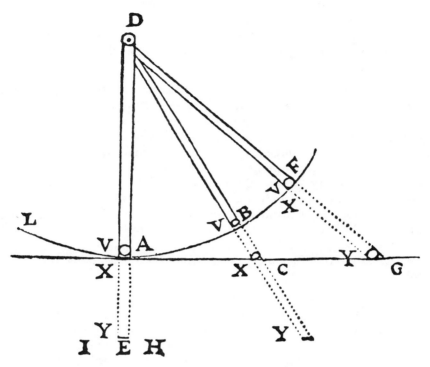

FIGURE 7.1 From [René Descartes,] *L'homme de René Descartes....A quoy l'on a ajouté Le monde, ou Traité de la lumière, du mesme Autheur* (Paris, 1677), p. 441.

another important fact about circular motion, that a body in circular motion tends away from the center of rotation. Though this conclusion is taken to be obvious in the context of his initial presentation of the laws of nature, both in *The World* and in the *Principles* (AT XI 44; Pr II 39), in the later application of the law of rectilinear tendency to the theory of light he attempts to argue for the conclusion. The strategy for establishing this centrifugal tendency is ingenious.

Descartes throughout works under the methodological assumption that to reason about tendencies, we must reason about the motions that would result were those tendencies to be realized. So, he begins by decomposing the rectilinear motion of a body along a tangent to a circle into two different motions (see fig. 7.1):

> But to understand this last point distinctly, imagine this inclina-
> tion that the stone has to move from A to C as if it were composed
> of two others, that is, one to turn around the circle AB, and the
> other to go straight along the line VXY, in such proportion that,

finding itself at the point of the sling marked V when the sling is at the point of the circle marked A, it ought to find itself later at the point marked X when the sling would be at B, and at the point marked Y when it would be at F, and thus it ought always to remain on the straight line ACG. (AT XI 85)[33]

In the case of refraction we examined in the previous chapter, the determination of the tennis ball is divided into two parts, a horizontal determination from left to right, and a vertical determination from up to down. Here it is a *tendency to motion* that is being divided. A rectilinear tendency along the tangent to a circle is divided into a tendency to revolve around a fixed point, together with a tendency to move in a straight line along a rotating radius of the circle. In presenting this argument in the *Principles* he makes this more graphic by imagining the rectilinear motion of an ant along the tangent to a circle as having been produced by the ant walking along a stick while that stick rotates around a center (Pr III 58; see fig. 7.2).

Descartes continues the passage from *The World* quoted above as follows:

> Knowing that one of the parts of its inclination, namely, that which carries it around the circle AB is in no way hindered by this sling, you can well see that the sling resists only the other part [of the inclination], namely, that which makes it move along the line DVXY, if it were not hindered. Consequently it tends (that is, it makes an effort) only to go directly away from the center D. (AT XI 85–86; see also Pr III 58–59)

The sling does not impede the circular component of the stone's motion, but it does impede the paracentric component, the tendency the ball has, on this analysis, to move away from the center. But in impeding the motion, the sling does not eliminate the tendency to motion (Pr II 57). If the sling were not impeding the stone, it would move out along the rotating radius. And so, Descartes concludes, the stone in the sling, indeed *any* body rotating around a center, has a tendency to move away from the center of rotation, a tendency deriving from the tendency the ball has to move in a straight line.

The argument, though very ingenious, is very problematic. Now, Descartes certainly intends the circular motion in question in the argument to be uniform circular motion around a center, motion of a constant angular velocity, and by the first law he is committed to the view that the tangential motion toward which the body tends is uniform rectilinear motion. But what Descartes does not realize is that as he sets the case up, the uniformity of the rotational motion of DVXY around

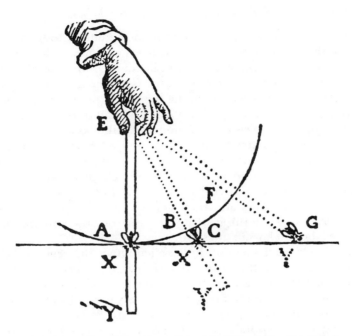

FIGURE 7.2 From *Renati Des-Cartes Principia Philosophiae* (Amsterdam, 1644), p. 99.

D is inconsistent with the uniformity of the motion of the stone along the tangent ACG. Suppose that the stone moving along the tangent ACG were to move with a uniform speed (see fig. 7.1). Then the rotating radius DVXY would have to move more and more slowly as the stone went further and further out the tangent; indeed, if the speed of the stone on the tangent were finite, DVXY would never become parallel with the tangent ACG. Suppose, now, that the radius DVXY were rotating uniformly around D, and that the stone, moving out along DVXY, were to remain on the tangent ACG. Then it is obvious that the motion of the stone along ACG would be accelerated as the angle ADF approached a right angle. (It is also obvious that in this circumstance, Descartes' ant in fig. 7.2 would have to run faster and faster to keep up, until the stick on which it was running became parallel to the original tangent, at which point it would have to run infinitely fast, no small feat even for an imaginary insect!) And so, I think, uniform rectilinear motion along the tangent cannot be broken down into components in the way Descartes tries to do; and without this decomposition, the argument collapses.

The Relations between Law 1 and Law 2

Descartes' first two laws are among the most familiar to modern ears; together they appear to coincide with what is now called Newton's principle of inertia, that a body in uniform rectilinear motion will remain in uniform rectilinear motion unless otherwise interfered with. Indeed, Descartes is certainly among the first, if not the first, to state in print that principle of inertia.[34] But it is no small puzzle on such a view that what appears in Newton and modern textbooks as one law is given in Descartes as two. Later in this chapter we shall have something to say about the different ways in which Newton and Descartes conceive of the persistence of rectilinear motion, and in chapter 8 we shall discuss the notion of inertia as Descartes treated it. But the account we have given of these two laws and their genesis in Descartes' program for physics gives us something of an account of how and why he distinguishes the two laws in question.

Descartes himself says nothing explicit about the relation between the laws in question. It is reasonably clear that the law that asserts the conditional persistence of rectilinear motion, is, from one point of view, a clarification of the first law, the persistence of states of body in general and motion in particular, or perhaps, a corollary of that law, something already contained in that law, though significant enough to warrant special mention. In the postil to Pr II 39, where he sets out law 2, the law is given as the claim "that every motion in and of itself is straight." Law 2, then, tells us what the motion God conserves in law 1 is *really* like: the motion that persists until interfered with, the motion God conserves is in and of itself straight. It is thus unsurprising that writing to Huygens on 18 or 19 February 1643, long after part II of the *Principles* was probably finished, he combined the two laws: "I consider that the nature of motion is such that once a body has started to move, that is sufficient to enable it to continue always, with the same speed and on the same straight line, until it is stopped or diverted by some other cause" (AT III 619). In this way, then, we can say that not only is *motion* a state of body, but so is uniform rectilinear motion, insofar as it is *that* that God will preserve in body, unless an external cause forces change.

But though the law of the persistence of rectilinear motion is, properly, a commentary on the law of the persistence of motion, it is not surprising that Descartes chose to separate it out and present it as an independent law. Historically, the two laws derive from quite different parts of his program. The law of the persistence of motion first arises in connection with the problem of free-fall, and receives its most important application in connection with the tennis ball analogies

Descartes uses in deriving the laws of reflection and refraction. The law of the persistence of rectilinear motion and its important consequence, that bodies in circular motion tend away from the centers of their rotation, first arose in quite a different context. Though there is no hard evidence, I suspect that the observations brought together in law C of *The World* and later in law 2 of the *Principles* date from the mid-1620s and are closely connected to the formulation of his theory of light in terms of the centrifugal pressure of celestial matter circulating around a central point in a giant vortex. Though I suspect that Descartes later came to see the connection between these two principles, their separate histories probably continued to influence the way he thought of them, and influenced him to keep them somewhat separate. Nor was this an entirely unreasonable way for him to proceed. The persistence of rectilinear motion, the principal claim of law 2 (law C), is a proper part of the law of the persistence of motion and, more generally, of all the states of body. But his real interest here was probably not in the conditional persistence of rectilinear motion; though all motion *tends* to go in straight lines, this is a tendency that can never be realized in the Cartesian world, full of matter. The real interest of the law for Descartes was probably in the consequence that he draws from the rectilinear tendency, the account of circular motion and the tendency that bodies in circular motion have to recede from the center. This is the principal, indeed the *only* use he makes of the law in developing his account of the world. Given the importance of this consequence, it is not surprising that Descartes would want to separate it and the special case of the principle of persistence on which it depends from the more general law of the persistence of all states of body.

Persistence in Historical Context

In the earlier sections of this chapter I discussed Descartes' laws of persistence. In this section I would like to put these laws into historical perspective. Two questions are relevant here, Descartes' relation to the scholastic tradition, and his relation to other seventeenth-century figures.

It is, of course, widely known that for the Aristotelian physicist, motion tends to come to rest, and one needs a special explanation for why a body in motion continues in motion. For Descartes and many of his contemporaries, though, motion as such continues, and one needs a special explanation for why it stops. It is hard to imagine a more basic transformation in the fundamental concepts of the theory of motion.

The problem the Aristotelian physicist faced is particularly acute for unnatural or violent motion in terrestrial physics. The Aristotelian physicist had something of an explanation for why heavy bodies fall to the center of the Earth and why light bodies rise to the sphere of the moon; it is their nature to do so (see chap. 4 above). There is also a reasonably compelling answer to the question of why a stone thrown or an arrow shot from a bow begins to move: it is being pushed by the hand or the bowstring. But, they held, everything moved needs a mover. What is it that moves the stone after it leaves the hand or the arrow after it leaves the bowstring? An answer widely discussed (and widely rejected, probably by Aristotle himself) was the theory of anti-peristasis, the view that it is the air surrounding the body which keeps it in motion when it leaves the hand or the bow; the initial impulse imposed directly on the violently moving body moves the surrounding air as well, which continues to cause the body to move.[35] Needless to say, later natural philosophers in the Aristotelian tradition were not altogether happy with this. By Descartes' day, Aristotelian natural philosophers commonly adopted what came to be called the impetus theory. While the impetus theory is a matter of great complexity and existed in a number of varieties, in brief the view was that what caused the body in motion to remain in motion was "a certain force or impetus, impressed by the hurling of the moved object, which inheres in it," as the Coimbrian Fathers put it in their commentary on the *Physics*.[36]

Descartes was certainly aware of this problem. Like most of his scholastic contemporaries, he thought that the account of the continuation of motion in terms of the surrounding air was scarcely plausible (AT I 117; AT III 12). And while he did not discuss the impetus theory explicitly, there is no reason to think that he thought it was any better. Writing in *The World*, after having presented his law A, Descartes refers explicitly to the problem: "Assuming [law A], we are free of the difficulty in which the learned find themselves when they want to explain the fact that a stone continues to move for some time after it leaves the hand of the person who threw it. We should ask instead, why doesn't it always continue to move?" (AT XI 41). Descartes also addresses this standard scholastic question in the *Principles*. Immediately after stating his law 1, he writes: "Nor is their any other reason why a projectile should remain in motion for some time after it is separated from the hand of the thrower than the fact that what is once moved continues to move, until it is slowed by bodies that are in the way" (Pr II 38). As he is aware, his law of the persistence of motion offers a solution to the problem of continued motion different from anything that the schoolmen offered. The arrow and the stone continue in motion not because of the surrounding air, and not because of any impressed force or impetus, but simply because motion is a state of body, and as a state, it persists in body until something causes it to change.[37] Or, to put it differently, for Descartes motion as such is natural and always persists; there is no important distinction to be drawn between natural motion, which needs no special explanation outside of the body, and violent motion, which does (AT III 39; AT V 404 [K 258]).

But despite Descartes' self-proclaimed opposition to the schoolmen on the question of the continuation of motion, in a deeper sense, he may well share some of the intuitions of the schoolmen. Motion is a state that persists, for Descartes. But as we shall see in more detail in chapter 9, it persists because God persists in *causing* motion, and were God to withdraw his continual impulsion, everything would stop. Henry More wrote to Descartes on 5 March 1649, asking if "matter, whether we imagine it to be eternal or created yesterday, left to itself, and receiving no impulse from anything else, would move or be at rest?" (AT V 316; see also AT V 381). He answered: "I consider 'matter left to itself and receiving no impulse from anything else' as plainly being at rest. But it is impelled by God, conserving the same amount of motion or transference in it as he put there from the first" (AT V 404 [K 258]). Like the scholastics, Descartes thinks that continued motion requires a continued cause, in a sense. But instead of air or impetus, he turns to God.

Though Descartes' account of the continuation of motion, his laws of the persistence of motion, may have solved one problem that the schoolmen faced, it raised another. Basic to the Aristotelian cosmology was a radical distinction between terrestrial and celestial physics. Bodies below the sphere of the moon are made up of combinations of the four elements, earth, water, air, and fire. These elements have two natural motions, either down (earth and water), or up (air and fire), motions that restore them to their proper places in the world. But the celestial bodies are not made up of the same elements as we are down here; they are made up of a fifth essence (quintessence) whose nature it is to move circularly.[38] Now, for Descartes all bodies are made up of material substance, substance whose whole nature it is to be extended. And this, of course, goes for both celestial and terrestrial bodies. As he writes in the *Principles,* "there is one and the same matter in the heavens and on earth . . . and so matter in the entire universe is one and the same thing, insofar as it is recognized by this one thing, that it is extended" (Pr II 22–23). But insofar as all body is the same, it will all satisfy the same laws. And insofar as all bodies in motion tend to move in straight lines, Descartes must explain something his teachers did not have to explain, that is, what it is that deflects the rectilinear motion of the heavenly bodies into the roughly circular motion that astronomers, both Ptolemaic and Copernican, agree that they have. It was to explain this that Descartes had to introduce his vortex theory of planetary motion.

On this theory, the initial chaos that God created, a plenum without any empty space, sorts itself out into whorls of turning fluid matter, like whorlpools. Each of these vortices constitutes a separate system, with a central sun and, under some circumstances, surrounding planets. It is because all the planets, including Earth, are embedded in such a system that their motion is deflected from the rectilinear path that they tend to move in by law 2 (law C), and they turn around a central sun. But, of course, the fluid matter that makes up the heavens and whose turning constrains the planets to turn in circular paths is also extended stuff, and as such, it too tends to move in rectilinear paths. The fluid matter in any given vortex is constrained, Descartes holds, by the vortices surrounding it.[39] Its rectilinear tendency is also important to Descartes' physics. As I noted earlier, it is the rectilinear tendency of the subtle matter that, by law 2 (law C) gives rise to the centrifugal tendency that Descartes identifies with light (sunlight; see AT XI 84ff, Pr III 55–64).

Descartes was not, of course, the only thinker in the period to challenge the scholastic account of local motion and to propose principles of the persistence of motion. A full account of Descartes' place in the seventeenth-century theory of motion would require a full history of the scientific revolution in the theory of motion. But something should be said about Descartes' relations to two other important figures, Galileo and Newton. (Descartes' accounts of persistence will be compared with Kepler's account of inertia below in chapter 8, after we discuss the problem of collision.) The history of science in the period is sometimes presented in terms of the progress toward *the* principle of inertia; Galileo, Descartes, and Newton are all supposed to have made contributions to this, as well as Beeckman, Huygens, Gassendi, and others.[40] Actually, the history is much more complicated. Many shared the insight Descartes had, that motion persists in and of itself, and that bodies tend to move in rectilinear paths. But beneath the apparently similar formulations are very significant differences.

Galileo, like Descartes after him, believed that a body in motion persists in motion; it is, for example, assumed in the account of projectile motion in the beginning of the fourth day of the *Two New Sciences*.[41] But, for Galileo, unlike Descartes, this was not a basic law of motion, a basic property of body as such; indeed Galileo never presents *anything* as a basic law of motion. Instead it is an apparently derivative fact about the behavior of naturally heavy bodies. Heavy bodies going down inclined planes are accelerated, and going up inclined planes are decelerated. But bodies moving on horizontal planes are neither accelerated nor decelerated; their motion, thus, must remain constant.[42] This does resemble the conclusion that Descartes reached, that motion as such persists. But the differences here are quite significant. First of all, for Descartes, the natural motion of bodies is rectilinear; bodies in motion tend to continue their motion in straight paths. But the "natural" motion Galileo attributes to bodies is *circular:* as Galileo notes on a number of occasions, the horizontal plane is an arc of a great circle.[43] This difference points to another important difference between the two. The fact that motion once impressed persists is used by Galileo *mainly* to diffuse certain standard objections to Copernicanism. Anti-Copernicans advanced a variety of arguments to show that things would be quite different on Earth if it really moved, as the Copernicans said it does; they claimed that balls dropped from towers would fall differently if the Earth were in motion, that cannon balls would fall at different distances if shot eastward rather than westward, and so on.[44] It is here that Galileo's appeal to the persistence of (circular) motion does some of its most important work for him. For, he reasoned, if horizontal motion persists as much as rest does, then there can be no experiment that will allow us to tell whether we are at rest or in uniform motion. As Galileo puts it in the *Two Great World Systems:*

> Shut yourself up with some friend in the main cabin below decks on some large ship, and have with you there some flies, butterflies, and other small flying animals. Have a large bowl of water with some fish in it; hang up a bottle that empties drop by drop into a wide vessel beneath it. With the ship standing still, observe carefully how the little animals fly with equal speed to all sides of the cabin. The fish swim indifferently in all directions; the drops fall into the vessel beneath; and, throwing something to your friend, you need throw it no more strongly in one direction than another, the distances being equal. . . . When you have observed all these things carefully, . . . have the ship proceed with any speed you like, so long as the motion is uniform and not fluctuating this way and that. You will discover not the least change in all the effects named, nor could you tell from any of them whether the ship was moving or standing still.[45]

And if uniform (circular) motion and rest are physically indistinguishable in this way, then the supposed experimental refutations of the Copernican cosmology must be wrong. The persistence of motion for Galileo is thus closely connected with the physical indistinguishability of motion and rest. But Descartes never contemplated any such thing. For Descartes, rest and motion are importantly different, indeed, physically different, as we shall see in more detail in the next chapter, and he was willing to go to some lengths to ground a genuine distinction between motion and rest in his definition of motion. Furthermore, it was important to Descartes that there be a real experimental difference between circular motion and rest. Bodies in motion around a center exert a centrifugal force, central to his physics insofar as it is that centrifugal force that will explain light in his system. For Descartes there *must be* a difference between rest and rotational motion, and *his* principles of persistence are intended, at least in part, to explain that difference, and ground the centrifugal force that arises out of it.

Descartes is in some ways closer to Newton. For both it is rectilinear motion that persists, and though Descartes' proofs of his principles from the immutability of God are far from anything Newton would have used, both see the persistence of rectilinear motion as one of the basic properties bodies have. And, as we shall later see, both think that there are forces associated with bodies that resist changes in their states of rest and motion. But again, there are important differences to note.[46] For Newton the principle of inertia he frames is an essential step in the account of planetary motion he gives. His interest is in the force that deflects a body from rectilinear motion, and in the mathematical description of that force; the persistence of rectilinear motion, Newton's first law of motion in his *Principia,* is the ground of a series of propositions that will, in book III, allow him to establish from astronomical observations of the trajectories of the planets and their moons the existence of an inverse-square-law force between all bodies which will explain why the planets and their moons move in the nonrectilinear paths that they are observed to follow. Descartes, however, does no calculation. He has neither the mathematical tools, nor the interest. His interest in the rectilinear persistence of motion is not in deriving the *cause* of circular motion, the *centripetal* force connecting the planets to the sun and the moons to their planets that Newton is interested in establishing; for Descartes it is enough to say that planets travel around the sun on a *vortex.* Descartes' main interest in the persistence of rectilinear motion is in grounding an effect of that rotation, the *centrifugal* tendency that will ground his theory of light. And this tendency is treated by Descartes in an entirely qualitative way. Nowhere does he attempt to calculate its magnitude; it is enough for Descartes that it is there.

MOTION AND ITS LAWS:
PART 2, THE LAW OF IMPACT

IN THE PREVIOUS chapter we discussed Descartes' conservation principle and the first two of his laws of motion. In this chapter we will continue the discussion with an examination of the third of the laws, the law of impact. Descartes' full treatment of impact was late in coming and, in the end, never fully worked out; the record shows him struggling with the problem, and never really arriving at a single satisfactory view on the problem. To appreciate his thinking on the question it will be very useful to work carefully through the sequence of stages in which he confronted the problem.[1]

IMPACT IN *THE WORLD*

The striking similarity between the laws of impact that Descartes finally arrives at in the *Principles* and those found in Isaac Beeckman's *Journal* suggests that Descartes must have discussed this problem with his mathematical mentor during his period of apprenticeship.[2] But the first written record of Descartes' thoughts on impact does not occur until much later. In chapter VII of *The World* he attempts to set out a law that specifies what happens to bodies in impact, what I called law B earlier, the second of the three laws that he presents there:

> [Law B:] When a body pushes another, it cannot give it any motion without at the same time losing as much of its own, nor can it take any of the other's away except if its motion is increased by just as much. (AT XI 41)

The law certainly bears on impact insofar as it imposes a constraint on what can and cannot happen in the collision of bodies. The rule as stated mandates that when two bodies collide, *if* one body transfers motion to another, then it must lose some of its own. But the law falls far short of a genuine law of impact. Descartes offers no numerical measure of the motion a body has before collision, and no numerical measure of the amount that it can lose to another body in collision. Nor, for that measure, does he deal with the question of whether or not every collision must result in a transfer of motion from one body to another. At best the supposed law of impact is a special case of the conservation principle that will be stated explicitly later in that chapter of *The World* and elevated to the status of the most general constraint on motion some years later in the *Principles*.

Descartes' purported impact law in *The World* is, thus, no impact law at all. But the Descartes of *The World* is not without some ideas on how to solve the problem of impact. In a passage immediately following the presentation of law B, he presents the kernel of a genuine analysis of impact in the guise of a commentary on law B. He is discussing the question of what it is that slows down a projectile in motion; if, by law A, motion continues until the moving body encounters another, then what can explain the apparently spontaneous loss of motion that we see in everyday life when, for example, we throw a feather or a stone? The answer he offers is the resistance of the air, whose parts collide with the projectile and slow it down, in accordance with rule B (AT XI 41). But Descartes recognizes that this explanation seems to violate common sense in a way; we can understand how a stone can slow another stone, perhaps, but how can the air, which seems not to resist the penetration of the projectile and seems to offer such little resistance to being set into motion, slow a stone? He answers:

> If, however, one fails to explain the effect of the air's resistance in accordance with our second rule [law B] and if one thinks that the more a body can resist the more it is capable of stopping the motion of others (as one can, perhaps, be persuaded at first), one will, in turn, have a great deal of trouble in explaining why the motion of this stone is weakened more in colliding with a soft body of middling resistance than it is when it collides with a harder body that resists it more. Or also why, as soon as it has made a little effort against the latter, it at once turns on its heels rather than stopping or interrupting the motion it has. Whereas supposing this rule [law B], there is no difficulty in all this. For it teaches us that the motion of a body is not retarded by collision with another in proportion to how much the latter resists it, but only in proportion to how much the latter's resistance is over-come, and to the extent that, in yielding to the former, the latter receives into itself the force of motion that the former loses. (AT XI 41–42)

The language here is obscure, and Descartes' ideas difficult to disentangle. But in presenting what he claims to be an illustration of law B, Descartes clarifies just how he thinks bodies behave in impact. The cases that he is considering are those of a projectile (a stone) colliding with a soft body of middling resistance (air, a pillow, etc.), and slowing down, and that same projectile colliding with a hard body of greater apparent resistance (a stone wall, say), and changing its direction but not losing any of its original speed. Descartes' reason for being interested in these cases is quite probably connected with his program for optics. For in the case of reflection of light, we are dealing precisely with a case in which a smaller body (a tennis ball on Descartes' model) changes its direction without changing its speed upon impact with a hard surface, while in refraction, the ball loses motion to a body of "middling" resistance, water or a permeable cloth.[3] The puzzle such cases raise is this: why should the lesser resistance of the softer body result in a loss of speed, while the greater resistance of the harder body result in no loss of speed? Descartes' answer seems to be that the motion is lost not to resistance per se, but only in actually imposing motion on another body. This much follows directly from law B. But the examples suggest a conception of resistance and its role in determining the outcome of impact that goes beyond anything found in that law. Descartes suggests that while resistance is not what slows the projectile down, it does have a role in determining the outcome of a collision: where resistance is relevant is in determining whether or not motion is to be transferred from one body to the other *at all.* If the resistance to the projectile is too great, then the projectile will rebound without transferring *any* of its motion to the other body. If, on the other hand, the resistance to the projectile is only "middling," then the projectile can impose motion onto the other body, and in doing so, lose some of its own. Thus Descartes writes that "the motion of a body is not retarded by collision with another in proportion to how much the latter resists it, but only in proportion to how much the latter's resistance is overcome, and to the extent that, in yielding to the former, the latter receives into itself the force of motion that the former loses."

What is suggested by this example is that Descartes regarded the analysis of impact as involving two separate stages. First, one must compare the resistance a body offers to a colliding body with the ability that other body has to overcome the resistance, whatever precisely that ability might consist in, and however it and the resistance might be measured. If the resistance is greater than the colliding body can overcome, then the colliding body rebounds without transferring any of its motion. If the resistance is sufficiently small, then an appropriate transfer of motion from one body to the other is effected. The picture that Descartes presents suggests a *contest* between two bodies. If A has the ability to overcome the resistance of a body B, then A wins and gets to impose its terms, that is, it gets to impose its motion on B (though, of course, in doing so A must lose some of its own motion, in accordance with law B). But if B's resistance is too great for A to overcome, then A loses and rebounds, without imposing any motion on A and thus, without losing any motion of its own.[4]

This is the state of the problem of impact in *The World*. Unfortunately, it leaves much to be desired, even if we leave questions of truth aside. The contest view of impact, the view that represents Descartes' deepest intuitions about the behavior of bodies in collision, seems not to be original with him; it can be found in many other authors of the early seventeenth century and seems to have been a common starting place for those who were attempting to solve the problem of collision.[5] But in Descartes' *World* it remains, at best, implicit in the discussion. The physical quantities it appeals to are barely listed, not to mention defined mathematically or otherwise, making it impossible to determine in any particular case which body wins the impact contest. But even if we could in general determine the winner of the contest, it would do us little good. In *The World* Descartes does not specify the postcollision state of affairs in any detail, what the winner wins, and what the loser loses. We know that sometimes there is rebound, and sometimes motion is transferred. But when motion is transferred, we have no idea of how much. In short, the reader of *The World* perceptive enough to discern the impact-contest view implicit in Descartes' discussion is still left very much in the dark about how to apply that analysis to any particular case, how to determine the winner of the contest, and how to describe in any detail the postcollision state of affairs. At the end of the chapter of *The World* that contains his discussion of impact, Descartes claims that he could fill in all of the details of his account of impact: "I could have set out here many additional rules for determining in detail when and how and by how much the motion of each body can be diverted and increased or decreased by colliding with others" (AT XI 47). But one suspects that this is idle boast. It will be some years before Descartes is at all clear about how exactly to handle the problem of impact.

IMPACT IN THE LATIN *PRINCIPLES*

Descartes' correspondence in the 1630s shows that he was not unconcerned with the problem of impact during that period.[6] But after the abandonment of *The World* in 1633, there seems to be no attempt at a systematic account of impact until the early 1640s, when Descartes was drafting the Latin *Principles*. There the impact law, law B of *The World*, appears as law 3 of the *Principles,* considerably changed from its initial statement. The contest view, at best implicit in the earlier discussion, becomes the heart of the law, now clearly distinguished from the conservation principle, and the bare law is supplemented with a series of seven rules in which Descartes attempts to show how the law applies in practice. The problem is by no means solved; as we shall later see, Descartes gives the problem of impact considerable attention in his later writings. But it is worth taking stock of the situation as it stands in 1644.

The law of impact is stated in the Latin *Principles* as follows:

> [Law 3:] When a moving body comes upon another, if it has less force for proceeding in a straight line than the other has to resist it, then it is deflected in another direction, and retaining its motion, changes only its determination. But if it has more, then it moves the other body with it, and gives the other as much of its motion as it itself loses. (Pr II 40L)[7]

Here the impact contest is quite clearly presented. Consider two bodies, B and C, where B is the "moving body" of law 3, and C is the other body B "comes upon." Descartes posits in B a force for proceeding in a straight line (a force for proceeding, for short) and in C a force resisting the motion of B (a force of resisting, for short). If the force for proceeding in B is less than the force of resisting in C, then B loses the impact contest and changes its determination while retaining its motion (quantity of motion); this is what Descartes calls the first part of the law in the argument that follows (Pr II 41). On the other hand, if the force for proceeding in B is greater than the force of resisting in C, then B wins the impact contest and retains the direction in which it moves, transferring some of its motion to the loser, C; this is the second part of the law (Pr II 42).

The law of impact is grounded in the fundamental distinction Descartes draws between motion and determination, a distinction discussed at some length above in chapter 6. He writes in explanation of law 3:

> *Motion is not contrary to motion, but to rest; and the determination in one direction is contrary to determination in the opposite direction.* And we must note that one motion is in no way contrary to another motion of equal speed; properly speaking, one finds only these two contrarieties. One is between motion and rest, or also between the speed and slowness of motion, to the extent that that slowness partakes of the nature of rest. The other is between the determination of motion in some direction and the collision with a body

at rest in its path *[in illa parte]* or in motion in a different way; and this contrariety is greater or lesser in proportion to the direction in which the body colliding with the other is moving. (Pr II 44L)[8]

These two contrarieties or oppositions, between motion and rest or motions of different speeds, and between a determination in some direction and something that constitutes an obstacle to that determination, are treated separately in law 3 and define the two parts of the law.

The first part of the law deals with the question of determination in some direction. If B's force for proceeding is less than C's force of resisting, then B's determination to go in the original direction is completely altered, while its motion (quantity of motion) remains unchanged. Descartes justifies this as follows:

> The first part of this law is demonstrated from the fact that there is a difference between motion regarded in itself and its determination in a certain direction, by which it happens that this determination can be changed, while leaving the motion complete. For, as was said above [in Pr II 37] one thing, not composite but simple, such as motion is, always continues to be as long as it is not destroyed by any external cause. And in collision with a hard body, there is a cause that prevents the motion of the other body with which it collides from remaining determined in the same direction. But there is no cause for the motion itself to be given up or diminished, since motion is not contrary to motion; from this it follows that it should not be diminished on that account. (Pr II 41L)[9]

Here, when dealing with the impact contest at the highest level, the comparison between the force for proceeding in B and the force of resisting in C, we are dealing only with the second of Descartes' two contrarieties, the opposition between the determination B has to move in a given direction; the potential opposition between motion and rest or motions of different speeds does not enter here. This, I think, is the sense in which he says here that motion is not opposed to motion: in *this* case (though not in every case), the contrarieties that can arise between motion and rest, or between motions of different speeds, do not enter. When the force for proceeding in B is less than the force of resisting in C, it is only determination that is in question.

But if B's force for proceeding is greater than C's force of resisting, then the question of the opposition between motions may enter. B continues to move in the same direction and, thus, maintains a determination in its original direction.[10] But if C is at rest or if B and C have different speeds, then we must resolve the opposition that can arise between a body in motion and one at rest, or between two bodies moving with different speeds, by transferring motion from one body to the other. This, in any case, is what the passage quoted earlier from Pr II 44 would seem to imply. As we shall see when we discuss the application Descartes makes of this law in Pr II 46–52, he does not allow for any transfer of motion between bodies moving with equal speeds; transfer only takes place when the force for proceeding is less than the force of resisting, *and* when there is an opposition between motion and either rest or motion of a different speed. The proof Descartes offers for this law (which we shall take up in more detail in chapter 9, after discussing Descartes' account of divine immutability) is addressed at justifying the claim that motion is transferred from one body to another, and that one body must lose motion for another to gain it. Descartes writes:

> For since everything is full of body, and nevertheless, the motion of each and every body tends in a straight line, it is obvious that from the beginning, God, in creating the world not only moved its different parts in different ways but at the same time also brought it about that some [parts] impel others and transfer their motions to the others. Therefore, now, conserving it by the same action and with the same laws with which he created it, conserves motion not always placed in the same parts of matter, but transferring it from some into others in accordance with how they collide with one another. (Pr II 42)

The impact law as outlined in Pr II 40–44 is very abstract, and it is not at all obvious how to interpret notions like 'force for proceeding' and 'force of resisting', how to use the machinery developed in the law to actually specify the outcome of any given collision. Immediately after stating and defending his third law, Descartes gives what he presents as examples of the law in application, examples that are supposed to illustrate how the law is meant to work in practice. To apply his law, Descartes tells the reader, "one needs only to calculate *[calculo subducere]* how much force there is in each [body], either force for moving or force for resisting motion." And this, he tells us, "can easily be calculated if only two bodies collide with one another, and they are perfectly hard, and separated from all remaining bodies so that their motion is neither hindered nor helped by any others that surround them" (Pr II 45).

What follows is a series of seven rules of impact, which purports to show what happens when two perfectly hard bodies collide in a straight line. These rules are *not* themselves part of the third law, but only examples of how the law applies under certain assumed conditions. The conditions that Descartes assumes in giving these rules of impact are highly idealized; he assumes that the bodies in question are perfectly hard, that their motions are neither aided nor impeded by the media in which they sit, and that the collisions in question are direct and not oblique. But idealized as these conditions are, the rules he gives are as close as Descartes comes to showing how his third law works in the world.[11] The seven rules are relatively straightforward, and are summarized in a table in the appendix to this chapter. Though Descartes does not present it exactly in this way, the rules divide naturally into three cases. In case I, he considers the situation in which the two bodies, B and C, are moving in opposite directions; this case includes three rules, one for the situation in which both size and speed are equal, one for the situation in which B is larger than C and the speeds are equal, and one for the situation in which the sizes are the same, but B moves faster than C. (Descartes does not consider the case in which both size and speed are different.) Case II considers the situation in which one of the two bodies, C, is at rest; the three rules given in this case deal with the situations in which B is smaller than, larger than, and equal to C. And finally, case III considers the situation in which B and C both move in the same direction, with B moving faster than C. Though there is only one rule given in the text, in the Latin *Principles* this rule is divided into two special cases, the situation in which the ratio of the speed of B to the speed of C is less than the ratio of the size of C to the size of B, and the situation in which it is greater. (The case of equality is added in the French *Principles*.)

The rules are, for the most part, quite wrong, and were recognized as such by many later seventeenth-century thinkers.[12] But rather than compare Descartes' account of impact with what later authors saw, I would like to use these rules to try to get a sense of how Descartes was thinking about impact.

Unfortunately, the seven rules Descartes gives in the Latin *Principles* are not as helpful as they might be. At the end of his exposition of the rules, Descartes remarks that "these [rules] require no proof since they are per se obvious" (Pr II 52L). And indeed, in the course of setting out his rules in the Latin *Principles,* he simply states the results and offers virtually no argument at all.[13] Most strikingly, there is virtually no reference to the elaborately developed law 3 that these rules are supposed to illustrate, and no attempt to show in any precise way how to calculate magnitudes like force for proceeding or force of resisting that are supposed to be crucial for determining the outcome of a collision.

But despite the brevity of exposition in the Latin *Principles,* and the various assumptions Descartes makes about initial conditions, it is possible to make some reasonably good conjectures about Descartes' understanding of law 3 from examining the specific rules of impact he offers. Consider rules 2 and 3, for example. These are circumstances in which the body B appears to win the impact contest and retain the original direction of its motion. B seems to win the contest in two circumstances, when it has the same speed as C but is larger than C (R2), and when it is the same size as C but is moving faster (R3). (See the table in the appendix.) This suggests that in moving bodies, both the force for proceeding and force of resisting are functions of size and speed alone, and are jointly proportional to them. It is important to recognize that Descartes does not say this explicitly, but it isn't too much to infer that the forces in question are measured by the product 'size times speed'. This seems to be what he has in mind in R7 as well, where he is also dealing with two moving bodies colliding, this time, though, moving initially in the same direction. In R7a, B seems to win and retain the original direction of its motion when $m(B)v(B) > m(C)v(C)$, and in R7b, B seems to lose the impact contest and rebound when $m(B)v(B) < m(C)v(C)$, suggesting, again, that size times speed is a measure of both force for proceeding and force of resisting for a body in motion. These cases (R2, R3, and R7) also shed some light on the question of how Descartes thought that incompatible speeds are to be reconciled. It is important to note that although he does not mention this, in all of the numerical examples he gives in these rules, indeed, in all of the rules, the quantity of motion, size times speed, is the same before and after the collision; there can be little doubt that he saw this as a fundamental constraint on the outcome of collisions. Even though the conservation principle is global, and holds for the world as a whole, it is clear since the first statement of the conservation principle in *The World*, where it was immediately applied to the laws governing the behavior of individual bodies, including law B, the ancestor of the collision law of the *Principles,* that the conservation principle was meant to hold for individual bodies and subsystems of bodies, like two bodies in collision. Indeed, it has been suggested that the conservation principle only becomes quantitative in the period of the *Principles,* and then largely to enable him to say something definite about the outcomes of collisions.[14] But even if we hold that the conservation principle holds in impact, this alone does not determine the postcollision speeds of the bodies; there is an infinity of ways in which the conservation principle can be satisfied. In R2, R3, and R7 he seems to hold that the winner of the impact contest will only impose enough of its speed on the loser to render their speeds compatible, and eliminate the opposition that he calls attention to in Pr II 44. And so in R3 and R7a, where the initial speeds are unequal and B wins the impact contest, B only imposes as much speed on C as is needed for them both to move with the same speed. And in R2, where the initial speeds are equal, the winner, B, imposes no motion at all on C.

These rules are reasonably clear. But there are complications in the account. The most obvious one, the one that puzzled many of Descartes' contemporaries and continues to puzzle his readers, concerns the notion of force as applied to a body at rest. Again, he says nothing explicit about how to measure such forces in the Latin *Principles*, but the rules of impact, particularly R4 and R5, are very instructive here. In R4, C is at rest and $m(B) < m(C)$; in this case B collides with C and rebounds with its original speed. (Here, no doubt, the case of reflection in optics loomed large in Descartes' mind.) In R5, C is again at rest, but here $m(B) > m(C)$; in this case B collides with C and imparts enough motion to it for the two bodies to be able to travel off in B's original direction at the same speed. Presumably, the force for proceeding that B exerts is proportional to the product of its size, $m(B)$, and its speed, $v(B)$, as in the other cases we examined. But how is the force of resisting that the resting C exerts to be measured? The fact that B loses the contest and rebounds when $m(B) < m(C)$ and wins the impact contest when $m(B) > m(C)$ suggests that size is a parameter here. But speed also seems to enter into calculating the force of resisting that C exerts, not the speed of C, which is zero, but the speed of B. In explaining R4 Descartes makes the following comment in the Latin *Principles*, the only such attempt at argument in the presentation of the seven rules there: "[A] resting body resists a greater speed more than it does a smaller one, and this in proportion to the excess of the one over the other. And therefore there would always be a greater force in C to resist, than there would be [a force] in B to impel" (Pr II 49L). While the passage is by no means lucid, Descartes seems to be saying that the force of resisting that C exerts against B is proportional to $v(B)$, the speed with which B approaches.[15] This suggests that on Descartes' conception, the force of resisting in C is to be measured by the product of $m(C)$ and $v(B)$. And so, however quickly B were to move, it would always be reflected by a larger C, and, conversely, however slowly B might move, it would always be able to set a smaller C into motion.[16]

Descartes' conception of the force of rest introduces what many have seen as a serious problem in his law of impact: his treatment of impact in R4 and R5 is a clear violation of the relativity of motion and rest. In R4 Descartes deals with the case in which C is at rest and $m(B) < m(C)$; in R5, C is still at rest but $m(B) > m(C)$. If the distinction between motion and rest were arbitrary, these two cases would be simple redescriptions of one another, depending upon whether we choose a frame in which the larger body is at rest, or choose one in which the smaller body is at rest. But the outcomes of these two cases are quite different for Descartes; when the larger body is taken to be at rest, the smaller one is reflected and the relative speed between the two bodies remains the same as before the collision, while when the smaller body is taken to be at rest, the two bodies travel off together after the collision. Many have made this observation, and assuming that Descartes' account of motion was meant to be purely relativistic, have seen an inconsistency here.[17] Descartes' account of impact was clearly wrong, and R4, in particular, is a gross misconception of what happens in nature. But inconsistent it isn't. As I noted in chapter 6, it was important to Descartes for there to be a genuine distinction between motion and rest. For Descartes, the case in which C is in motion is *physically distinct* from the case in which it is at rest. And so, for him, the situations described in R4 and R5 are not mere redescriptions of one another; one cannot arbitrarily designate which of two bodies in relative motion is in motion and which is at rest. This, I think, will come out more clearly in some of the later explanations Descartes gives of R4, where he calls upon the technical terminology introduced in Pr II 25f to specify the sense in which C is taken to be at rest in the context of R4–R6.

Another complication with the rules concerns the cases of equality. In large part the seven rules of impact Descartes gives can plausibly be construed as expansions on the abstract law 3, illustrations of how the quantities involved can be calculated and outcomes determined. But not so for the cases of equality. Only two of the rules deal with cases in which the two forces in the impact contest are equal; R1 deals with the case in which B and C of equal size and speed collide from opposite directions, and R6 deals with the case in which the resting C has the same size as the moving B, and thus B's force for proceeding—$m(B)v(B)$—equals C's force of resisting—$m(C)v(B)$. (There is another possible case of equality for the situation dealt with in R7, but this is not treated in the Latin *Principles*.) Here law 3 can give us no real guidance; law 3 deals only with the cases in which a given body either wins or loses a given impact contest, and leaves no provision for a tie. I can discern no principle in the solutions that Descartes offers in these two cases. In R1, the two bodies both rebound with the same speed they originally had; both apparently lose the impact contest. In R6, on the other hand, the moving B manages to set the resting C into motion, as it would if it won the impact contest, but it also rebounds after the collision, as it would if it lost. Because of the case of equality, and the consequent possibility of ties in the impact contest, at very least, it seems clear that law 3, by itself, cannot determine the outcome of collisions, even if we were to understand how force for proceeding and force of resisting are to be measured.[18]

An obvious question about the law of impact concerns the nature and status of the forces that Descartes posits in stating the law: what are these forces, the force for proceeding and force of resisting? Where do they come from, and what place do they occupy in Descartes' world of extended and thinking substances? This is a crucial question, of course, but I will put off any serious consideration of it until chapter 9, where we shall discuss divine immutability and the grounds of the laws of motion.

This completes my account of impact as treated in the Latin *Principles*. While the third law seems to have its roots in *The World*, much of it seems to have been newly worked out for the Latin *Principles*. This is particularly true of the seven rules of impact. At least one of the rules seems to clash with what Descartes held about impact throughout the 1630s; as early as 1632 and as late as 1640, Descartes wrote to correspondents that a small body with sufficient speed could move a larger body at rest, no matter how much larger the resting body was, suggesting that the crucial idea of force of resistance offered by a body at rest, as conceived in rules 4 and 5, may only come with the actual composition of the *Principles* in the early 1640s and the attempt to systematize his thought on motion (AT I 246–47; AT II 622–23, 627; AT III 210–11).[19] Indeed, Alan Gabbey has made a very convincing case for the claim that the seven rules are a late addition to the *Principles*, and were hastily added only after what is now Pr II 45 and Pr II 53 were fully drafted.[20] The account of impact Descartes hastily sketched in the Latin *Principles* confused his readers, and in the years immediately following the publication of the *Principles* in 1644, Descartes does his best to explain himself, to justify the rules of impact, if not the principles on which he claims they are based in the Latin *Principles*.

IMPACT IN 1645

One of the early readers of the *Principles* was Claude Clerselier, later to become the translator of the *Objections and Replies* to Descartes' *Meditations*, and, more importantly, Descartes' literary executor and editor of his correspondence, *The World*, and other texts. While Clerselier's original letter concerning the rules of impact does not survive, he seems to have been especially puzzled by Descartes' account of impact, particularly R4, and wrote Descartes asking for further elucidation. Clerselier's puzzlement elicited a response from Descartes, a letter probably written on 17 February 1645, while Descartes was in the midst of thinking about the French translation of the *Principles*, then in progress by the Abbé Picot.[21] Descartes' letter to Clerselier goes into considerable detail on the grounds for his account of impact in the *Principles*, and represents one of the most important documents we have on that question in Descartes' thought.

Descartes begins the letter by attempting to explain the reasoning behind R4. Paragraph 1 begins directly with what appears to be a paraphrase of the account given in the Latin *Principles*, particularly Pr II 49.

In paragraph 1 of that letter Descartes reviews the impact contest model, reminding Clerselier that if one body is to move another, it must have "more force to move it than the other has to resist it."[22] In the case in which one of the bodies is at rest, he notes that the force in the moving body can depend only on its size, for the resting body "has as many degrees of resistance as the other, which is moving, has of speed." He offers what he presents as a "reason" for this fact about the resistance of the resting body: "if it [the body at rest] is moved by a body which moves twice as fast as another, it ought to receive twice as much motion from it; but it resists twice as much this twice as much motion." It is difficult to see an argument in this statement; it looks little more than a simple restatement of what Descartes had said earlier in the paragraph and in the Latin *Principles,* that the force of resisting in a resting body is proportional to the speed of the moving body.

In paragraph 2 Descartes attempts to illustrate the purported argument of paragraph 1 with an example. He considers two bodies, B and C, with B in motion and C at rest. He begins by assuming that if B were to set C into motion, it would have to impart to C enough motion for both B and C to travel off in the same direction with the same speed, as happens in R3 and R7a. (He seems to forget the possibility that B might transfer some of its motion to C and rebound with the rest, as happens in R6. Presumably he is assuming that B and C are of unequal size, and so one of the two must win the impact contest.) Now, if B is smaller than C, then Descartes argues that it would have to transfer less than half of its speed to C for them to move at the same speed after collision, assuming the conservation principle, presumably. So, he reasons, "if B is to C as 5 is to 4, with B having 9 degrees of motion, it must transfer 4 of them to C to make it go as fast as it goes." He then asserts, "this is easy, for it has the force to transfer up to four and a half (that is, half of what it has) rather than reflecting its motion in the other direction." If we assume this principle, it is easy to show, as Descartes then goes on to do, that if B is smaller than C, it will not be able to move the resting C, for in order to do so, it would have to transfer more than half of its speed to C for them to be able to move off together. Descartes claims that "C resists more than B has the force to act."

It is difficult to see much more of an argument here than in the earlier discussions of the case. But there does seem to be something different going on here. In the account of R4 in Pr II 49 and in paragraph 1 of the Clerselier letter, Descartes simply takes it for granted that the force of resisting that C exerts increases in proportion to the speed of B. In paragraph 2, though, Descartes gives something of an account of this conception of the force of resisting. In impact, Descartes seems to hold, B is attempting to impose enough motion on

C to make it travel at the same speed and in the same direction as B would after the collision. Now, the faster B is going, the more motion it would have to transfer to C for the result to obtain. And so, if we think of the force of resisting in C as measured by the quantity of motion it *would* have if B *were* to succeed in pushing it after the collision, then the faster B goes, the more the resistance C offers.[23]

This is clearly something of an advance over the earlier discussions. In the Latin *Principles* the analysis of impact seems to be based on what might be called a simple impact contest (IC) model, where the outcome of a collision depends upon the balance of two forces, the force for proceeding and the force of resisting, where these forces are measured in terms of the modes a body has at the moment of impact; in a moving body, force seems to be measured by the product of size and speed, and in a body at rest, it seems to be measured by the product of the size of the resting body and the speed of the body colliding with it. But in the Clerselier letter, Descartes seems to present an interestingly different conception of the force of resisting that a body at rest offers a colliding body, a view not even hinted at in the Latin *Principles*. On this view, force is not proportional to *motion* but to *change of motion*, the quantity of motion that a resting body *would* have if it *were* set into motion by a collision. This conception of force is still firmly anchored in the impact contest model of impact, what we might call a sophisticated impact contest (IC) model in contrast to the simple IC model of the Latin *Principles;* we are still thinking in terms of first comparing the forces two bodies have to determine whether there will be rebound or a transfer of motion from the one to the other. But this conception of force of rest also looks forward to the Newtonian conception of force in terms of change of motion, as Alan Gabbey has perceptively pointed out.[24] Similarly, one can see in paragraph 2 a conception of the force for proceeding in a moving body somewhat different from the one found in the Latin *Principles*. On this conception, the force for proceeding would seem to be measured by the maximum amount of motion that a moving body can impose on another body, if it were to win the impact contest, and so would be proportional to *half* of its quantity of motion.[25] But all of Descartes' problems are not solved by these changes. While we may have a clearer conception of force of resisting in a resting body, and a better conception of why it should be proportional to the speed of the colliding body, there is still a deep arbitrariness in Descartes' account: I can see no argument for the claim he makes in paragraph 2 that a body in motion has force enough to impose only half of its motion on another; this is not obvious in itself, nor does it follow from anything he says either in this paragraph of the Clerselier letter or in the Latin *Principles*.[26]

This question is what Descartes seems to be addressing in a very important principle that is announced in the very next paragraph of the Clerselier letter. In paragraph 3 he tells Clerselier that he is pleased that he has found no other problems in the earlier sections of his *Principles,* and claims that any difficulties that Clerselier has with the rules of impact will dissolve as soon as he recognizes the "single principle" on which, he claims, they all rest. The principle is then given as follows:

> When two bodies having incompatible modes collide, there must really be some change in these modes, in order to render them compatible, but . . . this change is always the least possible, that is, if they can become compatible by changing a certain quantity of these modes, a greater quantity of them will not be changed.[27]

Descartes clarifies this principle, what might be called the principle of least modal change (or PLMC) by specifying the modes at issue in impact, speed and determination, the two modes that he had appealed to in explaining law 3 in the *Principles* (see Pr II 41, 44):[28] "And we must consider two different modes in motion *[mouuement]:* one is motion *[motion]* alone, or speed, and the other is the determination of this motion *[motion]* in a certain direction, which two modes change with equal difficulty."[29]

Paragraph 4 of the letter then goes on to apply the principle announced in paragraph 3 to the crucial rules R4, R5, and R6. When a moving B encounters a resting C, we must alter the incompatible speeds and determinations to make them compatible. Now, Descartes argues, if C is larger than B, as in R4, we would have to impose more than half B's speed onto C in order for B to drive C ahead of it, in accordance with the conservation principle. In this case, B would have to change more than half of its speed and more than half of its determination, "insofar as this determination is joined to its speed."[30] So, Descartes argues, "being reflected, without changing C, it changes only its entire determination, which is a lesser change than a change of more than half of this same determination and more than half of the speed." On the other hand, if C were smaller than B, as in R5, then the situation is reversed, and B would push C. The case of equality, R6, is dealt with by arguing that when B and C are equal in size, the bodies split the difference, as it were, and B reverses its direction while, at the same time, imposing some motion on C, half the motion it would impose if B were the larger body. This, in essence, implicitly introduces another principle, what might be called a principle of the mean, a kind of symmetry principle to deal with the cases of equality that don't really fit into the the strict impact contest framework of the Latin *Principles.*[31]

If we can accept the PLMC and the analysis of the modes of a body in motion on which it rests, then we have a kind of explanation for the key assumption of paragraph 3, that B can only transfer up to half its speed to a body C. For to transfer more than half its speed, and thereby more than half its determination joined to the speed is a greater change in B than to change its entire determination. And through the PLMC we can also directly derive others of Descartes' rules of impact, most obviously R2, R3, and R7a.

But as apparently attractive as the PLMC is, the analysis of impact that Descartes is proposing to Clerselier here is highly problematic. First of all, in the discussion of R4 and R5, Descartes' argument considers only the modal changes in B. But what of the modal changes in C? If, in R5, B is to move C, then won't it change C's *entire* motion and *entire* determination? And if so, then won't this be a greater overall modal change than B's reflecting? And so, it would appear, if we take the PLMC seriously, a resting body should *never* be moved, no matter how large the colliding body might be. Also, we must observe that the PLMC sits very awkwardly with others of the rules that Descartes holds, in particular, R1 and R7b. In the situation dealt with in R1 it would seem that it would constitute less of a modal change for either B or C alone to rebound than for both to do so. One might appeal to some sort of principle of symmetry to deal with this case of equality, but then one would be faced with the problem of the apparent incompatibility of the two principles, the PLMC leading us to one solution of the problem, and the principle of symmetry leading us to another. And in R7b, where B, following C, rebounds upon collision, it seems clear that there could in certain circumstances be less change if B were to impose motion on C without changing its whole direction, as in the case of R7a.[32] And finally, I find a basic conceptual unclarity with respect to the treatment of determination as a quantitative notion in the context of the PLMC. On the one hand, it is connected with speed, so a change of speed automatically results in a change of determination, and of the same magnitude as the change in speed; speed seems to be counted twice in the calculation. But when a body reverses itself, Descartes says that its determination is completely changed, treating determination now as a purely directional notion. And there is no obvious way of quantifying change of determination when *both* speed and direction are altered, or when the direction is changed by something other than a complete rebound, situations that come up in R3, R6, and R7b, and would come up if he were to try to extend his account of impact to handle oblique collisions. Descartes recognized the importance of both magnitude and direction, without knowing how exactly to combine them.[33]

Ultimately the PLMC is very problematic. But despite the enormous problems, it represents a very interesting step in Descartes' thinking about impact. It has been noticed that the PLMC is curiously distant from the impact contest model of collision articulated in Law 3 of the Latin *Principles,* and there are debates about which has priority, why he withheld it from the *Principles,* and so on.[34] It has been suggested, for example, that the fact that the PLMC is an obviously teleological principle (which it is) might have made Descartes uncomfortable about going public with it in a work in which final causes are explicitly rejected (Pr I 28). But I see no reason to assume that he had anything like the PLMC in mind when formulating the impact rules in the Latin *Principles.* What I find interesting about the PLMC is precisely that it seems to bear no connection to the impact contest model of law 3. Descartes begins the Clerselier letter clearly within the framework of the impact contest model of the *Principles,* attempting to show how it can deal with cases like those treated in R4 and R5. But by paragraph 3, he has introduced a completely different way of thinking about impact. The PLMC represents a way of deriving the outcome of a collision *directly,* without a detour through force for proceeding and force of resisting and the balance of forces in an impact contest. Indeed, it can be applied without any of the conceptual apparatus of force at all; all we need is a conception of the modes of a body in motion and at rest, and a way of talking about changes in those modes in a quantitative way. This turns out not to be so easy, I think. I don't know whether Descartes thought of what he was doing as presenting the foundations of his earlier view, or whether he was just reluctant to admit that his views had changed. But the answer he gives to Clerselier, though presented as an explication of what he had said before, represents an interesting step beyond the impact contest model of the Latin *Principles,* and may represent a new way of conceptualizing impact that he was trying out in this letter.

Before moving on, I would like to make one last observation about the treatment of impact in the Clerselier letter. Earlier, when talking about the treatment of impact in the Latin *Principles,* I noted that the rules of impact seem to assume a genuine distinction between motion and rest so that the situations Descartes deals with in R4 and R5 represent physically distinct states of the world. I suggested that Descartes had in mind here the sort of distinction I tried to explicate above in chapter 6, based on the careful definition of motion he offers in Pr II 25. This is what he seems to be saying quite explicitly in paragraph 5 of the Clerselier letter: "By a body which is without motion I understand a body which is not at all in the act *[en action]* of separating its surface from those of the other bodies which surround it, and which, as a consequence, makes up a part of another larger hard body." The appeal seems to be directly to the formal definition of motion in the *Principles,* and the account that follows of the difference between motion and rest.

At the same time Descartes was writing to Clerselier, he was also helping to prepare a French translation of his Latin *Principles.* He did not do the translation itself; the responsibility for that was in the hands of the Abbé Picot. But there is every reason to believe that he took the publication of the French translation as an opportunity to explain some things at greater length, particularly the rules of impact, which raised such questions (see AT IXB Xf; AT IV 187 note d; 396; AT V 168 [Descartes 1976, p. 35]). It would be wrong to suppose that every difference between the Latin original and the French translation had Descartes' approval; there are some obvious mistakes and mistranslations even in the sections having to do with impact.[35] But there are differences so significant between the two editions and so suggestive of his letter to Clerselier that it is very difficult to imagine that they are not the work of Descartes himself.

There is little in the way of significant addition in the French edition of Pr II 40–45, where Descartes sets out law 3, its explication and its defense. As in the Latin edition, a relatively straightforward impact contest view of the matter is presented and defended. There is reason to believe that Descartes hardly looked at the French translation of those sections. As Pierre Costabel has pointed out, there are some errors of translation in those sections that are so glaring that Descartes could hardly have failed to notice them, had he but read them over with any care.[36] Descartes' attention seems to have been squarely on the sections that follow, Pr II 46–52, where the seven rules of impact are presented. There the changes are quite interesting.[37]

Many of the changes are rather minor. The cases are generalized a bit, for one. In the first group of rules, R1–R3, instead of having B move from right to left and C move from left to right, as in the Latin, the French text has them simply coming from opposite directions. And in R7, the French text, unlike the Latin, acknowledges the possibility that B might be larger than C. The French edition also adds to R7 the case of equality, where $v(B)/v(C)=m(C)/m(B)$, a case that the Latin edition omits. In at least one case, R4, numerical examples missing in the Latin edition are added to the French. But more importantly, the French edition adds considerable explanation to these sections, quite brief in the Latin edition. In the original Latin edition Descartes had concluded the exposition of the seven rules by claiming that "these require no proof since they are per se obvious" (Pr II 52L). In the

French edition, though, Descartes ends by remarking on the certainty of the *demonstrations* that he has given of those rules (Pr II 52F).

In the Latin *Principles* the analysis of impact seems to be based on what I earlier called the simple impact contest (IC) model, where the outcome of a collision depends upon the balance of two forces, the force for proceeding and the force of resisting, where these forces are measured in terms of the modes a body has at the moment of impact. In the Clerselier letter this simple model is replaced with a sophisticated IC model, where those forces are measured in terms of a future hypothetical effect, the motion that they would have to have after the impact. In the Clerselier letter Descartes also adds a principle of the mean for dealing with one of the cases of equality, R6. But, as we saw, in addition to these elaborations on the conceptual framework of the Latin *Principles*, he also proposes a least change principle, what I called the PLMC, a way of conceptualizing the problem of impact that doesn't seem to involve forces at all, that seeks to determine the postcollision states of the bodies in impact in terms of the least change in the precollision modes of motion and determination. Between the Latin edition of the *Principles* and the various lines of thought in the letter to Clerselier, there are, thus, a number of different ways of thinking about the problem of impact that demand some sort of resolution, and one would think that the new edition of the *Principles* would have been an ideal time to tie up some loose ends. But when preparing the French version of the *Principles* Descartes, unfortunately, did not take the opportunity to rethink the problem of impact from the ground up. While he was clearly unhappy with the treatment of impact in the Latin *Principles*, he felt quite pressed by other matters, and claimed that he simply did not have the time to work on the question of impact; writing to Mersenne on 20 April 1646, Descartes asked his friend to tell Picot, should he see him, that he, Descartes, hasn't had as much as "a quarter of an hour in the entire last year" to think about the laws of motion (AT IV 396). One also suspects that Descartes wasn't quite sure how to proceed with the laws of impact. Thus, in the explanations added to the rules of impact in the French *Principles*, one finds a *mélange*, elements that clearly derive from the reflections Descartes sent to Clerselier, superimposed on a framework derived directly from the original Latin *Principles*.

Resting at the bottom of the pile, as it were, is the text of the Latin *Principles*, rendered into French; there are some stylistic changes introduced by the translation, of course, but relatively little is omitted. Some of the additions introduced in the French edition seem to continue directly the ideas implicit in the original edition. For example, the French edition of R2, Pr II 47, is virtually identical with the Latin edition. Descartes makes only one brief addition; dealing with the case in which B and C directly collide from opposite directions, with $v(B)=v(C)$ and $m(B)$ $m(C)$, Descartes notes that "B, having more force than C, could not be forced to rebound by it." This seems to be a straightforward evocation of the simple IC model implicit in the Latin *Principles;* the outcome seems to depend on the balance of forces, which are measured directly in terms of the size and speed of bodies.

But there are other changes that clearly show the thinking of the Clerselier letter. Unlike the Latin *Principles,* Descartes seems aware that the cases of equality pose a kind of problem for the impact contest model. In the treatment of R6 in the French edition, Pr II 51F, for example, Descartes reasons much as he implies we should in the Clerselier letter. There, where we are dealing with a moving B colliding with a resting C of the same size, he reasons that since the two bodies are equal in size, "there is no more reason why [B] should rebound than push C," and thus, "these two outcomes should be divided equally" (Pr II 51F). That is, the result in R6 is a compromise between the results in R4, where $m(B) < m(C)$, and R5, where $m(B) > m(C)$.[38] The reasoning can be construed in a similar way in the French version of R1, the case of direct impact between bodies of the same size, moving with the same speed. But what Descartes explicitly appeals to is a different kind of symmetry principle, the principle that since B and C are in exactly the same circumstances, the outcomes should be comparable. So, he reasons, B should not push C and C should not push B, but both should rebound (Pr II 46F).

But most interesting are the additions that seem to suggest the sophisticated IC model and the PLMC. The largest addition is to the account of R4 in Pr II 49, the case that Descartes centers on in the Clerselier letter, where a smaller B collides with a C at rest; the French version of this section runs almost four times the size of the original Latin. What is extremely interesting is that the PLMC, originally formulated in the Clerselier letter to deal with this case, as well as that of R5 and R6, is not mentioned in this context in the French edition. Instead, the explanation Descartes offers is squarely in terms of the impact contest model. The French reads that "it is certain that C should resist more the faster B goes toward it, and its resistance ought to prevail over the action of B because it is larger than it." As in the Latin, the rule is conceived as a contest between B and C, where the force of resisting in C is taken to be greater in proportion to the speed of B. But the justification that Descartes offers derives directly from paragraphs 1 and 2 of the Clerselier letter. Descartes writes:

> Thus, for example, if C is twice as large as B, and B has three
> degrees of motion, it cannot push C, which is at rest, without

transferring to it two degrees (namely, one for each of its halves) and without retaining only the third for itself, because it is not greater than each of the halves of C and afterwards, it cannot go faster than they go. (Pr II 49F)

While it is not quite as explicitly put as in the Clerselier letter, Descartes seems to be reasoning that the resistance offered by C is a function of the speed with which it would have to move, if it were set into motion, and that the force for proceeding in B is proportional to the amount of motion that it could impose on a body with which it collides. What we have, in essence, is the sophisticated IC model of impact.[39]

The PLMC is not summoned to elucidate R4, strangely enough. But it is not forgotten in the French edition of the rules. While not given explicitly, there is a clear reference to it in the explanation of R3, the case in which B and C directly collide from opposite directions, with $m(B)=m(C)$ and $v(B) > v(C)$. Discussing the case in which B moves with six degrees of speed and C with four, Descartes reasons that "afterwards both would go with five degrees of speed, since it is easier to communicate one of its degrees of speed to C than it is for C to change the entire course of the motion B has" (Pr II 48F). While this may possibly be interpreted in terms of the impact contest model, I see here a clear reflection of the PLMC of the Clerselier letter.

In the end, in the French edition of the *Principles,* the problem of impact is left a bit up in the air. It is not clear (to me, at least) just what Descartes ends up thinking; if there is anything at all clear in the treatment of impact, it is that Descartes himself has not managed to sort out his thinking on this problem. Nor is it much of a surprise that he hasn't, in a way. As I pointed out earlier, the first two laws of the *Principles,* law 1, which concerns the persistence of motion, and law 2, which concerns the persistence of determination and the tendency that bodies in curvilinear motion have to recede from the centers of their rotation, have deep roots in the Cartesian program for physics; law 1 is closely connected to Descartes' early studies of free-fall and his later studies of the reflection and refraction of light, and law 2 is closely connected to the account of what light is, on which Descartes attempted to ground his optical studies in the 1620s. On the other hand, the conservation principle, later to become so visible in the *Principles* and in later Cartesian physics, and the law of impact seem to emerge rather later. The conservation principle is important in its first appearance in *The World* only as an intermediate step in the proof of other laws. While the impact law makes an appearance of sorts in *The World* as law B, it plays little, if any, substantive role in the actual business of Cartesian physics, both in *The World* and afterward.[40] As late as 26

February 1649, Descartes wrote Chanut saying that "one need not" spend much time with the rules of impact, since "they are not necessary for understanding the rest" of the *Principles* (AT V 291 [K 246]). The conservation principle and the impact law with which it is closely connected thus seem to be rather later additions to the Cartesian framework. They arise, I think, not from the inner needs of the Cartesian program for physics, that is, from the particular questions that Descartes wanted to treat and could not treat without them, but from the need for completeness and system, what became, more and more, a ruling theme in Descartes' thought. It is not surprising, then, that they get their first careful and explicit statements only in the *Principles*, where Descartes is attempting to set out his thought in a careful, rigorous, and systematic way, under the influence of his recent reading of scholastic manuals, like that of Eustachius a Sancto Paulo, and with the intention of capturing the attention of the schools, and it is not surprising that the impact law and the numerical measure of quantity of motion arise at the same time, as Pierre Costabel has emphasized. The question of impact seems to remain an unsettled question up until the end of Descartes' life, work in progress that he never quite finished.

But though, as I suspect, Descartes never quite finished the account of impact, never quite settled even its basic analysis, I do think that the work Descartes did on the problem in the 1640s shows a very interesting trend. Descartes begins in the Latin *Principles* with a general principle to understand impact, the impact contest model, implicit in the brief discussion of impact in *The World* and widely shared by others of his contemporaries who had thought about the problem. This general principle is supplemented later in the composition of the Latin *Principles* by seven rules, attempts to apply the very general law 3 to a variety of specific cases. But what is interesting in the writings that follow is the increasing attention Descartes gives to these specific cases; in the Clerselier letter of 1645, Descartes discusses the examples only, and in the French *Principles* of 1647, only the rules of impact, and not the general law 3 seem to have had his attention. Descartes seems to be coming to grips with the problem of impact in a most uncartesian way, through understanding what is going on in individual special cases. In trying to justify the rules of impact to Clerselier, and to himself, no doubt, rules originally presented as "per se obvious," Descartes is led to a deeper understanding of the phenomenon of impact, from a simple IC model, to a more sophisticated IC model, and to a least change principle, the PLMC, a principle that seems to depart altogether from the impact contest analysis of impact, and seems to require no forces to enter into the determination of the outcomes. It is unlikely that in the end Descartes would have accepted a principle like the PLMC. As I have pointed out, that principle is, in the end, no less problematic in its application than the impact contest model is. Descartes may have come to realize this while pondering the impact laws for the French edition. Or, he may have been uncomfortable with such a teleological principle, given his public rejection of teleological arguments, though this doesn't seem to have prevented him from making implicit reference to the principle in the French version of R3.[41] But it is impossible to say just where Descartes would have ended, if he had the opportunity (and the inclination) to continue work on the problem.

Descartes on Inertia

Though I have completed what I have to say about the laws of impact, strictly speaking, there is another closely connected issue worth discussing, the question of inertia. What I have in mind is the question not of Descartes' relation to Newton's so-called law of inertia, but Descartes' views on inertia as understood by his contemporaries.

The Latin word "inertia" literally means laziness, and this is not far from the way the notion was understood by Descartes' contemporaries when applied to body. The word, and the view that bodies have a kind of innate laziness or natural inertia, is closely associated with the physical theories of Kepler in the early seventeenth century. Writing in the second edition of the *Mysterium Cosmographicum* (1621), Kepler notes:

> The bodies of planets . . . should not be considered as mathematical points, but as bodies endowed with matter, and as having a certain sort of weight, as it were, . . . that is, insofar as they are endowed with a faculty for resisting motion imposed from the outside, in proportion to the bulk of the body, and the density of the matter. For since all matter tends toward rest in the place in which it is, . . . it thus happens that the moving power of the sun fights against this inertia in matter.[42]

Kepler assumed, as a basic and inborn property of bodies, that matter as such resists motion and tends toward rest, which opposes the force he thought the sun exerts on the planets to keep them in motion.

Such a conception of matter is quite foreign to Descartes' conception of the world. It is precisely this sort of inertia that Descartes meant to deny when writing to Mersenne in December 1638:

> I don't recognize any inertia or natural sluggishness [*tartiueté*] in bodies, no more than M. Mydorge does, and I believe that when even one man walks, he makes the whole mass of the earth move

ever so little, because he steps now upon one part, and later upon another. But I don't disagree with M. Debeaune that the largest bodies, pushed by the same force (like the largest boats pushed by the same wind) move more slowly than the others. Perhaps this is sufficient to ground his arguments, without having recourse to this natural inertia, which can in no way be proven. (AT II 466–67)[43]

But though Descartes denies that bodies have any natural inertia, any natural tendency to resist motion, he does grant that it is harder to set a larger body in motion than a smaller one. In a letter written directly to Debeaune a few months later, on 30 April 1639, Descartes grants that in this precise sense, one might want to say that bodies have natural inertia. Immediately following a discussion of his conservation principle he notes:

> If two unequal bodies receive the same amount of motion as one another, this same quantity of motion doesn't give as much speed to the larger as it does to the smaller. Because of this one can say that the more matter a body has, the more natural inertia it has. To this one can add that a body which is large can better transfer its motion to other bodies than a small one can, and that it can be moved less by them. In this way there is a sort of inertia which depends on the quantity of matter, and another which depends on the extension of its surface. (AT II 543–44 [K 64–65])[44]

The position that Descartes is outlining in these few passages is an interesting one. Descartes quite clearly denies the existence of any inherent tendency bodies have to come to rest or to resist the acquisition of motion. But at the same time he grants that it is harder to set a large body in motion than a small one. Because motion is conserved, both in the world as a whole and in individual collisions between perfectly hard bodies, the same quantity of motion will give a lesser speed to a heavy body than it will to a smaller one; in this very limited sense Descartes is willing to grant that there is a natural inertia in bodies. What is basic here is not the supposed force of inertia but the *law*, the constraint on motion that derives not from matter itself but directly from God's constant activity on the world, keeping the same quantity of motion in the world as he created in the beginning. Inertia emerges on this view as a kind of "imaginary" force; while bodies behave *as if* there were some kind of internal resistance to being set into motion, all there *really* is is bare extended substance, behaving in accordance with the laws that an immutable God's continual sustenance imposes on it.

APPENDIX

DESCARTES' IMPACT RULES:
PRINCIPLES, II 46–52

Consider bodies B and C, where v(B) and v(C) are the speeds B and C have before impact, v(B)′ and v(C)′ are their speeds after impact, and m(B) and m(C) are their respective sizes.

Case I: B is moving from right to left, and C is moving from left to right.

R1. If m(B) = m(C), and v(B) = v(C), then after the collision, v(B)′= v(C)′= v(B) = v(C), B moves from left to right, and C moves from right to left (i.e., B and C are reflected in opposite directions). [Pr II 46]

R2. If m(B) > m(C), and v(B) = v(C), then after the collision, v(B)′= v(C)′= v(B) = v(C), B and C move together from left to right (i.e., B continues its motion and C is reflected in the opposite direction). [Pr II 47]

R3. If m(B) = m(C), and v(B) > v(C), then after the collision, B and C move together from right to left (i.e., B continues its motion and C is reflected in the opposite direction) and v(B)′= v(C)′= {[v(B) + v(C)]/2}. [Pr II 48]

Case II: C is at rest and B collides with it.

R4. If m(B) < m(C), then after the collision, C remains at rest and B rebounds (i.e., B moves off in the opposite direction) with v(B)′= v(B). [Pr II 49]

R5. If m(B) > m(C), then after the collision, B and C move together in the direction in which B was moving before the collision, with v(B)′= v(C)′= {m(B)v(B)/[m(B) + m(C)]}. [The formula is inferred from the example in Pr II 50 using the conservation principle.] [Pr II 50]

R6. If m(B) = m(C), then after the collision, C moves in the direction B originally moved with v(C)′= (1/4)v(B) and B would be reflected in the opposite direction, with v(B)′= (3/4)v(B). [Pr II 51]

Case III: B and C move in the same direction, with v(B) Ç> v(C)

R7a. If m(B) < m(C) and "the excess of speed in B is greater than the excess of size in C," i.e., v(B)/v(C) > m(C)/m(B), then after the collision, B transfers to C enough motion for both to be able to move equally fast and in the same direction. I.e., v(B)′= v(C)′= [m(B)v(B) + m(C)v(C)]/[m(B) + m(C)]. [The formula is inferred from the example in Pr II 52 using the conservation principle. In the French version, Descartes drops the condition that m(B) < m(C), though he keeps the condition that v(B)/v(C) > m(C)/m(B).] [Pr II 52]

R7b. If m(B) < m(C) and "the excess of speed in B" is less than "the excess of size in C," i.e., v(B)/v(C) < m(C)/m(B), then after the collision, B is reflected in the opposite direction, retaining all of its motion, and C continues moving in the same direction as before, with v(B) = v(B)′and v(C) = v(C)′. [Pr II 52]

R7c. If $m(B) < m(C)$ and $v(B)/v(C) = m(C)/m(B)$, then B transfers "one part of its motion to the other" and rebounds with the rest. [This rule is only in the French edition. There is no example in the text from which one can infer a formula, but perhaps Descartes means that B would transfer half of its speed to C, so that by the conservation principle, $v(B)' = v(B)/2$ and $v(C)' = (3/2)v(C)$.] [Pr II 52F]

PRINCIPLES, II 46–52

Latin Version (1644)	French Version (1647)
46. First [rule]. First, if these two bodies, say, B and C, were exactly equal, and moved with the same speed, B from right to left and C in a straight line from left to right, then when they encountered one another, they would be reflected, and afterwards, each would proceed to move, B to the right and C to the left, without losing any part of its speed.	**46. First [rule].** The first [rule] is that, if these two bodies, say, B and C, were exactly equal and moved with the same speed in a straight line toward one another, then when they encountered one another, they would both equally well be reflected, and each would return in the direction from which it had come, without losing any part of its speed, *since in this circumstance there is no cause that can take it [i.e., speed] away, but there is a very evident cause that should force them to rebound; and because the cause would be the same for both, they would both rebound in the same way.*
47. Second [rule]. Secondly, if B were the slightest bit larger than C, assuming everything else as before, then only C would be reflected, and both would move to the left with the same speed.	**47. Second [rule].** The second [rule] is that, if B were the slightest bit larger than C, and they encountered one another with the same speed, then only C would be reflected in the direction from which it came, and afterward both would continue their motion together in the same direction. *For B, having more force than C, could not be forced to rebound by it.*
48. Third [rule]. Thirdly, if they were equal in bulk *[mole]*, but B moved the slightest bit faster than C, then not only would both proceed to move to the left, but also half of the speed by which the one exceeded the other would be transferred from B to C. That is, if previously there were six de-	**48. Third [rule].** The third [rule] is that if these two bodies were of the same size, but B had the slightest bit more speed than C, then not only would it happen that after the encounter, C alone would rebound and both would move off together, as before [i.e., in Rule 2], but also it would be

256

grees of speed in B and only four in C, then after their encounter with one another both would tend *[tendere]* to the left with five degrees of speed.

necessary for B to transfer to C half of the speed by which the one exceeded the other, *because [C] being in front of it, it could not go any faster than it [C].* That is, if before their encounter, B had had, for example, six degrees of speed and C had had only four, *it [B] would transfer one of its excess degrees of speed to it [C],* and thus afterwards both would go *[iroient]* with five degrees of speed, *since it is easier for B to communicate one of its degrees of speed to C than it is for C to change the entire course [cours]* of the motion B has.

49. **Fourth [rule].** Fourthly, if body C were entirely at rest, and were just a bit larger than B, then whatever the speed with which B moved toward C, *it would never move C,* but would be repelled by it in the opposite direction, since a resting body resists a greater speed more than it does a smaller one, *and this in proportion to the excess of the one over the other.* And therefore there would always be a greater force in C to resist, than there would be [a force] in B to impel.

49. **Fourth [rule].** The fourth [rule] is that if body C were just a bit larger than B, and were entirely at rest, *that is, not only is there no apparent motion, but also it is not surrounded by air nor by any other fluid bodies, which, as I shall discuss below [in Pr II 59] dispose the hard bodies they surround to be more easily moved,* then whatever the speed with which B could go toward it [C], *it would never have the force to move it,* but it would be forced to rebound in the same direction from which it had come, since *seeing that B could not push C without making it go as fast as it itself would go afterwards,* it is certain that C should resist more the faster B goes toward it, *and its resistance ought to prevail over the action of B because it is larger than it.* Thus, *for example, if C is twice as large as B, and B has three degrees of motion, it cannot push C, which is at rest, without transferring to it two degrees (namely, one for each of its halves) and without retaining only the third for itself, because it is not greater than each of the halves of C and afterwards, it cannot go faster than they go. In the same way, if B has thirty degrees of speed, it must communicate twenty of them to C; if it has three hundred, it must communicate two hundred; and so [it] always [communicates]*

twice what it retains for itself. But since C is at rest, it resists the reception of twenty degrees ten times more than it resists the reception of two, and it resists the reception of two hundred one hundred times more, so that the more speed B has, the more resistance it finds in C. And since each half of C has as much force to remain in its rest as B has to push it, and since both resist at the same time, it is evident that they ought to succeed in forcing it [i.e., B] to rebound. Consequently, whatever the speed with which B goes toward C, at rest and larger than it, it will never have the force to move it.

50. **Fifth [rule].** Fifthly, if the resting body C were smaller than B, then however slowly B moved toward C it would move it [C] with it [B], by transferring part of its motion to it in such a way that afterwards, both would move equally fast. Namely, if B were twice as large as C, it would transfer to it a third of its motion, since that one third would move the body C as fast as the two remaining ones would move the body B, being twice as large. And so, after B encountered C, it would move slower by one third than it moved before, that is, it would need as much time to move through a space of two feet as it previously needed for it to move through a space of three. In the same way, if B were three times larger than C, it would transfer to it a fourth of its motion, and so on.

50. **Fifth [rule].** The fifth [rule] is, if, on the other hand, the body C were ever so slightly smaller than B, the latter could not go so slowly toward the other (which, again, I assume to be perfectly at rest) that *it would not have the force* to push it and transfer to it the part of its motion necessary to bring it about that they would afterwards go at the same speed. Namely, if B were twice as large as C, it would only transfer to it a third of its motion, since that one third would move C as fast as the two remaining ones would move B, since it is assumed to be twice as large. And so, after B encountered C, it would go slower by one third than it went before, that is, in the time that it could have traversed three spaces, it could only traverse two. In the same way, if B were three times larger than C, it would only transfer to it a fourth of its motion, and so on. *B could not have so little force that it would not always suffice to move C. For it is certain that the weakest motions should follow the same laws, and have, in proportion, the same effects as the strongest, although we often think we notice the opposite on this earth, because of the air and the other fluids which always surround hard bodies that move and*

which can greatly increase or decrease their speed, as we shall see below. [Pr II 59]

51. Sixth [rule]. Sixthly, if the resting body C were exactly equal to body B, which was moving toward it, then it [C] would in part be impelled by it, and in part it [C] would repel it in the opposite direction. Namely, if B went toward C with four degrees of speed, it would communicate one degree to C, and with the three remaining [degrees] it would be reflected in the opposite direction.

51. Sixth [rule]. The sixth [rule] is that if the body C were at rest, and exactly equal in size to body B, which was moving toward it, then it would be necessary that it would in part be impelled by B, and in part it will make it rebound, so that if B went toward C with four degrees of speed, it would be necessary that it transfer one to it, and with the three remaining [degrees] would return in the direction from which it had come. *Since it is necessary that either B push C without rebounding and thus that it [B] transfer two degrees of its motion; or that it [B] rebound without pushing it, and consequently, that it retain these two degrees of speed along with the other two which cannot be taken from it; or, finally, that it rebound and retain one part of these two degrees and push it [C] by transferring to it the other part: it is thus evident that since they are equal, and thus there is no more reason why it should rebound than push C, these two outcomes should be divided equally. That is, B should transfer to C one of these two degrees of speed, and rebound with the other.*

52. Seventh [rule]. Finally, if B and C were moving in the same direction, C more slowly and B following it more quickly, so that at length, [B] would hit it [C], and if C were larger than B, but the excess of speed in B were greater than the excess of size in C, then B would transfer to C enough of its motion for both of them to move afterwards equally fast and in the same direction. Moreover, if, on the other hand, the excess of speed in B were less than the excess of size in C, B would be reflected in the opposite direction, and would retain all of its mo-

52. Seventh [rule]. The seventh and last rule is that if B and C go in the same direction, and C precedes but goes more slowly than B, so that at length, [B] would hit it [C], then it could happen that B would transfer a part of its speed to C in order to push it [C] in front of it [B], and it could also happen that it could transfer none at all, but rebound with all of its motion in the direction from which it came. Indeed, [these outcomes can happen] *not only when C is smaller than B, but also when it is larger; as long as that with which the size of C surpasses*

tion. This excess is calculated as follows. If C were twice as large as B, and B did not move twice as fast as C, it [B] would not impel it [C], but would be reflected in the opposite direction. But if it [B] moved more than twice as fast, it would impel it [C]. Namely, if C had only two degrees of speed, and B had five, then two degrees would be taken from B, which transferred into C would bring about only one degree [of motion] since C is twice as big as B. From this it would happen that the two bodies, B and C would afterwards move with three degrees of speed. *And we must settle the rest in the same way. These require no proof since they are per se obvious.*

that of B is less than that with which the speed of B surpasses that of C, B should never rebound, but push C, transferring into it one part of its speed. On the other hand, when that with which the size of C surpasses that of B is greater than that with which the speed of B surpasses that of C, B must rebound, without communicating any of its motion to C. *And finally, when the excess of size which C has is perfectly equal to the excess of speed which B has, the latter ought to transfer one part of its motion to the other, and rebound with the rest.* This can be calculated as follows. If C were exactly twice as large as B, and B didnot move twice as fast as C, but lacked something of it, then B should rebound without increasing the speed of C, and if B moved more than twice as fast as C, then it should not rebound, *but should transfer as much of its motion to C as is needed to bring it about that they move together afterwards with the same speed.* For example, if C had only two degrees of speed, and B had five (which is more than double), then it should communicate two of the five, which two being in C would bring about only one [degree of motion] since C is twice as big as B, and thus, afterwards they would both go with three degrees of speed. *And the demonstrations of all of these are so certain, that even were experience to seem to make us see the contrary, we would nevertheless be obligated to put more faith in our reason than in our sense.*

I have tried to give rather literal translations of the two texts. The differences are too numerous to note; I have put into italic those passages that seem to me to be the the most significant differences between the two versions.

DESCARTES TO CLERSELIER:

17 February 1645
Sir,
[1.] The reason I say that a body without motion would never be moved by another [184] smaller than it, whatever the speed with which this smaller body

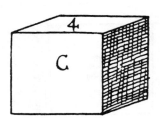 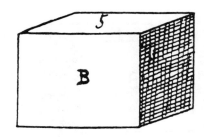

FIGURE 8.1 From *Lettres de M. Descartes*, vol. 1 (Paris, 1657), p. 530.

could move, is that it is a law of nature that a body that moves another must have more force to move it than the other has to resist it. But this greater [force for moving] can depend only on its size, for that which is without motion has as many degrees of resistance as the other, which is moving, has of speed. The reason for this is that if it is moved by a body which moves twice as fast as another, it ought to receive twice as much motion from it; but it resists twice as much this twice as much motion.

[2.] For example, body B cannot push body C without making it move as fast as it itself would move after having pushed it. [See fig. 8.1.] In particular, if B is to C as 5 is to 4, with B having 9 degrees of motion, it must transfer 4 of them to C to make it go as fast as it goes. This is easy, for it has the force to transfer up to four and a half (that is, half of what it has) rather than reflecting its motion in the other direction. But if B is to C as 4 is to 5, B cannot move C, unless of its nine degrees of motion it transfers 5 to C, which is more than half of what it has, and which, as a consequence, C resists [185] more than B has the force to act. This is why B ought to rebound in the other direction instead of moving C. And without this, no body would ever be reflected by encountering another. [See fig. 8.2.]

[3.] Moreover, I am glad that the first and principal difficulty that you have found in my *Principles* concerns the rules in accordance with which the motion of bodies in impact changes, since I conclude from that that you have not found any [difficulties] in what precedes them and that you will not find any more of them in what follows, nor any more in the rules [of impact] either, when you notice that they depend only on a single principle, which is that *when two bodies having incompatible modes collide, there must really be some change in these modes, in order to render them compatible, but that this change is always the least possible,* that is, *if they can become compatible by changing a certain quantity of these modes, a greater quantity of them will not be changed.* And we must consider two different modes in motion *[mouuement]:* one is motion *[motion]* alone, or speed, and the other is the determination of this motion *[motion]* in a certain direction, which two modes change with equal difficulty. [186]

[4.] Therefore, to understand the fourth, fifth, and sixth rules, where the motion of a body B and the rest of a body C are incompatible, we must notice that they can become compatible in two ways, namely, *if B changes the entire determination of its motion,* or, *if it changes body C's rest, transferring to it such part of its motion*

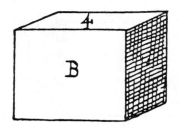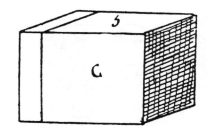

FIGURE 8.2 From *Lettres de M. Descartes,* vol. 1 (Paris, 1657). p. 531.

as will enable it [B] to push C in front of it as fast as it will itself go. And I have said nothing in these three rules but this, that when C is larger than B, it is the first of these two ways which takes place, and when it is smaller, it is the second, and finally, when they are equal, this change is made half in the one way and half in the other. For when C is the largest, B could push it ahead of itself only if it were to transfer to C more than half its speed, and at the same time more than half of its determination to go from right to left, insofar as this determination is joined to its speed. So instead, being reflected, without changing C, it changes only its entire determination, which is a lesser change than a change of more than half of this same determination and more than half of the speed. On the other hand, if C were smaller than B, it should be pushed by it, for then B gives it less than half of its own speed and less than half of the determination which is joined to it. This is less than changing the entire determination, which would be changed if it were reflected.

[5.] This is not at all in contradiction with experience, since, in [187] these rules, by a body which is without motion I understand a body which is not at all in the act *[en action]* of separating its surface from those of the other bodies which surround it, and which, as a consequence, makes up a part of another larger hard body. For I have said elsewhere [Pr II 30] that when the surfaces of two bodies separate, all that is positive in the nature of motion will be as much in the one which is commonly said not to move at all as in the one which is said to move; and I have also explained afterwards why a body suspended in the air can be moved by the least force. [Pr II 56]

[6.] But yet I must acknowledge that these rules are not without difficulty, and I shall try to clarify these things more, if I am now capable of doing so. But because my mind is now occupied with other thoughts, I will wait for another time, if you please, to send you my views at greater length. . . . [AT IV 183–87]

The paragraphs have been numbered for convenient reference in the text. The page numbers in AT IV are given in brackets at the start of every new page. The last paragraph of the letter, which deals with other matters, has been omitted.

GOD AND THE GROUNDS OF THE
LAWS OF MOTION: IMMUTABILITY,
FORCE, AND FINITE CAUSES

IN THE LAST TWO chapters we have been discussing the laws of motion, the conservation principle and the three laws that follow it. In the course of that discussion, we have made reference to the fact that these laws are grounded, for Descartes, in the activity of God, in the immutable and constant way in which God keeps the world in existence. In this chapter we shall deal directly with this aspect of Descartes' position. We shall begin with an exposition of his views on God as the creator and continual sustainer of the world, including a brief digression on the question of whether or not his account of God commits him to a kind of temporal atomism, as is often assumed it does. Then we shall discuss the ways in which this view grounds the laws of motion, and the closely related question of the ontological status of force in Cartesian physics. We shall close with a discussion of the role of finite causes in Descartes' system, both bodies and minds, and consider the extent to which Descartes can be regarded as an occasionalist.

DIVINE SUSTENANCE

We saw in chapter 6 that duration has, for Descartes, the status of an attribute of created substance, both body and mind; as he writes in the *Principles,* "since any substance that ceases to endure ceases to be, there is only a distinction of reason between it [i.e., a substance] and duration" (Pr I 62). But just as he distinguishes motion, a mode of body, from the action that is its cause, he also distinguishes duration, an attribute of finite substance, from its cause; though duration may be an attribute of all finite substance, finite substances cannot themselves be the causes of their continued duration. And consequently, he argues, God, the infinite substance, must continually sustain them and maintain their existence.

While the doctrine goes back to at least the time of the composition of *The World* (AT XI 37, 44–45) and figures in the summary of the system Descartes gave in the *Discourse* (AT VI 35–36, 45), this doctrine receives its most extensive and careful development in the early 1640s. He writes in Meditation III:

> All of the time of my life can be divided into innumerable parts, each of which is entirely independent of the others, so that from the fact that I existed a short time ago, it does not follow that I ought to exist now, unless some cause as it were creates me again in this moment, that is, conserves me. (AT VII 48–49)

Now, he argues, "plainly the same force and action is needed to conserve any thing for the individual moments in which it endures as was needed for creating it anew, had it not existed" (AT VII 49). Clearly such a power is not in us; if it were, then, Descartes reasons, I would also have been able to give myself all of the perfections I clearly lack (AT VII 48, 168–69). And so, he concludes, it must be God that both creates and sustains us (AT VII 49–50, 111, 165, 168–69, 369–70; Pr I 21). This conclusion, of course, holds for bodies as well as it does for us. It is not just *minds,* but *all* finite things that require some cause for their continued existence. And as with the idea of ourselves, "when I examine the idea of body, I perceive that it has no power *[vis]* in itself through which it can produce or conserve itself" (AT VII 118; see also p. 110). And so, we must conclude that the duration of bodies, too, must be caused by God, who sustains the physical world he created in the beginning.

It is in this sense that "the nature of time or of enduring things" is "such that their parts are mutually independent" (Pr I 21); just as God could have created this mind without this body, or this body without that body, he could have created the world up to a given time without creating anything that follows, and he could create any individual portion of time without creating any others (AT VII 369–70). And were God to withdraw his sustenance, the world would cease to exist. Writing in August 1641 to Hyperaspistes, Descartes noted: "Nor should we doubt that if God ceased his concourse, everything he created would vanish immediately into nothing, since before he created them and provided them with his concourse, they were nothing" (AT III 429 [K 116]).[1]

Descartes conceives of God's continual sustenance of his creatures as their *efficient* cause; writing in the *First Replies,* he notes, "I should not hesitate to call the cause that sustains me an efficient cause" (AT VII 109). And, as I shall later argue in connection with the derivation of the laws of motion, he might also be considered as an efficient cause of motion insofar as he sustains the impulses that are the causes of particular motions. But God's causality here is in one respect importantly different from other efficient causes that we are familiar with from our experience. In reply to Gassendi's *Fifth Objections,* Descartes distinguishes between two sorts of efficient causes, a *causa secundum fieri,* a cause of becoming, and a *causa secundum esse,* a cause of being. An architect is the cause of becoming with respect to a house, as is a father with respect to his son. But, he claims, "the sun is the cause of the light proceeding from it, and God is the cause of created things, not only as a cause of becoming, but as a cause of being, and therefore must always flow into the effect in the same way, in order to conserve it" (AT VII 369). And so, just as the sun must continue its illumination for daylight to persist, so must God continue his activity in order for the world and its motion to be sustained.[2] This continual sustenance is also unlike the more ordinary efficient causes insofar as it requires a kind of power beyond the capacities of created things. While finite things may be able to stand as the efficient causes *secundum fieri* of things in the world, only God can stand as their cause *secundum esse.* As we noted earlier, in Meditation III Descartes declares that "plainly the same force and action is needed to conserve any thing for the individual moments in which it endures as was needed for creating it anew, had it not existed." From this Descartes infers that "it is also one of those things obvious by the light of nature that conservation differs from creation only in reason" (AT VII 49). That is, the activity and power needed to sustain a thing in its existence is identical to the activity and power necessary to create anything from nothing (see also AT VII 165, 166). But it is not clear just how literally we are expected to take this doctrine, whether we are supposed to think of God as sustaining the world through a probably infinite series of mini-creations. Though he sometimes says that "conservation" is to be understood as the "continual production of a thing" (AT VII 243; see also Pr II 42), elsewhere he is a bit more guarded, suggesting that God *as it were* continually reproduces his creatures *[continuo veluti reproducat]* (Pr I 21; see also AT VII 110).

Despite Descartes' clear commitment to the doctrine that God must continually support his creation, it is not at all clear just how he thinks God actually performs this most remarkable of his feats. Nor, I suspect, did Descartes think that there was any particular need for detailed explanation or defense. As far as he was concerned, he was appealing to an old and widely accepted doctrine, a doctrine with which his audience could be expected to be both familiar and sympathetic. When in the *Fifth Objections* Gassendi challenged his appeal to a conserving God, Descartes responded: "When you deny that to be conserved we require the continual influx of a first cause, you deny something that all metaphysicians affirm as obvious" (AT VII 369).[3] And in defending himself against Gassendi's criticisms he seems to have turned directly to his copy of St. Thomas.[4] God's sustenance of this world of created things is explicitly discussed in the *Summa Theologiae* I q104 a1, and this passage is the apparent source of Descartes' answer to Gassendi. Like Descartes, Thomas distinguishes between causes *secundum fieri* and *secundum esse,* and appeals to the same examples Descartes does—the builder of a house, the parent of a child, and the sun as illuminator—to clarify the sense in which God is the cause of the world as enduring.[5] While he does not say explicitly that God's activity in sustaining the world is identical with his activity in creating it, Thomas does say that "God's conservation of things is not through any new action, but through the continuation of the action by which he gives them being."[6] Thomas' views were widely shared by scholastic thinkers in the sixteenth and seventeenth centuries, many of whom would be happy to agree with Descartes that "this conservation is the very same thing as creation, differing only in reason."[7] It was not always clear what the doctrine meant to Descartes' more theologically minded contemporaries. Antonius Walaeus, writing in 1640, for example, noted that "the way in which this happens is incomprehensible . . . to humans."[8] But, as we shall later see, when worried about deriving the laws of motion it was not important for Descartes to specify *how* God sustains the world: what will be important is *that* he does, and that he, the sustainer, is immutable and constant in his operation and in that way preserves motion and certain of its features. As for the incomprehensibility of the means by which God sustains the world, I suspect that Descartes might have given any questioner an answer he often gave when questioned about mind-body interaction, that it is unfair to hold him to any higher standards of intelligibility than those that apply to those from whom the doctrine was borrowed.[9]

A Digression:
Continual Recreation and Temporal Atomism

God sustains the world through his continual recreation, and continually sustains the motion he placed in it in the beginning. Surrounding this indisputably Cartesian doctrine is an important disagreement among his commentators over the nature of duration in the world that God sustains, a controversy that, it is claimed, has serious consequences for our understanding of Descartes' conception of time, motion, and his derivation of the laws of motion.

What might be considered the received view is argued by a variety of commentators, but it receives its most careful development in Gueroult.[10] For Gueroult and the other commentators who hold this view, the independence of all portions of time from one another, the claim that every moment requires God's sustenance, together with the claim that this sustenance constitutes a continual recreation, leads Descartes to the position that God sustains the world through the successive creation of discrete, independent, and atemporal instants, like the motionless frames of a movie. Gueroult writes: "[From] the concrete and real point of view, which is that of the creation of existence . . . we are concerned with a repetition of indivisible discontinuous creative instants."[11] And so, on this view, duration comes out as radically discontinuous, composed of a collection of indivisible temporal atoms (1:196). Motion emerges on this view as discontinuous as well, the successive recreation of a body in different places in the successive discrete instants that constitute duration:

> All movement is accomplished in time, not in an instant, and since the different moments of movement are instantaneous, they are not movements; each of them is a set of geometric relations defining statically at each instant the different situations that bodies occupy with respect to one another. The radical instantaneousness of change at each moment is merely the *creation* at each time of a new *state* of things or of a new universe in which the geometric relations of distance between bodies *are* different. . . . From the concrete point of view of creation . . . all temporal movement arises from the repetition of instants of movements that, since they are instantaneous, cannot be movements but indivisible and discontinuous actions or creations. (1:196, 199)

Gueroult does acknowledge that Descartes sometimes seems to hold duration and motion to be continuous, divisible to infinity, and, thus, not made up of any elementary parts. But, Gueroult argues, this is not the way the Cartesian world *really* is. The continuity we attribute to the enduring world is found only when we view the world from what he calls the "abstract point of view" (1:198):

> Our mind, in conformity with the law that imposes regularity, as if from the outside, on the intrinsically independent acts of creation, translates the continuous repetition of the free and discontinuous creation of independent and self-sufficient states, into the continuity of one and the same movement, which is proposed to our eyes as accomplishing a path, developing and engendering itself in time. (1:197)

Just as when watching a movie, our mind creates continuous action from a sequence of frozen frames, Gueroult claims that the Cartesian continuum of duration and motion is imposed by us onto a world of discrete and timeless instants.

While the received view is widely held, it has not been without its critics. Though not the first to hold this view, J.-M. Beyssade has recently made the question a central issue in his reevaluation of Descartes' first philosophy.[12] In opposition to Gueroult, Beyssade argues for the continuity of duration in Descartes, for the view that every part of duration is, itself, a finite portion of time, and that duration is not a succession of durationless instants. To understand time and duration, Beyssade argues, we must start not with God and his creation, but with ourselves and our experience of them; the soul is better known than body, Beyssade reminds us, and it is through our understanding of the notion of duration as it applies to us as thinking substances that we can understand how it applies to the world of bodies.[13] But, Beyssade notes, thought for Descartes is never instantaneous, and always takes place in time.[14] Insofar as thinking, that which constitutes our essence, must take place in time, our existence cannot be a succession of timeless instants. The situation is similar for bodies, Beyssade thinks. Though when we consider bodies in a purely geometrical fashion we can *imagine* their duration as being made up of timeless instants, in the real world of bodies, we must consider them as enduring. Just as all thought takes time, all operation does as well, and so, Beyssade concludes, what God sustains is not body at an instant, but body extended in time.[15] For Beyssade, Gueroult's abstract point of view is the *real* point of view, and his concrete point of view the mere abstraction.

Beyssade does not deny, of course, that Descartes is concerned with timeless instants in a number of important contexts, and, indeed, that he even talks about God conserving bodies as they exist at a given instant (see, e.g., AT XI 45–46; Pr II 39). But, Beyssade argues, such instants are not *parts* of duration, strictly speaking. A hunk of extended substance can be divided into innumerable parts. But for these divisions to be genuine *parts,* they must be extended as well. Points, lines, surfaces, geometrical objects that lack extension in length, width, and breadth, are not *parts* of a body, but *limits* or *boundaries.* Beyssade argues:

> Every duration or part of duration contains a before and after; . . .
> the instant is its limit or boundary. If we are not mistaken,
> Descartes always takes this word ['instant'] and its Latin original
> *'instans'* in the strict sense of a limit, for a privation or negation of
> duration, which excludes all temporal priority, and which cannot
> be made smaller.[16]

When Descartes means to speak of a very small *part* of time, Beyssade claims, he uses the term 'moment': "An instant is a boundary [of duration], the moment is a part, like a century, a year, a day or an hour, considered provisionally as undivided."[17]

The distinction is an important one. Durations, no matter how small, can be parts of an enduring world, and thus candidates for God's sustaining activity. But while there may be instants *in* duration, as boundaries of finite durations, instants, Beyssade claims, cannot be parts of an enduring world; they cannot compose durations nor can we intelligibly talk about God creating a single instant by itself, without creating the finite duration it serves to bound, any more than we can talk about God creating a two-dimensional surface without the body that it bounds.

The issues separating the two positions are reasonably clear. The received view, as articulated by Gueroult, what I shall call the discontinuity thesis (D-thesis) asserts the following:

D-thesis. The created world God sustains through his continual recreation is a world of durationless instants (temporal atoms), each created by God in succession.

Beyssade's alternative, what I shall call the continuity thesis (C-thesis) asserts the following:

C-thesis. The created world God sustains through his continual recreation is a world in which any portion of time is divisible into portions smaller still, and is not made up of any ultimate elements, with or without duration; durationless instants are not *parts* of the duration God sustains, but only mark the boundaries between one part and another.

I shall argue that there is no strong reason for attributing *either* view to Descartes.

Now, temporal atomism, the view that duration is composed of indivisible temporal elements, is a view by no means unheard of in the history of philosophy; Epicurus and his followers probably held such a view, and in the medieval Islamic tradition, it is a view not unconnected with the way the world depends upon God's continual recreation.[18] But, so far as I can see, there is no particularly good reason for thinking that Descartes held such a view; all of the statements Descartes makes about the way the world depends upon God for its sustenance can quite straightforwardly be interpreted in a way consistent with the C-thesis.

Consider first, the statement of the doctrine of independence of portions of time Descartes gives in the *Meditations:*

All of the time of my life can be divided into innumerable parts, each of which is entirely independent of the others, so that from the fact that I existed a short time ago, it does not follow that I ought to exist now, unless some cause as it were creates me again in this moment, that is, conserves me. (AT VII 48–49)

In this passage Descartes is clear that it is the *parts (partes)* of time that are independent of one another and can exist without any others. But nothing in the passage tells us what counts as a part of time, and whether or not atemporal instants constitute such parts. If, as Beyssade plausibly maintains, atemporal instants are boundaries rather than parts, then the independence of the parts of time does not in any way entail the possibility that *instants* may exist independently of one another. Descartes often formulates the independence claim as he does in the *Meditations,* in terms of *parts* of time (see, e.g., AT VII 109, 110, 369–70; Pr I 21). But even when he uses other formulations, there is no good reason to attribute to him the D-thesis. For example, he sometimes says that "all of the moments of the duration [of the world] are independent of one another" (AT V 53 [K 222]; see also AT VII 370; AT V 193 [K 232]) or that we cannot endure without God beyond "a single moment" (AT VI 36; see also AT VII 111). Elsewhere he talks about the independence of the "present time" from other times, and the need for a cause to sustain me "in the present time" (AT VII 165; AT VII 50). But, as Beyssade suggests, a moment may well simply be a very short period of time; at very least, we can say that nothing in the passages cited forced us to understand the term differently. And so when he says that we require a sustaining cause to exist beyond a moment, he may mean simply that to exist beyond the portion of time in which God does sustain us requires God's recreation. It is obvious that the "present time" can be understood similarly, not as an instant, but as a short temporally extended portion of time.

The doctrine of the independence of parts of time or moments does not force us to attribute a temporal atomism to Descartes. But other aspects of his conception of the way God sustains the world might be thought to be more suggestive of the D-thesis. Descartes, of course, holds the thesis which he regards as traditional, that creation and conservation are, in God, the same, and that there is only a distinction of reason between creation and conservation, that no less of a cause is required to conserve something than was required to create it in the first place (AT VI 45; AT VII 49; AT VII 165, 166). And so, he holds that God conserves the world by a continual reproduction, and that God, as it were, creates me anew in every moment (AT VII 49, 109, 110, 243; Pr I 21). Following Beyssade one might try to dismiss these statements as arguments for the D-thesis by claiming that here a moment is thought of as a small but enduring portion of time. But the use to which Descartes puts the principle in deriving his laws of motion makes reasonably plausible that the moments at which the world is conserved are timeless instants. For example, in the proof of law C from *The World* he says that God conserves a body "as it is in the very instant when he conserves it," where he makes clear that the instant excludes *all* motion, that is, that it is durationless (AT XI 44, 45). In the parallel passage of the *Principles,* the proof of law 2, Descartes says that God conserves a body as it is "in the very moment of time in which he conserves it." But, as in *The World,* he goes on to note that "no motion takes place in an instant," suggesting that the moment of time in question is without duration (Pr II 39). One might infer from this that since God recreates the world at every durationless instant, that duration is just this sequence of durationless instants, created individually by God. But this does not at all follow. To say that God recreates the world at every instant is to say that *every* instant can be regarded as the *beginning,* as the *boundary* of a newly created world. But though every instant can be regarded as a moment of creation, the beginning of a newly created world, it does not *follow* that what is being created is a bare instant or a sequence of bare instants, or that God could create an atemporal instant without creating a finite duration for that instant to bound. The view that God continually recreates the world at every instant is, of course, *consistent* with the D-thesis; but the doctrine of continual recreation is *also* consistent with its contrary.

The idea that duration is made up of discrete timeless units is also suggested by other passages. For example, in a passage of the *Principles* Descartes talks about a ball in a rotating tube "at the first moment" of rotation, in comparison with its state "at the second moment" (Pr III 59). Similarly he talks about the possibility that he might not exist "in the next moment" or "in the next time," and claims that the present time is independent of the "immediately preceding time," and that an individual moment can be separated from "neighboring [moments]" (AT V 193 [K 232]; Pr I 21; AT VII 165; AT VII 370). But these passages are also indecisive on the question at hand. While Descartes can be read as holding a theory of discrete temporal atoms, he can just as easily be interpreted in terms of finite moments of time; finite stretches of time have temporal orders of before and after just as much as timeless instants do. Indeed, one could argue, these passages are more naturally interpreted in terms of the C-thesis than as evidences of the discontinuity of time for Descartes. For he surely realized that if there were temporal atoms, then between any two temporal atoms there must be a third, if time is to be indefinitely divisible, as he sometimes implies (AT III 36). But if time is indefinitely divisible, then there will be no such thing as the next instant or the preceding instant. However, if we construe him as talking about brief but finite moments, then there is no particular problem here, since we can quite easily divide things up in such a way that there is a unique preceding and a unique succeeding moment, for the purposes of making some particular point. Descartes' discussion of the ball in the rotating tube suggests that he has this in mind, at least in that context. For, he says, in the "first moment" of rotation, the ball "will advance only very slowly" toward the end of the tube, while in the second, "it will proceed a bit more quickly" (Pr III 59). But, as he had himself noted earlier in the *Principles* (Pr II 39), no motion can take place in an instant. So, it is plausible to infer that the moments in question in this context are not durationless instants but only very short periods of time.

I have argued that nothing Descartes says suggests that he held the discontinuity thesis. But while the case in favor of the continuity thesis is somewhat better, it too is not altogether convincing.

All of the passages I have discussed can be interpreted in a way consistent with the C-thesis, but I see little that *forces* us to read them in that way. Beyssade's argument that psychological time, in which we find no durationless instants, is basic and that duration in the physical world is understood in those terms is interesting, but hardly decisive. As Descartes often remarks with respect to infinity, there are many things we *know* God can do that go beyond our comprehension, like dividing a finite body into an infinity of extended parts (Pr II 34). To suppose that since all of *our* actions must take place in time, all divine actions must as well, as Beyssade would have Descartes reason, would seem to be an unwarranted limitation of God's omnipotence. As we saw in connection with the vacuum, Descartes is very reluctant to put limits on what God can and cannot do, and I suspect that he would have been very hesitant to adopt the view that God cannot recreate the world in an instant as it is at that instant, independently of all other times in which the world exists. Similarly, though it is certain that Descartes did not regard surfaces and points as *parts* of extended bodies, it is not so clear that he thinks of instants as mere boundaries, incapable of being created independently of a finite duration, as Beyssade would have it. While the position is an interesting and plausible one, and certainly was available to him, I know of no passage where such a view is seriously discussed or even touched upon.

The best reason for attributing the C-thesis to Descartes that I can think of is an argument that Beyssade does not consider. Now one of the arguments that Descartes uses against the view that bodies are made up of points is the argument that since extension is an attribute of body and points are not extended, points cannot be bodies or parts of bodies, and, consequently, bodies cannot be composed of points (see, e.g., AT III 213–14 [K 80]). But, as we noted in chapter 6, *duration* is an attribute of all creatures, both thinking and extended substances. And so, it would seem, God cannot create a finite substance without duration. The state of the created world at an instant, extension or thought without duration would thus, it seems, have to be a fiction, an abstraction from what is real, enduring finite substances, just as bare space without body is an abstraction from extended substances. But, one must admit, however compelling this argument might be, and however much Descartes might be forced to accept it on his own terms, there is no evidence that would suggest that he ever saw it.

In the end, it seems, both the C-thesis and the D-thesis are, for the most part, consistent with what Descartes says about duration and time, and how he uses these notions in his system. This suggests that the issue of the continuity or discontinuity of time was not an issue that worried him greatly. Nor should it have worried him, at least in connection with the foundations of his physics. Later, when we discuss in detail the derivation of his laws of nature, I will show that the issue of the continuity of time is irrelevant to Descartes' arguments. He would, I think, have been somewhat surprised at the issues his commentators chose to emphasize.

GOD AND MOTION

Descartes turns directly to God in order to derive the laws bodies satisfy. In this section we must begin our exploration of just how the derivation is supposed to work.

It is tempting to suppose that Descartes' God is a law-giver, who decides what laws best suit the world and then imposes them on the bodies he created. Such a view is, to be sure, suggested by some of the language Descartes uses in his writings. Writing in *The World*, for example, he does talk of "the laws which God has imposed on [nature]" (AT XI 36). But, I think, we should not read too much into such turns of phrase. Were we to try to figure out what laws a perfect God would impose on nature, as Leibniz, for example, will later try to do, then the project would come to grief, Descartes thinks, since we would never be able to determine which laws God would have chosen; to do so would require us to know God's ends, his reasons for choosing one set of laws over another. But, he thinks, all of God's ends "are equally hidden in the inscrutable abyss of his wisdom," as he told Gassendi in the *Fifth Replies* (AT VII 375). Descartes thus writes in the *Principles:*

> And so . . . we should not take any reasons for natural things from the ends which God or nature proposed for themselves in making them, since we should not glorify ourselves to such an extent that we think that we are privy to their counsel. But considering him [i.e., God] as the efficient cause of all things, we shall see what it appears we must conclude by the light of reason he gave us, from

those of his attributes which he wanted us to have some notion of, about those effects of his which appear to our senses. (Pr I 28)

We must then proceed from the consideration of God as the efficient cause of motion, as well as everything else, and derive the laws bodies in motion obey from his nature. The laws, then, are not formulated by God and then imposed on nature. Rather, they follow directly out of the way the God who exists with the nature he has causes motion in the world he first created and now sustains. But to understand how this grounds the laws, we must understand how it is that God causes motion in the world, and to understand this, we must return to the doctrine of God as sustainer of the universe.

The doctrine of God as the continual and sustaining cause *secundum esse* of a world of enduring things that Descartes holds to was, I think, nothing out of the ordinary in the intellectual context in which he worked. But there was an important way in which Descartes departed from the tradition on this question.

After arguing that the world of creatures depends on God for their sustenance, Thomas turns to a closely related question, "whether God immediately conserves every creature."[19] Now, Thomas had established that "the being of every creature depends on God so that unless it were conserved by the operation of divine power, it could not subsist beyond a moment, but would be reduced to nothingness."[20] But, Thomas claims, God does not always do this *directly;* "he conserves things in their being through certain mediating causes."[21] Thomas' position is that "God immediately created everything. But he instituted in this creation of things an order among things, so that certain things depend on others through which they are secondarily conserved in their being."[22] Thomas appeals to an analogy from Aristotelean cosmology to explain himself. The diurnal motion (the first motion) of the heavens is the cause of the continuing of things generated, while the motion of planets through the Zodiac (the second motion) is the cause of their diversity. God causes the motions, but the motions are themselves causes of various effects. Were God to withdraw his concourse, then all would fall into nothingness; in that sense God is the primary cause of everything. But though God is the primary cause, he works through mediating causes.[23]

Descartes seems to depart from the view Thomas and his many later followers held. For the schoolmen, the world God sustains is a world of matter and substantial forms. These forms are active principles that constitute an important class of the mediating causes of change that the schools recognized. But in a physical world whose only constituents are extended bodies, a world without forms (at least if we set aside human souls), then this class of mediating causes is not available to press into service.[24] What Descartes chooses in their place is God, who will act not only as the general conservator of the world, but as the *direct* cause of motion and change in that world.

But Descartes is not at all clear about how God accomplishes this remarkable feat. From the earliest statements of this view in *The World,* the way God causes motion is intimately intertwined with Descartes' account of God as the moment-by-moment sustainer of the world. For example, in *The World,* he grounds law C in the fact that "God conserves everything through a continual action," and as a consequence, he sustains motion in a body as it exists in the body at the moment it is being sustained, as we shall see in more detail later in this chapter (AT XI 44). In introducing God as the general cause of motion in the *Principles,* he announces that God "created matter in the beginning, at the same time as motion and rest, and now through his ordinary concourse alone, preserves as much motion and rest in it as he created then" (Pr II 36). This connection between the divine conservation of the world and the preservation of motion continues throughout the proofs of the laws in the *Principles.* God as cause of motion is closely linked to God as sustainer of the world in the correspondence as well. Descartes wrote Newcastle in October 1645, summarizing the conservation principle, "now, through the same action by which he conserves this matter, he also conserves in it just as much motion as he had put there" (AT IV 328 [K 183]). And writing in response to Henry More in August 1649, he noted that aside from the force God gave to creatures, a theme to which we shall return later in this chapter, "the force moving [a body] can be that of God, conserving as much transference in matter as he placed in it at the first moment of creation" (AT V 403–4 [K 257]). The close link between God as sustainer of the world and God as cause of motion has suggested that conservation of body and conservation of motion are, in *some* sense, the very same thing. A plausible suggestion is that for Descartes, motion is quite simply the divine recreation of bodies in different places with respect to one another in different moments; God, on this view, moves bodies not by shove or impulse, but by his recreation alone. In this way, God is conceived of as creating motion in the same way that a cartoonist creates the illusion of motion on the movie screen.[25]

This view, what we might call the "cinematic view" of God as creator of motion, is certainly consistent with what Descartes actually says about God in this connection. Furthermore, it is a view that is quite explicitly found among a number of Descartes' followers.[26] But there is no place, at least no place I know of, where Descartes says anything explicitly that would commit himself to this view; while it may be a natural interpretation for us and his followers to make, particularly if one is inclined to see Descartes as a temporal atomist, it is not anything that he straightforwardly *says* he believes. And secondly, there is a different strain in his writing, one that suggests a very different view of how God causes motion in the world. Descartes may well not have been entirely clear on this through much of his career, and it is quite possible that the cinematic view is a correct representation of his thought in *The World* or the *Principles*. But at the end of his life, particularly in the letters to Henry More, he articulates a somewhat different view.

In order to appreciate the view, we must return to the point I made that what God sustains for Descartes is not a world of substantial forms and matter, the substances of the schools, but a world of extended substances; God enters the physical world as a cause of motion precisely because there are no such forms in the world to serve as the source of change. To understand how God causes motion for Descartes we must, I think, understand how the forms God replaces were thought to cause motion; and to understand this, we must understand how Descartes thought the soul, the only substantial form he recognizes, causes motion.

Now, Descartes thought that it was evident that the human soul, the form of its body, can cause that body to move, and despite the complaints of his readers, he felt that this was relatively unproblematic. Writing to Arnauld on 29 July 1648, Descartes noted:

> That the mind, which is incorporeal, can impel a body, is not shown to us by any reasoning or comparison with other things, but is shown daily by the most certain and most evident experience. For this one thing is among the things known per se, which we obscure when we try to explain it through other things. (AT V 222 [K 235])

The mind can cause motion in a body, Descartes holds; it is something we know through experience, directly, something we cannot explain in other terms. Insofar as we comprehend it, it is because "we have within us certain primitive notions . . . on the model of which we form all our other knowledge," Descartes explained to Elisabeth a few years earlier, in 1643, in particular, an idea of the union of mind and body, "on which depends [the idea] of the power *[force]* the soul has to move the body" (AT III 665 [K 138]). And, insofar as we understand the schoolmen's forms through this notion we have for our own mind and its union with the body, as we discussed above in chapter 4, we understand the substantial forms of the schools to cause motion in exactly the same way.

Descartes isn't very informative about just how mind (or form) moves body; on his view there isn't much that can be said, other than to direct our attention to the experience we all have that is supposed to make it all clear. But though there isn't much that we can say, there is no confusing the sense in which mind causes motion in a body with the way God sustains the body that mind supposedly moves. One of the axioms Descartes uses in the geometrical presentation of his arguments appended to the *Second Replies* reads thus: "It is greater to create or conserve a substance, than it is to create or conserve the attributes or properties of a substance" (AT VII 166). This passage is not without its difficulties.[27] But the clear sense is that Descartes wants to distinguish causes that relate to the modes or properties of a thing (*modal causes,* as I shall call them) from causes that create or sustain the very being of a substance (what I shall call *substantial causes*). God, sustaining the world, is clearly a substantial cause. But minds are clearly not; insofar as they cause changes in the motion of bodies, they can at best count as modal causes. And insofar as substantial forms are understood on the model of souls acting on bodies, Descartes would have had little trouble classifying them, with minds, as modal causes; they are causes of motion, a mode in bodies, bodies which are assumed to be sustained by the Divine Sustainer, who is the unique substantial cause.

God enters Descartes' physics to do the business substantial forms did in the Aristotelian system, as he understood it, to cause bodies to behave in their characteristic ways. And, I think, by the late 1640s, if not before, Descartes held the view that when God causes motion, he causes it in just the way forms were taken to do it, that is, on his account, in just the way that we do it, by way of an impulse that moves matter in a way that we can comprehend only through immediate experience. In April 1649 Descartes wrote to Henry More:

> Although I believe that no mode of acting belongs univocally to God and to his creatures, I confess, nevertheless, that I can find no idea in my mind which represents the way in which God or an angel can move matter, which is different from the idea that shows me the way in which I am conscious that I can move my own body through my thought. (AT V 347 [K 252])

And so, Descartes suggests to More, God is conceived to move bodies in just the way we do, using the same primitive notion we use to understand how we move our own bodies.

If this is how we conceive of God as a cause of motion, then, it would seem, we are conceiving of him as a modal cause when it comes to motion. Conceived as such, there would appear to be a distinction between God as sustainer of the world, a substantial cause, keeping things in existence, and God as cause of motion, a modal cause, causing bodies to have the particular motion they have, determining, at least in part, their modes. The difference between these two roles God plays for

Descartes comes out, again, in the letters with More. Writing to Descartes on 5 March 1649, More asked if "matter, whether we imagine it to be eternal or created yesterday, left to itself, and receiving no impulse from anything else, would move or be at rest?" (AT V 316). Descartes at first ignored the question, and so More repeated it again in his letter of 23 July 1649 (AT V 381). Descartes' answer appears in August 1649: "I consider 'matter left to itself and receiving no impulse from anything else' as plainly being at rest. But it is impelled by God, conserving the same amount of motion or transference in it as he put there from the first" (AT V 404 [K 258]).

The picture that comes through seems to be a relatively simple one. Bodies can be conserved with or without the divine impulse. Without the impulse, they are at rest, with it, in motion.[28] This is further suggested in the following paragraph of the same letter, where Descartes notes that "when . . . I said that the same amount of motion always remains in matter, I understood this as concerning the force [vis] impelling its parts, which attaches itself at one time to one part of matter, and at another time to another part" (AT V 405 [K 258]). This suggests that what God directly conserves is *not* motion itself, the effect, but its *cause,* the *impulsion* he introduced into the world at its creation, the cause that *results* in motion. As the earlier passage from the letter of April 1649 suggests, this impulsion is understood as a *real shove,* something that can only be understood on analogy with the way in which *we* cause motion in bodies. God's conservation of body, thus, seems separable from his role as cause of motion, and as cause of motion, he seems to act as *we* would act in the circumstances; God's motion seems to result from a divine act of will, a Divine Shove. This is what we might call the "divine impulse view" of how God causes motion, in contrast to the cinematic view, outlined earlier.

But there is a question with respect to rest that must be addressed on the divine impulse view. Descartes' answer to More suggests a very simple view of the difference between motion and rest, *sub specie causalitatis,* that is, *sub specie Dei:* as long as God is pushing, a body moves, and when God stops pushing, the body stops. But matters are a bit more complicated. For there is a sense in which rest, too, involves activity, divine activity, presumably, and is not a mere consequence of the lack of divine impulse.

In a number of contexts Descartes emphasizes that rest is as real as motion, and requires as much activity as motion does. Writing in *The World,* for example, Descartes notes:

> They [i.e., the scholastics] attribute to the least of their motions a being much more solid and true than they do to rest, which they

call only a privation. But me, I conceive of rest as also a quality, which ought to be attributed to matter while it remains in one place, just as motion ought to be attributed to it when it changes place. (AT XI 40)

And in the *Principles* Descartes emphasizes that "no more action is required for motion than for rest," and argues that it is a prejudice of the senses to think otherwise, that is, to think that "in all motion there is action, while in rest there is a cessation of action" (Pr II 26, 24).

Much of what Descartes means to assert about rest here is not relevant to the issue at hand, the apparent claim that a body at rest is a body without divine impulse. One thing Descartes means to assert is a point that we discussed in some detail in chapter 6, that rest is a mode of body, and just as much a real mode of body as motion is; this is clearly what he has in mind in the passage quoted from *The World*. Even when Descartes discusses the activity connected with rest, the question of the divine shove or lack thereof is not always at issue. For example, in writing to More on 15 April 1649, Descartes considers the case of a boat in which one man stands, pushing off from the shore, while another stands on the shore pushing the boat into the river. In this example Descartes concludes that "the action by which the boat recedes from the ground is not any less in the ground itself than it is in the boat" (AT V 346; see also p. 348). Descartes' point here is that when we consider two bodies in motion with respect to one another, the body that we consider to be at rest (the bank of the river) is just as active as the body said to be in motion (the boat), at least at the moment when the motion is being caused. But here we are dealing with rest only in the vulgar sense. As Descartes emphasized in the *Principles,* there is as much motion in the bank, commonly said to be at rest, as there is in the boat said to be in motion (Pr II 29–30). The boat example he uses with More is just to emphasize that the distinction *commonly* drawn between motion and rest is a distinction of no real use to the serious student of physics.

But there is a sense in which rest, rest *properly* understood, involves activity that does lead us to a view somewhat more complex than the simple picture Descartes suggested to More in August 1649, the view on which rest involves no divine impulse at all. When Descartes says that rest involves as much activity as motion does, part of what he means is that a body in motion will not spontaneously come to rest, and so rest itself requires a cause as much as motion does; indeed, it requires as much activity to stop a body in motion, that is, cause its rest, as it did to put it into motion in the first place. In the *Principles* Descartes writes: "Not only do we need to make an effort [*conatus*] to move external bodies, but often we also need one to stop their motion, when it isn't done by heaviness or by some other cause" (Pr II 26). Furthermore, Descartes holds that it takes a certain effort to put a resting body in motion, that a body will not only remain at rest until an external cause changes its state, but that in that state, the resting body will *resist* change. As we saw in our examination of the laws of impact in chapter 8, Descartes thinks of the force of resisting a resting body exerts as a kind of *reactive* force, in the sense that it arises in the resting body *only* during collision, and has a magnitude that is not a constant, but is proportional to the speed of the colliding body. In his letter to More of August 1649, Descartes emphasizes that this force is distinct from the state of rest. In his letter of 23 July 1649 More had attributed to Descartes the view that "rest is an action, indeed a certain opposition or resistance" (AT V 380). Descartes replied: "Even if a resting thing has that opposition by virtue of the fact that it is at rest, that opposition is not, by virtue of that, rest" (AT V 403). This suggests that on a more complete account of the divine impluse view of God as the direct cause of motion and rest, the body at rest will not *always* be without an impulse from God. Though without the divine impulse the world would be at rest, in a world with some bodies in motion, the others, those at rest, will receive their own impulses. These resistive impulses, though, unlike those that cause motion, are in resting bodies *only* at the moment when they are being collided with by bodies in motion; they are intermittent impulses, not the constant impulse that seems required to keep moving bodies in motion.[29]

DIVINE IMMUTABILITY AND THE LAWS OF MOTION

We have been exploring how God, for Descartes, sustains the world, and in sustaining the world, how he causes bodies to move. Now we can turn to the main question at hand. The conservation principle and the three laws of motion Descartes proposes are supposed to follow out of the fact that God is immutable and operates in a constant way in sustaining his creation and the motion that he put into it. In this section we shall explore just how these arguments are supposed to work. We shall begin by considering the arguments for the conservation principle, law 1 and law 2 of the *Principles;* these three arguments all deal in a fairly direct way with the persistence of some feature of the world. Then we shall turn to the way divine immutability supports the collision law, a somewhat different sort of argument that involves the question of the consistency of divine immutability and fact that what God sustains is a world in flux.

Let us begin by considering the conservation principle. Descartes is clear from the beginning that the principle is supposed to follow in *some* way from the immutability of the God who is responsible for keeping the world in existence. The argument is given very briefly in *The World:*

> Now, these two rules [i.e., laws A and B] follow in an obvious way from this alone, that God is immutable, and acting always in the same way, he always produces the same effect. Thus, assuming that he had placed a certain quantity of motions in the totality of matter from the first instant that he had created it, we must admit that he always conserves in it just as much, or we would not believe that he always acts in the same way. (AT XI 43)

The account of the argument in the *Principles* is somewhat more extended, but it seems to be basically the same argument. Descartes begins by noting that "although that motion is nothing in moving matter but its mode, it has a certain determinate quantity," which we can imagine to be conserved, size times speed, he goes on to suggest through example, as we have seen. The argument then continues:

> We also understand that there is perfection in God not only because he is in himself immutable, but also because he works in the most constant and immutable way. Therefore, with the exception of those changes which evident experience or divine revelation render certain, and which we perceive or believe happen without any change in the creator, we should suppose no other changes in his works, so as not to argue for an inconstancy in him. Whence it follows that is most in agreement with reason for us to think that from this fact alone, that God moved the parts of matter in different ways when he first created them, and now conserves the whole of that matter in the same way and with the same laws *[eademque ratione]* with which he created them earlier, he also always conserves it with the same amount of motion. (Pr II 36)[30]

As Descartes summarizes the view earlier in the same section, in the beginning God "created matter at the same time as he created motion and rest, and now, through his ordinary concourse alone he conserves as much motion and rest in the totality as he then put there."

The argument here is a difficult one. Later we shall deal with the apparent exception Descartes introduces into the account in the *Principles*. But putting that aside for now, let us try to unravel the main claim. Now, because God is immutable, he works in an immutable way, Descartes asserts. One thing that this should mean is that any hunk of matter God creates, he will continue to sustain; once created, a hunk of matter will not simply pass out of existence. So far Descartes is not saying anything surprising to his contemporaries. Thomas, for example, taught quite clearly that while God has the ability to annihilate things, he does not exercise that ability.[31] But Descartes wants to go farther than this; he wants to say that God not only conserves *things* in the world, but also at least some of their properties. Motion is a mode, which admits of a certain definite quantity. Descartes' claim is that God conserves not only the material substances he created in the beginning, but the exact quantity of motion he puts in them at the beginning.

The argument appears to be problematic in a number of respects. Descartes thinks that in sustaining the world, God maintains the same amount of motion. But there are many other modes of body that can be treated quantitatively. In creating the world of extended substance with motion and rest God created a specific number of discrete bodies, one would think, and a specific number of triangles, spheres, cubes, and so on. Now it seems unlikely that Descartes thought that these particular quantities remain constant in the world. However, if the immutability of God entails the conservation of quantity of motion, then, why doesn't it entail the conservation of quantity of bodies or quantity of spheres, or the quantity of *anything* that we can specify? Connected to this is the problem of the measure of quantity of motion. Even if we hold that an immutable God conserves *some* quantity connected to motion, why size times speed? An even more basic problem concerns the very appeal to divine immutability. If God really is immutable, then, it would seem, the world should not change *at all,* and one state should be exactly like the one that preceded it; a genuinely immutable God should create an immutable world, it would seem.[32]

I think that these problems are particularly acute if we understand Descartes to hold the cinematic view of God as a cause of motion. Certainly, on that view, there is no particular reason why quantity of motion should be conserved, as opposed to any other quantitative aspect of body, or why size times speed should be chosen as a measure of question. But particularly puzzling on this view is why an immutable God should create motion at all. If God is immutable and acts immutably, then why should he sustain (create) bodies in different places at different moments of time? And if motion is nothing but God's sustaining bodies in different places at different times, then, it would seem, a world sustained by an immutable God should be a world without motion.

This, I think, is not by itself a reason for denying that Descartes held the cinematic view; Descartes was fully capable of being confused and not following out the consequences of what he believes. But it is interesting to note that the divine impulse view of God as cause of motion is able to make somewhat better sense of the argument for the conservation principle. As we discussed earlier, Descartes tells More that matter left to itself would remain at rest. But, he tells More, "it is impelled by God, conserving the same amount of motion or transference in it as he put there from the first" (AT V 404 [K 258]). And then, in the following paragraph of that same letter Descartes notes that "when . . . I said that the same amount of motion always remains in matter, I understood this as concerning the force *[vis]* impelling its parts, which attaches itself at one time to one part of matter, and at another time to another part" (AT V 405 [K 258]). This suggests that what God directly conserves is not the quantity of motion itself, the effect, but its *cause*, the *impulsion* he introduced into the world at its creation, the cause that results in motion. The conservation law, understood in this way, rests on the fact that God conserves the same amount of impulsion, the amount of *shove* with which he started things in motion. And, in conserving the same amount of cause, he conserves the same amount of the effect, the same quantity of motion in the world. This is how he seems to be thinking of conservation in August 1649. But it is not out of the question that this idea was what he had in mind in earlier formulations as well. In *The World*, for example, he notes that God "acting always in the same way, always produces the same effect" (AT XI 43). Similarly, in the *Principles*, Descartes argues from the fact that God "works in the most constant and immutable way" (Pr II 36). Both of these suggest that the constancy of motion may follow from the constancy of the impulsion that is its cause.

On the divine impulse view, the impulsion that causes motion comes directly from God, and, so, it is his direct shove, so to speak, that keeps things in motion. Understood in this way, the conservation law is a direct consequence of God's commitment to keep pushing, as it were, and to keep pushing as hard as he did when he set the world into motion; it is a direct expression of God's commitment to continue the work of creation by sustaining bodies and by sustaining the effort necessary to keep them in motion. But, it should be pointed out, while Descartes does seem to hold that God is the immediate cause of motion in the world, the argument for the conservation principle need *not* depend upon the doctrine of God's *direct* impulsion. What is important is that God, having created an impulsion that causes motion in bodies, *sustains* that impulsion, whether it be in him or in one of his creatures.

If the argument for the conservation principle is understood in this way, a constancy of effect derived from a sustained cause, then it is comprehensible why a numerical quantity of one mode is sustained, while others are not. God creates motion in the world in a way in which he doesn't create a particular number of things, or a particular number of things of a particular shape. Individual bodies with particular shapes are, as it were, side-products of the motion God creates in the world. As motion gets redistributed in the world in accordance with the collision laws, the number of objects and their shapes will change. But, Descartes claims, insofar as God sustains the activity with which he originally created the world, the quantity of motion will remain unchanged. Furthermore, on the divine impulse view, it is comprehensible how an immutable God can create a world in motion. On the divine impulse view, God's causing motion is distinct from his sustenance of bodies. In sustaining bodies and in sustaining the divine shove that is the cause of their motion, motion, the effect of that divine shove, should follow. But something very arbitrary does remain, even on this view. While Descartes may have good reason to assert, on the basis of his argument, that *some* quantity connected with motion must persist, no justification is offered for the particular quantity Descartes assumes, size times speed. One suspects that the measure was just obvious to him.[33] But in this, clear and distinct perceptions failed Descartes. For, as we noted above in chapter 7, Leibniz was later to show that in a world governed by Descartes' conservation law, perpetual motion is possible.

So far we have been discussing only the way divine immutability supports the conservation principle. But the arguments Descartes advances for the other two principles of persistence, laws 1 and 2 of the *Principles,* work in very much the same way.

Descartes is not very forthcoming about how the persistence of the state of motion or rest in a given body, law A of *The World* and law 1 of the *Principles,* is supposed to be grounded in divine immutability. In *The World,* the argument Descartes appears to advance for law A seems to be an argument not for the persistence of motion or any other quality in a particular body, but for the conservation of quantity of motion in general in the world:

> Thus, assuming that he had placed a certain quantity of motions in the totality of matter from the first instant that he had created it, we must admit that he always conserves in it just as much, or we would not believe that he always acts in the same way. (AT XI 43)

And in the *Principles,* while Descartes precedes the statement of law 1 with the claim that this and all of the other laws follow "from the same immutability of God" from which he derived the conservation principle, no specific argument is offered (Pr II 37).

But despite the lack of specific argument, it is not difficult to see how Descartes could argue from divine immutability for this law. If God sustains his creation at all times, then he will sustain not only motion in general, as the conservation principle holds, but motion as it is found in specific bodies, insofar as he can. That is, he will sustain it insofar as sustaining motion in one body is consistent with sustaining it in another; when it isn't, the impact law will determine what is to happen. Leaving the impact law aside for the moment, this argument is relatively unproblematic for Descartes, and seems to follow out of divine immutability, whether we take the cinematic view or the divine impulse view of how God causes motion. If, on the cinematic view, it is coherent to talk about God conserving motion at all, then it is coherent to talk about God conserving motion in a given body, just as he would conserve the other states of a body. And if God conserves motion by maintaining his divine impulse, then his immutability would seem to entail that he should conserve it in a given body, other things being equal.[34]

While Descartes is relatively silent on how God grounds law 1, his account of law 2 is quite explicit; indeed, it gives us perhaps the best view of how he thinks divine immutability grounds his laws. In *The World* Descartes argues as follows:

> This rule [law C] is based on the same foundation as the two others, and depends only on the fact that God conserves each thing by a continual action and consequently, he does not conserve it as it was some time ago, but precisely as it is at the same instant in which he is conserving it. Now, of all motions, only the rectilinear is entirely simple, and has a nature that can completely be grasped in an instant. For to conceive it, it is sufficient to think that a body is in the process of moving in a certain direction, which is found in each of the instants that can be determined during the time in which it is moving. On the other hand, to conceive circular motion or any other there might be, we must at least consider two of these instants, or rather, two of its parts, and the relations between them. . . . I am not saying here that rectilinear motion can take place in an instant, but only that everything required to produce it is found in bodies in every instant that can be determined while they move, but not everything required to produce circular motion. (AT XI 44–45)

Immediately after giving this argument, Descartes illustrates it with the example of a stone rotating in a sling. While it is in the sling, it goes in a circular path, of course. But, he continues, "supposing that the stone then begins to leave the sling, and God continues to conserve it such as it is at that moment, it is certain that he will not conserve it with the inclination to go circularly . . . but with the inclination to go rectilinearly" (AT XI 46). The argument in the *Principles* is very similar, though somewhat more succinct:

> The reason *[causa]* for this rule [law 2] is the same as that of the

previous one, namely, the immutability and simplicity of the operation through which God conserves motion in matter. For he
conserves it precisely as it is in the very moment of time in which
he conserves it, without taking into account the way it might have
been a bit earlier. And although no motion takes place in an
instant, it is obvious that in the individual instants that can be
designated while it is moving, everything that moves is determined to continue its motion in some direction, following a
straight line, and never following a curved line. (Pr II 39)

Central to the argument is the consideration of the moving body as
it is in any instant of its motion. By instant here Descartes seems to
mean an atemporal stage of the body in motion, a stage in which there
is no duration. Considered in an instant, the body is not actually in
motion, he grants; motion can only take place in time. But, the claim
is, in such an instant the body has "everything required to produce
motion." If we understand Descartes here on the cinematic view of how
God causes motion, then I think that he is, once again, stuck with God
continually recreating (sustaining) a motionless universe; on the
cinematic view I cannot understand the momentary motion to be anything but God's determination to create the body at a different place
in a succeeding time. If this is what the momentary state of a body in
motion is, than in conserving it, God conserves the momentary intention, as it were, an intention that so far as I can see can never be
realized. But I think matters are somewhat more intelligible on the
divine impulse view. On that view it is plausible to suppose that what
Descartes has in mind by the momentary state of the body in motion is
the impulse, the shove that will result in an actual motion in the succeeding time; considering the moving body as it is in an instant allows
us to set aside motion, the effect, and examine its cause. And when we
examine its cause, we see that at any instant, a body is determined to
move in only one direction. That is, at any instant, the shove that produces the motion in time can only be a shove in some one determinate
direction. A succession of shoves can move the body in a curvilinear
path, but any individual shove at a particular moment can push it in
only one direction. It is in this sense that only rectilinear motion can be
comprised in an instant, and in an instant only that required to produce rectilinear motion can be found.
    At this point in the argument the immutable God enters. God con-
tinually conserves the world, and every instant can be regarded as the
beginning of a new creation. And so, Descartes claims, God in his immutability conserves the world as it is at the moment he conserves it;
insofar as each moment can be regarded as the beginning of a new

creation, God does not look to the world as it was before, but only as it is at the moment in which he is conserving it. Now, if at a given moment a body is being shoved in a particular direction, God the conserver will conserve that shove, at least as long as it is not opposed by some contrary cause. But as long as the cause, the shove, is conserved in the timeless instant, then the effect, rectilinear motion, will result in time. And so, a body in motion will move in a rectilinear path unless otherwise interfered with. That is to say, a body in motion has a tendency to move in a rectilinear path. Again, I am not sure that this establishes that Descartes' view on God as cause of motion was, indeed, what I have called the divine impulse view, but understanding Descartes in this way does allow us to make reasonably good sense of the argument he may have had in mind for this law.

In the context of this argument it is crucial to distinguish the tendency to move that Descartes argues for as the main conclusion of law 2 (law C) from that which is found in the moving body at any instant, a momentary tendency that appears in the premises of the argument. That which is found in the body at an instant is something that *determines* the body to move, that which is *required* for the body to move, in short, a *cause* of motion, an impulse, a shove, a force, on at least one understanding of Descartes' argument. The tendency to motion that Descartes *establishes* as a conclusion of the argument is not a cause, but is the *effect* of the continuation of that cause; it is simply the property the body has by virtue of which it would move in a certain direction if there were no impediments.[35] Were the instantaneous impulse identical with the tendency Descartes is attempting to establish, the argument would be trivial; it is because they are not identical that Descartes must appeal to God the immutable conservator to connect the two.

It is interesting to note here how weak the appeal to God is. There is *nothing* in the argument that requires God to be the *source* of the impulse that keeps bodies in motion; as long as the immutable God *sustains* the impulse, the cause of motion, then it matters not at all where it comes from. Though, as a matter of fact, I think, Descartes does believe that the impulse that moves bodies comes directly from God, as I argued earlier in this chapter, the argument would proceed in *exactly* the same way if there were genuinely immanent causes of motion in bodies. And secondly, the argument is completely neutral on the question of whether the world God creates is a succession of independent, atemporal instants, or whether instants exist only as boundaries of temporally extended durations. In regarding a moving body at an instant, Descartes is not committed to the view that God could have created that instant without having created anything else, any more than looking at a two-dimensional surface of a cubical body entails that God could have created that surface by itself. And though Descartes certainly holds that God sustains the cause that results in motion, it does not really matter *how* he does it, whether by creating a succession of independent instants, or by creating a temporally extended moment which begins with the instant in question.

Now, finally, we come to the law of impact. In arguing for his conservation principle and the other two laws he offers, Descartes stresses the immutability of God. This seems plausible enough when dealing with the persistence of motion and directionality, which is what those laws deal with. But with the impact law, we are dealing with a different sort of problem, not the persistence of any feature of bodies, but change, the change of direction and speed that a body can undergo in colliding with another. How can one call on the immutability of God to explain the change of motion, the transfer of motion from one body into another? This is the problem Descartes faces in justifying his law of impact.

One possible solution is to say that while God preserves motion, it is matter that causes its change; God maintains bodies in rectilinear motion at a constant speed, but that it is other bodies that actually cause the speed or direction of the motion to be changed. Such an account is suggested in a celebrated and often quoted passage in *The World*. Law C of *The World* (corresponding to law 2 of the *Principles*) asserts that motion in itself is rectilinear, and that all curvilinear motion must thus result from the constraint of other bodies. Following the statement and defense of this law, Descartes notes:

> Thus, in accordance with this rule, one must say that God alone is the author of all of the motions in the world insofar as they exist and insofar as they are straight, but that it is the different dispositions of matter which render them irregular and curved. In the same way the theologians teach us that God is also the author of all of our actions insofar as they exist and insofar as they have some goodness, but that it is the different dispositions of our wills which can render them evil. (AT XI 46–47)[36]

This account, as attractive as it might be as a way of saving divine immutability in the face of a changing world, will not really do. Descartes surely recognized the theological problems raised by this account; it is one thing to introduce the free will humans have as a causal agent that can act independently of God, in a sense, but it is quite another thing to introduce matter in that role. He would surely have seen the shadow of Manichaeism in such a view, and it is difficult to imagine that the same philosopher who shortly before rejected the doctrine of uncreated eternal truths as an unwarranted restriction on God's freedom would have seriously advanced a view of this sort. It seems quite clear to me that if ever Descartes meant this passage seriously, it surely did not represent his final thoughts on the matter.

While there is the suggestion that change of motion is due to matter rather than to the immutability God, even in *The World* there is another account in the works, an attempt to understand how even change can derive from divine immutability. After presenting his impact law in *The World*, Descartes offers a justification for it and law A, the law that precedes it, a justification that, he claims, is grounded in the fact that "God is immutable, and acting always in the same way, he always produces the same effect" (AT XI 43). He first notes that from divine immutability it follows that the same "quantité de mouvemens" is always preserved in the world. In dealing with impact in law B of *The World*, what he emphasizes is the fact that when one body transfers motion to another through impact, it must lose as much motion as it gives to the other. Though he does not seem to have a definite measure of motion in mind here, he does seem to think, incorrectly, that this follows directly from the conservation of motion in general. But after introducing this argument, he goes on to offer an argument to show not only that the transference of motion from one body to another in collision is not inconsistent with divine immutability, but that from divine immutability one can show that bodies *must* transfer motion from one to another in impact:

> And supposing together with this [i.e., that God placed a certain quantity of motion(s) in matter at creation] that from the first instant, the different parts of matter, in which these motions are found unequally dispersed, began to retain them or transfer them from one to another, according as they had the force to do so, we must necessarily think that he always makes them continue this same thing. (AT XI 43)

Descartes' idea here seems to be that already in the first instant of creation, the bodies God created were exchanging motion "according as they had the force to do so," and an immutable God would continue to sustain them in this way. It is interesting to note here what this justifies and what it doesn't. The appeal to conservation is supposed to account for why one body moving another in impact must lose some of its own motion, and the argument that follows is supposed to account for why there is any transfer of motion at all by an immutable God. But it is interesting that there is no justification at all for the impact contest model that stands hidden behind the discussion. The idea that one body moves another only "according as [it has] the force to do so" is something that Descartes seems to take for granted here.

The justification of the law of impact comes up again in the *Principles,* of course. As I argued in chapter 8, law 3 of the *Principles* shows considerable advance over the corresponding law B of *The World.* In the *Principles,* Descartes makes no reference to the view suggested in *The World,* that matter itself is in some sense directly responsible for changes in motion in the world; this view seems altogether abandoned. But the arguments he offers in the *Principles* clearly have their roots in the earlier reflections of *The World.*

In the *Principles,* Descartes divides the proof of the law into two parts, corresponding to the two parts of the law as stated there. The first part of law 3 concerns the case in which force for proceeding is less than force of resisting, and so the advancing body rebounds without losing any of its speed; the second part concerns the case in which force for proceeding is greater than force of resisting, in which case the body continues to advance, and transfers some of its motion. The proof of the first part implicitly appeals to law 1, the law that a simple thing will remain in the same state unless an "external cause" changes it (Pr II 37L). Now, Descartes argues:

> in a collision with a hard body, there appears to be a cause that prevents the motion of the other body with which it collides from remaining determined in the same direction, but there is no cause for the motion itself to be taken away or diminish, since a motion is not contrary to a motion, and thus it follows that it should not diminish. (Pr II 41L)[37]

This may explain how a change in determination can be rendered consistent with law 1, but it doesn't really justify law 3, so far as I can see. This simply constitutes a restatement of the impact contest model; Descartes is simply assuming here that bodies do have forces, force for proceeding and force of resisting, and that the one force being less than the other constitutes a reason (cause) for the change in direction in the one of them.

The proof of the second part of law 3 is, in essence, the very same argument that Descartes offered in *The World.* He writes:

> The other part [of law 3] is demonstrated from the immutability of the operations of God, now continually conserving the world by the same action by which he once created it. For, since everything is full of bodies, and nevertheless, the motion of each body tends in a straight line, it is obvious that from the beginning, God, in creating the world, not only moved its parts in different ways, but at the same time also brought things about in such a way that some would impel others, and transfer their motion to them.

Consequently, now conserving it with the same action and by the same laws with which he created it, he conserves motion not always fixed in the same parts of matter, but passing from some into others as they collide with one another. And this very change among his creatures is an argument for the immutability of God. (Pr II 42)

This version of the argument offers little more than it did in *The World*. The only addition seems to be the observation that since there are no empty spots in Descartes' world, if God creates motion then when he creates it he must, Descartes claims, also create it in such a way that already at that very moment it is being transferred from one body to the next. In this way, he claims, change is not only consistent with divine immutability, but it is, indeed, an argument for it.

The argument Descartes offers in *The World* and in the *Principles* is a very strange one in a number of respects. First of all, it takes for granted, without argument, that the world God created was a world in which there is and must be change from the very first instant. Once he can establish that there is such change, he can, perhaps, show that the change that God introduced in the beginning will continue as God sustains his creation. Perhaps, he might argue, we know empirically that there must have been such change at the beginning, for if there weren't, then there could be no change now, for God sustains the world as he finds it at every given instant. Or, perhaps, the argument is that there must have been change at the first instant if there was motion then, for without an empty space to move in, the rectilinear tendencies bodies have by law 2 (law C) would immediately begin to interfere with one another. A second strange feature of the argument is the very idea of transfer at an initial moment. Descartes certainly recognizes that there is no motion at an instant; this is an explicit remark that he makes in the context of his proof of law 2 (law C). But if there is no motion in an instant, how can there be a *transfer* of motion in an instant? And finally, even should the argument work, I think that it is fair to say that Descartes simply fails to reconcile the transfer of motion from one body to another with the immutability of God. In the end it remains a puzzle why an immutable God should create a world in which, from the beginning, change is such an integral part. Once he does it, he must of course sustain it. But, one might ask, why did he do it in the first place?

On top of the problems with this argument, it must be emphasized that Descartes seeks a justification for the impact law in only the most limited of senses. He takes it to be necessary to justify the claim that motion is ever exchanged, and, in the *Principles*, to justify the claim that at least sometimes, the direction of a motion can be changed without an exchange of motion. But the law that actually governs the exchange of motion, the law that determines when motion is exchanged and when it isn't, that determines what the actual postcollision states of body are to be, this substantive law is offered no justification at all. The actual law, and the impact contest model on which it rests, seem to be taken pretty much for granted. It is true that Descartes probably borrowed this basic model of impact from others,[38] but unlike other things he borrowed, the principle of the persistence of motion, for example, he sees no particular need to offer any justification.

So far we have been talking about the law of impact as it is explicitly stated in law B of *The World* and law 3 of the *Principles*. But, as I suggested in chapter 8, in working out the details of the particular rules of impact in the Clerselier letter of 1645, Descartes does suggest a different sort of principle, what I called the principle of least modal change (PLMC). Descartes, of course, never presents this as a law of impact, as an explicit alternative to the impact contest law that is presented in the *Principles,* and never attempts to justify its application. But it is interesting to note that such a law would have fit nicely into his conceptual framework. It is often emphasized in the literature that the PLMC is a kind of teleological principle, that to justify using it in physics would force us to speculate about God's ends in creating the world, and attribute to God a desire to minimize certain physical magnitudes. That it would. But it should also be pointed out that such a minimal change principle might fit nicely with Descartes' view of God as immutable; surely an immutable God would cause the least possible change, were he to admit change at all. A principle like the PLMC will not help us to explain why Descartes' God admits change into his world at the first or any other moment. But as a law governing how that change takes place, it would seem more promising than the impact contest model, which, it would seem, must be at best a brute intuition.

BEFORE ENDING THIS discussion of God and the laws of motion I would like to make one last point. Descartes is sometimes credited with having made physics mathematical. It is true that the objects of physics for Descartes are simply the objects of geometry made real. Furthermore, Descartes himself sometimes talks as if his physics were just a branch of mathematics. For example, at the end of part II of the *Principles,* concluding that section and introducing the discussion of cosmology and terrestrial physics, Descartes writes that "I admit no other principles in physics but those in geometry or abstract mathematics" (Pr II 64; see also AT I 410–11 [K 41]; AT I 420–21 [K 38]). But this cannot be right. While the objects of physics are the objects of geometry made real, in making them real they take on properties that they did not have in Euclid. Because they are created and sustained by God, and, perhaps, subject to his continual push, they satisfy laws of motion that are entirely foreign to the objects of pure mathematics. In trying to link Descartes' physics closely to mathematics, one forgets the crucial connection between Descartes' physics and his metaphysics; it is a crucial feature of his physics that it is grounded in God, and without that grounding there could be no Cartesian physics.

THE ONTOLOGICAL STATUS OF FORCE

In chapter 3 we discussed Descartes' conception of the nature of body. Bodies, for Descartes, are of course the objects of geometry made real, extended things all of whose properties are in a broad sense geometric, size, shape, and motion. But in discussing the laws of motion that Descartes proposes in *The World* and the *Principles,* there are a number of notions that Descartes makes reference to that hardly look as if they fit into this rigid and strict conception of the physical world. Most obvious here is the notion of force. In framing his law 3 in the *Principles,* both in the Latin and in the French versions of the text, Descartes talks quite explicitly about the force for proceeding that one body has, which must be balanced against the force of resisting that another body exerts against a body that collides with it. What place could such forces have in Descartes' spare and geometrical ontology?

Descartes does make some attempt to address the question at hand. In Pr II 43, after his statement and defense of the law of impact, and before presenting his seven rules, he writes:

> *What the force each body has to act or resist consists in.* Here we must carefully note that the force each body has to act on another or to resist the action of another consists in this one thing, that each and every thing tends, insofar as it can *[quantum in se est]* to remain in the same state in which it is, in accordance with the law posited earlier [Pr II 37]. Hence that which is joined to something else has some force to impede its being separated; that which is apart has some force for remaining separated; that which is at rest has some force for remaining at rest, and as a consequence has some force for resisting all those things which can change that; that which moves has some force for persevering in its motion, that is, in a motion with the same speed and toward the same direction. (Pr II 43)

The forces that enter into the calculation on the impact contest model, what we might call impact-contest forces, derive, then from law 1; because bodies remain in their states of rest or motion in a particular direction with a particular speed, they exert forces that keep them in their states, and resist change.[39]

But despite the fact that the impact-contest forces derive from the laws of persistence, and are simply an expression of the tendency every body has to remain in its own state, it still is unclear what status these law 1 forces might have in the Cartesian world of God, souls, and extended bodies. As with the impact-contest forces, Descartes often talks as if the forces arising out of law 1 are to be attributed to bodies themselves; this is certainly implied by the language he uses in the passage from Pr II 43 quoted above. Furthermore, writing to Mersenne on 28 October 1640, Descartes offers the following interesting paraphrase of his first law:

> It is very wrong to admit as a principle that no body moves itself. For it is certain that from the sole fact that a body has started to move, it has in itself the force to continue to move; in the same way, from the sole fact that it is stopped in some place, it has the force to continue to stay there. (AT III 213)

This passage would certainly seem to suggest that the law 1 forces bodies have to remain in their states, the forces that ground the initially problematic impact-contest forces of law 3, are found in the bodies themselves. But, of course, this is no less problematic than the original claim, that the impact-contest forces are found in Descartes' geometrical bodies.[40]

Descartes' commentators have offered a number of solutions to the problem of the ontological status of force. On the one hand, there are those, like Gary Hatfield, who would argue that force, for Descartes, is literally in God and not in bodies at all.[41] This would certainly solve the apparent ontological problem neatly, insofar as we would not have to deal with the problem of how forces or counterfactual properties like tendencies can be in things merely extended, as Descartes' bodies are supposed to be. But it sits very poorly with the explicit attributions Descartes makes of force to bodies, the impact-contest forces he makes reference to in stating and elucidating law 3, and the force to persist in motion that Descartes attributes explicitly to bodies themselves in writing to Mersenne; it seems absurd to say that it is God himself who literally *has* the force for proceeding or force of resisting that appear as parameters in a particular case of collision.

Others, though, have attempted to understand senses in which the extended bodies of Descartes' physical world can properly be said to have forces. Martial Gueroult, for example, grants that all force is ultimately grounded in God for Descartes. But yet, he claims, the characteristic of these forces [i.e., force of rest and force of motion], in contradistinction to the Divine will that they manifest, is that they are immanent in 'nature' or extension and . . . can be calculated at each instant for each body, according to the formula mv.[42]

Gueroult takes seriously the identification of the forces bodies exert with their tendency to remain in the same state, a tendency that is grounded in the way in which God preserves his creation, and causes it to endure. He writes:

> The principle of continuous creation implies that no created thing can exist unless it is sustained by a creative force, and that every force that inheres in a thing is nothing other than that by which God puts it in existence at each instant. Consequently, as distinct from rest and motion (which are modes of extension), force—whether it be the force of rest or the force of motion—cannot be a mode of extension any more than can duration or existence. In reality, force, duration, and existence are one and the same thing *(conatus)* under three different aspects, and the three notions are identified in the instantaneous action in virtue of which corporeal substance *exists* and endures, that is, possesses the force which puts it into existence and duration.[43]

Gueroult's position is quite ingenious. As I noted in chapter 3 above, the notions of existence and duration pertain to every substance as such, both thinking substance and extended substance; though not modes of extension, they pertain to extended substance insofar as it is substance. Now, Gueroult argues, insofar as the forces that bodies exert are identified with their very existence and duration as substances in the world, they depend on God, yet are in bodies; though such forces are not modes of extension, they are genuine features of extended *substances,* in just the way that existence and duration are in bodies.

Alan Gabbey argues for a similar position, which he represents as an extension of Gueroult's.[44] Gabbey appeals to the scholastic distinction between *causae secundum esse* and *causae secundum fieri,* the cause of being and the cause of becoming discussed earlier in this chapter to explain the ontology of force in Descartes' physics. Gabbey writes:

> Since the force of motion and the force to remain at rest are modally distinct expressions of God's action maintaining the body's existence in its modal disposition, both forces *per se,* that is as causes *simpliciter* of modal dispositions are *causae secundum esse,* and as such constitute the divine *causa universalis & primaria* described in *Principia Philosophiae,* II, art 36. When, however, the

forces of motion and of rest are viewed not *per se,* but as quantifi-able causes of change in the corporeal world, or as reasons . . . explaining absence of change of a certain kind in particular in-stances, they are *causae secundum fieri,* and are the causal agents at work in the . . . three Laws of Nature. . . . Force as *casua secundum esse* is also an attribute of body, in the sense that *qua* cause it is necessarily entailed by a body's duration, viewed *simpliciter* and irrespective of mode. On the other hand, forces as *causae secun-dum fieri* are clearly in body *diverso modo,* so they are modes of body, rather than attributes. . . . In this subtle and complex ontol-ogy of force Descartes does not exclude force from body or its actions. . . . Strictly speaking God is the ultimate real cause and the only true substance, but speaking at the 'practical' level of physical investigation, forces—whether of motion or of rest—are real causes in their own right and distinct from motion and rest.[45]

On Gabbey's rather complex view, then, force belongs to both God and body, in different senses. Insofar as forces are to be identified with the force that sustains bodies in their existence, the forces of motion and rest are *causae secundum esse* and belong both to God, the cause of the continued existence of bodies, and to bodies as attributes that have the same status as existence and duration do for Descartes. But considered as *causae secundum fieri,* the forces that enter into the laws of motion and determine the outcome of collisions, say, forces which can take on greater or lesser values in a given body at a given time, are also modes of body, Gabbey claims.

The Gueroult/Gabbey view is certainly attractive in many ways. It clearly links the forces that bodies exert to their continued existence in a particular state, as Descartes does in Pr II 43; it allows us to under-stand how, though grounded in God, the force can really be in body, in just the way that existence and duration are really in the bodies that exist and endure. But puzzles remain. Though the forces of motion and rest that enter into the impact law as force for proceeding and force of resisting can, in a sense, be accommodated in bodies as con-nected with their existence and duration, they cannot simply be identi-fied with existence and duration, as Gueroult seems to want to do; unlike existence and duration, the forces of motion and rest are vari-able, existing sometimes with one value, and sometimes with another in a given body. Gabbey recognizes this, and contrasts the notion of force which has the status of an unvariable attribute of extended sub-stance, and the variable notion of force which can take on different values and enter into physical calculations, and which is a mode in body. But once it is recognized that there is a sense of force that is not identified with the attributes of existence and duration in body, then the ontological problem resurfaces; for once it is granted that force is in body as a mode, what becomes of Descartes' commitment to the position that everything in body must be conceived as a mode of extended substance? Furthermore, the view Gueroult and Gabbey present and defend is predicated, it would seem, on the identification of God as sustainer of body with God as cause of motion. But, as I pointed out, there is at least one other strand in Descartes' thinking, what I called the divine impulse view, on which God causes motion by sustaining a divine shove, an action quite distinct from the action by which he sustains bodies in their existence. If that is the way God causes motion in the world, the way he sustains bodies in their motion and in their rest, then the force of motion and force of rest cannot be identified with the existence and duration of bodies in any direct way, as Gabbey and Gueroult do.

There may not be an altogether satisfactory view of the ontology of force in Descartes, one that is coherent and sensible, and is consistent with what he says about force in all of his writings and what he commits himself to in other contexts. But there is, I think, a way of conceiving force in Descartes' physics that while it is nowhere explicitly expressed, may capture the spirit of the way in which Descartes conceives of it; neither in God (as Hatfield would have it), nor in bodies (as Gueroult would have it), nor in both at once (as Gabbey would have it), Cartesian force is nowhere at all, I shall argue.

Let me begin by recalling the discussion of the notion of inertia in chapter 8 above. As I pointed out there, Descartes begins his discussion of inertia by denying any such thing, in the sense of a force or tendency in bodies to come to rest, such as Kepler assumes. But he is led to admit such a force in a limited sense. For, by his laws of motion, when a body B collides with another C and transfers motion, B must lose as much motion as it imparts to C. And Descartes holds, the larger C is, the more of its motion B must impart to it. In this sense Descartes is willing to grant that C has a kind of natural inertia, a kind of innate resistance to motion. But, as I suggested above, this inertia is a kind of imaginary force for Descartes. Bodies behaving in accordance with the laws of motion God's continual activity imposes on them will behave *as if* they resisted the acquisition of new motion in proportion to their sizes. But in reality, all we really have is two bodies operating in accordance with the laws God imposes; strictly speaking, the force of inertia is nowhere in Descartes' world.

My suggestion is that we read the forces that Descartes posits in the laws of motion in a similar way. As with the case of inertia, the underlying story is reasonably clear. There is God, the cause of motion in the world, either by divine shove or by continually recreating bodies at different places at different times, and there are bodies which have the modes of either motion or rest. By law 1, God will act on the world in such a way as to keep moving bodies moving, and resting bodies at rest. This can be *described* by saying that bodies, *as it were,* have a force to continue their motion, or exert a force to maintain their rest. But this is not to attribute anything real to bodies over and above the fact that God maintains their motion and as a consequence they obey a law of the persistence of motion. One can give a similar treatment to the force for proceeding and force of resisting that appear in the impact law. Again, the underlying causal story is reasonably clear. When faced with two bodies with conflicting states of motion or rest, both of which tend to persist, law 3 tells us how God acts; it tells us what God will preserve, and what he will change in order to render the conflicting states compatible. This can be *described* by attributing impact-contest forces to the two bodies in question, and specifying which of these forces is to prevail over the other. But again, the forces that enter into the discussion can be regarded simply as ways of talking about how God acts, resulting in the lawlike behavior of bodies; force for proceeding and force of resisting are ways of talking about how, on the impact-contest model, God balances the persistence of the state of one body with that of another. With the forces that arise in law 3, as with the so-called law of inertia, there is no need to attribute some new kind of property to bodies; all there really are in this case as in the other are bodies with different degrees of motion and rest, and God, who sustains that motion and rest, and reconciles incompatible modes in different bodies. And so, on this view, force is *nowhere,* strictly speaking, not in God, who is the real cause of all motion in the inanimate world, and not in bodies, which are the recipients of the motion that God causes. This, I think, is an attractive way of dealing with the ontological problems that the discussions of force in the *Principles* raise. It is precisely because the underlying causal story was so clear to Descartes, I think, that he never saw fit to worry much about giving a coherent and consistent account of the ontological status of the notions of force that he made reference to in his discussions of motion; in an important sense, they really weren't really there.[46]

Despite the attractiveness of this solution to the problem of force in a world of extended substance, I should emphasize that this view is not found in Descartes in any explicit way; what I am offering here is a reading, an interpretation of the collision law as given in the *Principles,* a way of understanding how Descartes' explicit talk of force in the *Principles* can be reconciled with his radical mechanist ontology. But while Descartes is not explicit about taking such a line in the *Principles,* it is interesting to note that the view of impact Descartes takes in the

Clerselier letter of 17 February 1645, discussed in chapter 8 above, is rather more suggestive of the sort of line that I am attributing to him here. In that letter, Descartes introduces what I called a principle of least modal change (PLMC) to determine the outcome of an impact. On that principle, that outcome results which least changes the modes of determination and speed in the two bodies involved in the collision. And so, as I pointed out, on that view the outcome of a collision is determined by the speed and determination directly, without any appeal to the force for proceeding and force of resisting that are at the heart of the impact contest model. In its apparent teleology, this does appear to step away from orthodox Cartesian doctrine. But in another sense it is a more truly Cartesian conception of impact than that which Descartes presents in the *Principles,* a theory of impact that doesn't make even apparent appeal to the notion of force that is so problematic for Descartes' account of body.

Descartes and Occasionalism

In the earlier sections of this chapter I have emphasized the role that God plays in Descartes' natural philosophy; in his continual sustenance of the world, God preserves motion and rest, and in doing so, causes bodies to behave in accordance with the laws of motion. However, God sustains not only bodies but also the minds united to some of those bodies, and so, it would seem, he must be considered the ultimate ground of everything that happens in the world. Should we then regard Descartes as an occasionalist?

Briefly, occasionalism, as understood by many of Descartes' followers, was the view that God is the only real cause in the world, at least in the world of bodies. On this view, the changes that one body appears to cause in another upon impact, the changes that a body can cause in a mind in producing a sensation or a mind can cause in a body in producing a voluntary action, are all due directly to God, moving bodies or producing sensations in minds on the occasion of other appropriate events. And so, on this view, the tickling of the retina and subsequent changes in the brain are only the "occasional causes" of the sensory idea I have of a friend in the distance; the real cause is God who directly moves my sense organs when the light approaches them, moves the parts of the brain when the sensory organs are moved, and then produces the sensory idea of I have of another person's face in my mind when my sense organs and brain are in an appropriate state. Similarly, it is God who is the actual cause of my arm's movement when I decide to raise it to wave; my volition is only an occasional cause. One follower, Malebranche, went so far as to say that in an important sense God is even the real cause of the volitions that I have. Occasionalism was widely held among many of Descartes' followers; it can be found in various forms in Clauberg, Clerselier, Cordemoy, de la Forge, Geulinx, and most notably, in Malebranche.[47] And throughout its seventeenth-century career it is closely associated with Descartes' followers.[48] But to what extent is it really Descartes' own view?

It seems to me quite clear and evident that Descartes agrees at least in part with his followers; I think that it is evident that the laws of motion that Descartes develops in both *The World* and the *Principles* are grounded in the view that God is the direct cause of motion in the world, at least in the inanimate part of the world.[49] It is because God creates and sustains motion directly in bodies that they obey the laws that they do.

Though God is a direct cause of motion in the world, "the universal and primary cause" Descartes identifies in his physics (Pr II 36), is he the *only* cause of change in Descartes' world? There are certain passages in Descartes' writings and features of his system that might well have led his followers to think so. When introducing his laws of motion in the *Principles,* he does identify God as "the general cause of all the motions there are in the world" (Pr II 36). Descartes seems to extend this theme even farther in an important letter he wrote to Elisabeth on 6 October 1645:

> All of the reasons which prove the existence of God and that he is the first and immutable cause of all of the effects which do not depend on the free will of men, prove in the same way, it seems to me, that he is also the cause of all of them that depend on it [i.e., free will]. For one can only prove that he exists by considering him as a supremely perfect being, and he would not be supremely perfect if something could happen in the world that did not derive entirely from him. . . . God is the universal cause of everything in such a way that he is in the same way the total cause of everything, and thus nothing can happen without his will. (AT IV 313–14 [K 180])[50]

In addition to these textual arguments, Descartes' views on God's continual sustenance of the world would seem to commit him to the view that God can be the only cause of change in the world. The argument is formulated neatly by Louis de la Forge:

> I hold that there is no creature, spiritual or corporeal, that can change [the position of a body] or that of any of its parts in the second instant of its creation if the creator does not do it himself, since it is he who had produced this part of matter in place A. For

example, not only is it necessary that he continue to produce it if he wants it to continue to exist, but also, since he cannot create it everywhere, nor can he create it outside of every place, he must himself put it in place B, if he wants it there, for if he were to have put it somewhere else, there is no force capable of removing it from there.[51]

The argument goes from the doctrine of continual recreation, authentically Cartesian, to the conclusion that God can be the only cause of motion in the world. When God sustains a body, he must sustain it *somewhere*, and in sustaining it where he does, he causes it to move or be at rest. And so, it seems, there is no room for any other causes of motion in the Cartesian world. Though the argument concerns motion, states of body and their causes, it would seem to hold for the causes of states of mind as well, insofar as the divine sustainer must sustain minds with the states that they have as much as he must sustain bodies in the places that they occupy.[52]

These are serious arguments, but I think that they are not decisive reasons for maintaining that Descartes held God to be the only genuine cause in his world.

Let us look more closely at the passage from the letter to Elisabeth. When reading this, it is very important to place it in context, and understand what exactly Descartes was addressing in this passage. In this series of letters, he is trying to console Elisabeth in her personal and familial troubles. In a letter written on 30 September 1645, Elisabeth wrote:

> [The fact] of the existence of God and his attributes can console us in the misfortunes that come to us from the ordinary course of nature and from the order which he has established there (as when we lose some good through a storm, or when we lose our health through an infection in the air, or our friends through death) but not in those [misfortunes] which are imposed on us by men, whose will appears to us to be entirely free. (AT IV 302)

Descartes' reply, as quoted above, is that all things, including human beings acting freely, are under the ultimate control of an omniscient, omnipotent, and benevolent God. In saying this, Descartes doesn't take himself to be saying anything particularly original; it is, indeed, a theological commonplace. While these kinds of theological issues have led thinkers in various theological traditions to take the issue of occasionalism seriously,[53] it is not appropriate to infer the full-blown metaphysical doctrine of occasionalism from this commonplace observation, and to conclude that Descartes held that God is the only real cause in nature.

The argument to occasionalism from the doctrine of continual recreation is also not decisive, I think. First of all, however good an argument it might be, I see no reason to believe that Descartes ever saw such consequences as following out of his doctrine of continual recreation. But more than that, I don't think that the argument is necessarily binding on Descartes. It is certainly persuasive, particularly if one takes what I called the cinematic view of God as a cause of motion, the view in which God causes motion by recreating a body in different places in different instants of time. But the argument is considerably less persuasive if one takes what I earlier called the divine impulse view of God as a cause of motion. On that view, God causes motion by providing an impulse, much as we take ourselves to move bodies by our own impulses. But if this is how God causes motion, then God's activity in sustaining bodies is distinct from his activity in causing motion, and there is no reason why there cannot be causes of motion distinct from God, other sources of the impulses necessary to move bodies.[54] It is not clear how this reply might work in the case of the causation of sensory ideas in minds by bodies. But even if it didn't, it seems to me sufficient to establish that there can, for Descartes, be genuine causes in the world other than God.

These arguments that purport to commit Descartes to the view that God is the only cause thus fail, I think. But the best reasons for questioning the claim that Descartes holds God to be the only genuine cause in the world are a number of passages in which he seems to allow other causes of motion in nature. The issue comes up quite explicitly in Descartes' last response to Henry More. He writes:

> That transference that I call motion is a thing of no less entity than shape is, namely, it is a mode in body. However the force [vis] moving a [body] can be that of God conserving as much transference in matter as he placed in it at the first moment of creation or also that of a created substance, like our mind, or something else to which [God] gave the power [vis] of moving a body. (AT V 403–4 [K 257])

Descartes is here quite clear that some created substances, at very least our minds, have the ability to cause motion. Furthermore, there is no suggestion in this passage that minds can cause motion in bodies only with God's direct help, as the occasionalists would hold. Indeed, as I pointed out earlier in this chapter, our ability to cause motion in the world of bodies is the very model on which we understand how God does it. It would then be quite strange if he held that minds are only the occasional causes of motion in the world. At least two passages in the *Principles* also suggest that Descartes meant to leave open the possibility that in addition to God, minds could cause motion in the world. In defending the conservation principle, for example, he argues that we should not admit any changes in nature "with the exception of those changes which evident experience or divine revelation render certain, and which we perceive or believe happen without any change in the creator" (Pr II 36). Such a proviso would certainly leave open the possibility that finite substances like our minds can be genuine causes of motion. Similarly, in presenting his impact law, law 3 in the *Principles,* he claims that the law covers the causes of all changes that can happen in bodies, "at least those that are corporeal, for we are not now inquiring into whether and how human minds and angels have the power *[vis]* for moving bodies, but we reserve this for our treatise *On Man*" (Pr II 40). Again, Descartes is leaving open the possibility that there may be incorporeal causes of bodily change, that is to say, motion.[55] And so, I think, we should take him completely at his word when on 29 July 1648 he writes to Arnauld: "That the mind, which is incorporeal, can set a body in motion is shown to us every day by the most certain and most evident experience, without the need of any reasoning or comparison with anything else" (AT V 222 [K 235]).

Minds can cause motion in Descartes' world. But what is the "something else to which [God] gave the power *[vis]* of moving a body" to which Descartes refers in his letter to More? Angels are certainly included, the passage from Pr II 40 suggests; angels are also a lively topic of conversation in the earlier letters between Descartes and More. Indeed, when he is discussing with More how we can comprehend God as a cause of motion through the way we conceive of ourselves as causes of motion, Descartes explicitly includes angels as creatures also capable of causing motion, like us and like God (AT V 347 [K 252]). It is not *absolutely* impossible that he meant to include bodies among the finite substances that can cause motion.[56] But I think that it is highly unlikely. If Descartes really thought that bodies could be causes of motion like God, us, and probably angels, I suspect that he would have included them *explicitly* in the answer to More; if bodies could be genuine causes of motion, this would be too important a fact to pass unmentioned. Furthermore, Descartes' whole strategy for deriving the laws of motion from the immutability of God presupposes that God is the real cause of motion and change of motion in the inanimate world of bodies knocking up against one another. Somewhat more difficult to determine is whether or not bodies can be genuine causes of the states of sensation or imagination. Though he persists in holding that mind can cause motion in bodies, he is somewhat more guarded about the causal link in the opposite direction. The argument for the existence of the external world presented in Meditation VI, where bodies are said to contain the "active faculty" that causes sensory ideas in us, would suggest that bodies are the real causes of our sensations. But, as I pointed out in chapter 3, later versions of the argument found in the Latin and French versions of the *Principles* don't make use of the notion of an active faculty in bodies, and seem to posit a progressively weaker conception of the relation between bodies and the sensory ideas that we have of them.[57] While there is room for disagreement, it seems to me that all of the important signs lead to the view that bodies (inanimate bodies, at least) have no real causal efficacy and lack the ability to cause either changes in motion in other bodies or sensations in minds.

And so, even though Descartes posits causes of change in the world in addition to God, finite minds, and angels at very least, he does seem to agree with his occasionalist followers in denying that bodies are genuine causes of motion, and may well agree with them in denying that bodies can cause sensations as well. Can we say, on the basis of this that Descartes is a quasi-occasionalist, an occasionalist when it comes to the inanimate world, though not in the world of bodies connected to minds? The doctrine of occasionalism is certainly flexible enough to allow this. But even if we choose to view him in this way, we mustn't lose sight of an important difference between Descartes and his occasionalist followers.

For many of Descartes' later followers, what is central to the doctrine of occasionalism is the denial of the efficacy of finite causes simply by virtue of their finitude. Clerselier, for example, argues for occasionalism by first establishing that only an incorporeal substance can cause motion in body. But, he claims, only an infinite substance, like God, can imprint new motion in the world "because the infinite distance there is between nothingness and being can only be surmounted by a power which is actually infinite."[58] Cordemoy argues similarly. Like Clerselier, he argues that only an incorporeal substance can be the cause of motion in a body, and that this incorporeal substance can only be infinite; he concludes by saying that "our weakness informs us that it is not our mind which makes [a body] move," and so he concludes that what imparts motion to bodies and conserves it can only be "another Mind, to which nothing is lacking, [which] does it [i.e., causes motion] through its will."[59] And finally, the infinitude of God is central to the main argument that Malebranche offers for occasionalism in his central work, *De la recherche de la vérité*. The title of the chapter in which Malebranche presents his main arguments for the doctrine is "The most dangerous error in the philosophy of the ancients."[60] And the most dangerous error he is referring to is their belief that finite things can be genuine causes of the effects that they appear to produce, an error that, Malebranche claims, causes people to love and fear things other than God in the belief that they are the genuine causes of their happiness or unhappiness.[61] But why is it an error to believe that finite things can be genuine causes? Malebranche argues as follows:

> A true cause is one in which the mind perceives a necessary connection between the cause and its effect. . . . Now, it is only in an infinitely perfect being that the mind perceives a necessary connection between its will and its effects. Thus God is the only true cause, and only he truly has the power to move bodies. I further say that it is not conceivable that God could communicate to men or angels the power he has to move bodies.[62]

For these occasionalists, then, God must be the cause of motion in the world because only an infinite substance can be a genuine cause of anything at all.

But, as I understand it, Descartes' motivation is quite different. Descartes seems to have no particular worries about finite causes as such. If I am right, he is quite happy to admit our minds and angels as finite causes of motion in the world of bodies. Indeed, it is through our own ability to cause motion in our bodies that we have the understanding we do of God and angels as causes of motion. When God enters as a cause of motion, it is simply to replace a certain set of finite causes, the substantial forms of the schoolmen, which, Descartes thinks, are unavailable to do the job. As we discussed in chapter 4, Descartes argued that the substantial forms of scholastic philosophy were improper impositions of mind onto matter, and must, as such, be rejected. But, one might ask, if there are no forms, what can account for the motion bodies have, for their characteristic behavior? What Descartes turns to is God. In this way Descartes seems less a precursor of later occasionalism than the last of the schoolmen, using God to do what substantial forms did for his teachers.

AFTERWORD

IN THE PREVIOUS chapters we have explored the foundations of Descartes' program for natural philosophy in some detail. In concluding I would like to make a few general remarks about the "big picture."

The piece of Descartes' system we have been looking at is an integral part of a larger program. Below it is the metaphysics from which it grows, and above it are cosmology and terrestrial physics, the account Descartes offers of light, gravity, magnets, plants and animals, as well as the sciences of morals, medicine, and mechanics that are supposed to flow from the physics. This is the Cartesian program, and I think that it is fair to say that it is this larger program that was historically so important among Descartes' contemporaries. Descartes was attempting to replace the school philosophy with his own; the *Principles* can be regarded as a *Summa philosophiae,* an account of everything there was to know, grounded not in Aristotle, but in the new mechanical philosophy, attempting to replace the increasingly shopworn medieval synthesis of God, metaphysics, and natural philosophy with a new and more up-to-date version. In this he had many followers. Numerous identified themselves as Cartesians, and numerous, like Spinoza, Leibniz, and Malebranche, were deeply influenced by the Cartesian idea of a mechanist system of metaphysics and natural philosophy, while significantly altering the details.

But there was another important trend in seventeenth-century thought, a nonmetaphysical and problem-oriented conception of natural philosophy. This is found in Descartes' near contemporary, Galileo, and in his successor, Newton. Galileo shunned the idea of a whole system and attacked individual problems in physics without giving undue attention to questions like the real inner essence of body, or the role of God in grounding the laws of physics.[1] And while Newton was certainly interested in such questions, he was quite willing to postpone their definitive solution to a later stage of scientific investigation, and explore nature only as far as the phenomena would allow him to penetrate. These thinkers seem to have broken more fully with the scholastic past, and it is their programs that seem to have had the more important influence on later generations of scientists. Though he tried as hard as he could to break with his teachers, the very conception of Descartes' project shows the extent to which he could not. (It is ironic, but in a way, the approach Descartes took to problems in physics in his earlier years comes much closer to that of moderns like Galileo and Newton than his later approach did.) In the introduction to the French edition of the *Principles,* Descartes wrote:

> Those who have not followed Aristotle...have nevertheless been saturated with his opinions in their youth...and this has so dominated their outlook that they have been unable to arrive at knowledge of true principles. (AT IXB 7)

This, in a way, is true of Descartes himself.

NOTES

Prologue

1. For one view of the philosophical significance of the history of philosophy, understood in this way, see Garber 1988c.

2. Not that it is always easy to determine which figures and doctrines are historically relevant to Descartes, even when we have done all of the necessary studies. The job of tracing out the proper context for Descartes is made particularly difficult by the fact that much of the thrust of Descartes' thought involves the rejection of external authority and, in fact, the rejection of external influence altogether; the philosophy of clear and distinct perceptions, grounded in the *Cogito* is intended precisely to build a system of thought grounded in the individual's immediate perception of the truth. See, for example, the comments Descartes made in connection with the supposed influence Beeckman claimed on his thought, AT I 159 (K 17); see also Garber 1986 for a further development of this point. As a consequence, Descartes hardly refers to other authors in his published writings, and in his correspondence, refers to them only when a correspondent (usually Mersenne) asks for his opinion on something specific.

Chapter One

1. On the history of La Flèche, see esp. Rochemonteix 1899. The dates Descartes actually attended the college are not altogether certain; see, e.g., Gouhier 1958, pp. 19–20; Rodis-Lewis 1971, pp. 18–20; Rodis-Lewis 1983.

2. See Gilson 1967, p. 119; Rodis-Lewis 1983, p. 617; Rodis-Lewis 1984, p. 35.

3. All we have is a charming but uninformative letter that may have been written by Descartes to his grandmother Jeanne Sain sometime between 1606 and 1609, and some records of his two degrees from Poitiers. For the letter, see Descartes 1936–63, 1:473–74. J.-R. Armogathe and V. Carraud have recently discovered a copy of the placard containing the theses Descartes defended in 1616, when he received his degrees in law. See Armogathe and Carraud 1987 and Armogathe, Carraud, and Feenstra 1988.

4. In general, the account Descartes gives of his intellectual development in the *Discourse* is the most complete such account found in the corpus. And though Descartes' memory is selective on certain points, for example his contact with

Isaac Beeckman, it accords reasonably well with the surviving documents. However, one should keep in mind that though it may be based on earlier sources (see, e.g., Gouhier 1973, pp. 11–14, Gadoffre 1987, and Curley 1987 for recent views on that issue), the *Discourse* was composed twenty years after Descartes left school. Furthermore, even within the context of the work, Descartes, writing anonymously, tells the reader that he can read the events related as a story *(histoire)* or fable (AT VI 4); the point of the *Discourse* is not autobiography, but to provide the reader with the example of how one person came upon the truth and a way to find it, an example that the reader may or may not want to follow in certain of its particulars (AT VI 4). However, on the specific issue at hand, the content of Descartes' education, I see no reason to doubt that it is modeled on Descartes' own experiences at La Flèche. For a general discussion of the reliability of the *Discourse* as history, see Gouhier 1973, pp. 283–86.

5. For general background on education in the period, see de Dainville 1940 and Brockliss 1987. On pedagogy in Jesuit schools in general and La Flèche in particular, see Rochemonteix 1899 and Mesnard 1956. The curriculum at La Flèche, as with all other Jesuit schools of the period, was governed by the *Ratio Studiorum,* a document giving the rules and regulations to be followed in Jesuit schools, and outlining in considerable detail the curriculum to be followed. The documents and drafts leading up to the *Ratio Studiorum* of 1599, the one in effect while Descartes was in school, are found in Pachtler 1887 and, more recently, in Lukács 1986; the *Ratio* of 1599 together with selected background documents are available in English translation in Fitzpatrick 1933. Gilson 1967 is an indispensable source of information on Descartes' school years; see his commentary on Part I of the *Discourse.*

6. On teaching of languages in Jesuit schools, see Lukács 1986, pp. 434–41; Gilson 1967, p. 111.

7. Lukács 1986, p. 98; cf. Gilson 1967, pp. 117–18. The quotation is actually from the *Ratio* of 1586, and does not appear in the *Ratio* of 1599. But the 1586 text is itself quoting from the *Constitutiones* of the Society, and the philosophy curriculum set out in the 1599 document clearly follows the earlier recommendation; see Lukács 1986, pp. 397–400. On Noël, see Rodis-Lewis 1971, p. 19. Noël was an interesting character; he turns up later, in 1647 and 1648, arguing against Pascal (and with Descartes) and for an Aristotelian view of the Torricelli experiments, that they do not establish that there can be a vacuum in nature. See Pascal 1963, pp. 199–221.

8. See Lukács 1986, pp. 397–400.

9. On the theological curriculum, see Lukács 1986, pp. 386–94. Professors of philosophy were required not only to have completed the course in theology, but to review it every two years; see Lukács 1986, p. 359. Furthermore, the goal of instruction in physics was for the philosophy teacher at the Jesuit school to prepare the student for the study of theology; see Lukács 1986, p. 397. On the connection between Aristotelian philosophy and theology, see, e.g., Descartes' remarks to Mersenne in 1629, AT I 85–86.

10. See Lukács 1986, p. 430; Gilson 1967, p. 101.

11. See Gilson 1967, pp. 113–14.

12. See Gilson 1967, pp. 116–17; Rodis-Lewis 1971, p. 427, n. 7.

13. See AT VI 9; Gilson 1967, pp. 120–21; Rodis-Lewis 1971, pp. 22–23.

14. See AT III 194–96 for a bibliography of editions of these works.

15. See Schmitt 1983, pp. 39, 41–43, 81, 86–87.

16. See Schmitt 1983 for a general account of Aristotelianism in the sixteenth and early seventeenth centuries.

17. See Rodis-Lewis 1971, p. 428, on the possible episode. On the early reception of Galileo's observations among the Jesuits, see Wallace 1984, pp. 281–84.

18. Gilson 1967, pp. 126–27; Rodis-Lewis 1971, pp. 20–21. Little is known of the views of Father François at the time he taught the young René; François published only after Descartes' death, in the 1650s. Among the books he published are pedagogical works, including *L'Arithmétique, ou l'Art de compter toute sorte de nombres avec la plume et les jettons* (Rennes: 1653), works on water and fountains, including *L'Art des fontaines* (Rennes: 1665) and *La Science des eaux* (Rennes: 1653), a book on astrology, the *Traité des influences célestes* (Rennes: 1660), and a work that seems to be about pure mathematics, the *Traité de la quantité considérée absolument et en elle-mesme . . . relativement et en ses rapports, matériellement et en ses plus nobles sujets, pour servir d'introduction aux sciences et arts mathématiques et aux disputes philosophiques de la quantité* (Rennes: 1665). But it is very difficult to say how much of what he published later was in his head forty years earlier.

19. Rodis-Lewis 1971, p. 21; Gilson 1967, pp. 127–28. Recent scholars have emphasized the extent to which Clavius' curriculum attempted to combine mathematics and physics. See, e.g., Wallace 1984, pp. 136–41; Schmitt 1983, pp. 104–5; Cosentino 1970; Cosentino 1971; Dear 1988, chaps. 3–4 passim. It is by no means clear what exactly this meant in practice nor how exactly Clavius or other Jesuit teachers of mathematics and physics saw the connection between the two disciplines; Schmitt suggests eclecticism, the two taught side by side. But given Descartes' later ambition to relate mathematics to physics, and to bring mathematical standards of certainty and proof to physics and, more generally, to philosophy as a whole, this feature of Clavius' thought is highly suggestive.

20. Quoted in Gilson 1967, p. 128.

21. On Descartes' mathematical education, see especially Rodis-Lewis 1987a. Descartes clearly knew some Clavius as a student. John Pell, writing in 1646, reported that Descartes had told him that "he had no other instructor for Algrebra [sic] than ye reading of Clavy Algebra above 30 yeares agoe" (i.e., before 1616) (AT IV 730–31). Furthermore, Descartes' early mathematical writings show clear evidence of having been directly acquainted with Clavius' mathematical works, as Costabel has ingeniously demonstrated; see Costabel 1983, pp. 640ff. There is also a reference to Clavius' edition of Euclid in a letter to Mersenne, 13 November 1629, AT I 71; but this tells us nothing about what he may have learned at school.

22. See also Costabel 1983, p. 638.

23. Descartes also sent copies of his *Principles of Philosophy* in 1644; see AT IV 143–44.

24. Daniel Lipstorp, *Specimina Philosophiae Cartesianae* (1653), quoted in AT X 47.

25. On the meeting between Descartes and Beeckman, see the excerpts from

Lipstorp and Baillet given in AT X 47–50. While the date is probably secure (it is found in Beeckman's notes, JB I 237 [AT X 46]), other traditional details of the meeting are somewhat suspect. See Rodis-Lewis 1971, p. 435, n. 47. Descartes emphasizes that his meeting with Beeckman was by chance in a letter to Beeckman 17 October 1630, AT I 167. In general, when citing passages from Beeckman's journals, I shall give the reference in JB, followed by the transcription given in AT X, when there is one.

26. The journals are given in JB; on their history, see AT X 17–18 and JB I xxx–xxxiii.

27. The 1619 letters can be found in AT X 154–69. On 1628 meeting, see AT X 331ff. On Beeckman's life and works, see Berkel 1983b; Dijksterhuis 1961, pp. 329–33; JB I i–xxiv; see also Sirven 1928, and Gouhier 1958, pp. 20–30.

28. Descartes' own record of some of those conversations can be found in a fragment preserved in a later copy by Leibniz, that Gouhier identifies as a fragment from the notebook *Parnassus*. See AT X 219ff and Gouhier 1958, p. 15.

29. For the *Compendium Musicae,* see AT X 89–141 or de Buzon 1987. The two small treatises can be found in JB IV 49–55 (AT X 67–78). On Descartes and Beeckman on music, see Berkel 1983a and Rodis-Lewis 1971, pp. 31–37; on hydrostatics, see Schuster 1980, pp. 48–49; on free fall, see Koyré 1978, pp. 79–89.

30. In physics, for example, Beeckman, unlike Descartes, accepted the existence of atoms and the void; see JB I 132. Berkel 1983a further emphasizes their different interests and points of view.

31. See Koyré 1978, pp. 116f, n. 61.

32. See JB I 24–25; Gabbey 1980, pp. 244–45. On the impetus theory, see Maier 1982, chap. 4; Grant 1971, pp. 48–54; Clagett 1959, chaps. 8, 9.

33. See Wallace 1981, chap. 15; Clagett 1959, pp. 658–59.

34. Beeckman's record of a conversation he had with Descartes (JB I 263; AT X 60) suggests that he discussed the principle with Descartes; he reports Descartes appealing to his principle ("mea fundamenta") "quod semel movetur, semper movetur, in vacuo" in the course of a discussion of free fall; see also Descartes' record, JB IV 51; AT X 78.

35. Beeckman did not unambiguously hold a conservation principle; indeed, in his collision laws of 1618, motion seems to be lost in certain circumstances. But the question as to why, then, the world does not wind down, a question intimately connected with that of a conservation principle, is one that clearly worried Beeckman. On this issue, see the helpful discussion of Beeckman on collision and conservation in Gabbey 1973, pp. 381–84. On gravity, see JB I 25.

36. See, e.g., JB I 132–34, 152–53, 201–2. Beeckman's view seems to have been a variety of Epicurean atomism: "physica res duplex est, inane et corpus" (JB I 132).

37. See, e.g., the appeal to corpuscular principles in JB I 244 (AT X 52) and JB IV 52 (AT X 68).

38. The fragment in question bears the marginal note: "physico-mathematici paucissimi." Note also the title Beeckman's brother Abraham gave the posthumously published collection of his selected notes, *D. Isaaci Beeckmanni . . .*

mathematico-physicarum meditationum, quaestionum, solutionum centuria (Utrecht, 1644); see JB I iii.

39. See Berkel 1983a.

40. See Descartes to Beeckman, 17 October 1630, AT I 157–67, excerpted in K 16–17. For an account of the quarrel, overly sympathetic to Descartes, perhaps, see Baillet 1691, 1:205ff. See also Gabbey 1970.

41. See Rodis-Lewis 1971, pp. 43–45 and p. 448, n. 108.

42. Descartes seems to have kept a notebook in this period, which he divided into smaller sections, entitled *Parnassus, Praeambula, Experimenta, Democritica,* and *Olympica;* see the inventory of Descartes' papers given in AT X 7–8. Though the notebook seems to have survived until the end of the seventeenth century, it is now lost. However, some of its content was saved in Baillet's biography and in excerpts that Leibniz had copied out in Paris in 1675–76; see AT X 173–77. Due to the ingenious work of Henri Gouhier, it is now possible to talk with some confidence about when these notes were written, and what they show. Gouhier (1958) analyzes the Leibniz manuscript and shows where the different fragments come from. Recent studies have also suggested that portions of the *Rules for the Direction of the Mind* may date from this early period as well. See Weber 1964; Schuster 1980, pp. 41–55.

43. See Rodis-Lewis 1971, pp. 58–59, 454 n. 1.

44. Given in AT X 179; see also pp. 181, 216.

45. The dreams are given in detail in Baillet 1691, 1:81–86, and in AT X 181–88; a translation can be found in Cottingham 1986, pp. 161–64. For convenience, I will cite the text as given in AT.

46. See, e.g., Rodis-Lewis 1971, pp. 47–55, and her notes for a discussion of the complexities.

47. This also appears in a brief reference to the dream in the Leibniz transcriptions, AT X 216.

48. See also note a on that page, where the editors discuss the reaction to this interpretation among Baillet's contemporaries.

49. See, e.g., Rodis-Lewis 1971, pp. 51–52.

50. See also the fragments on pp. 204, 230 (discussed in Gouhier 1958, pp. 68–69), and 255.

51. This fragment is not dated, but very likely comes from this period.

52. The conception of the unity of knowledge that Descartes was working out was a commonplace in sixteenth-century thought; see, e.g., Rodis-Lewis 1971, p. 444, n. 93. The same can be said about method; see, e.g., Jardine 1974, Gilbert 1960, and Ong 1958. On the vagueness of Descartes' vision of method at this point, see Schuster 1980 and Schuster 1986.

53. See also Rodis-Lewis 1971, pp. 59–63, and Gouhier 1958, pp. 16–17, 71ff.

54. On the geometry, see Schuster 1980, pp. 55, 88, n. 68; on the optics, see Gouhier 1958, pp. 77–78, and Rodis-Lewis 1971, pp. 70–72.

55. See Weber 1964, pp. 199–201, and, on the *Studium,* Rodis-Lewis 1971, pp. 72–80. Baillet clearly distinguishes between the *Studium* and the *Rules;* see AT X 203.

56. On Descartes' whereabouts during these years, see Rodis-Lewis 1971, pp. 80–105.

57. On Mersenne's program mainly in the 1620s, see esp. Dear 1988. Mersenne is, of course, associated with the new mathematical science then emerging, and his interest in it is clear throughout most of his career. But the 1620s was also a period of intense debate over skepticism, atheism, libertinism, and Mersenne was very much in the fray, defending religion, the Church, and even the teaching of Aristotle in works like *L'impiété des deistes* (1624), whose title accurately indicates its contents, and *La vérité des sciences* (1625), a defense of a Christian philosophy against skepticism and alchemy. It is often assumed that Descartes was involved with this movement, and attempts have been made to read his later philosophy as a response to the intellectual crisis he found in Paris. The classic attempt to do so is Popkin 1979. See also Schuster 1980, esp. pp. 55–57, 89, n. 72. Schuster reads Descartes as involved in a program like Mersenne's mathematisation of knowledge on a mitigated skeptical basis. Rodis-Lewis (1971, pp. 82–83, 467, n. 54, 55), on the other hand, rejects this view, arguing that Descartes' relations with Mersenne at this time were more personal than intellectual, at least as regards Mersenne's apologetic program, and the attack on skepticism and atheism.

58. Cornier to Mersenne, 22 March 1626, CM I 429.

59. Descartes to Villebressieu, Summer 1631?, AT I 213 (K 20); cf. Rodis-Lewis 1971, pp. 98–100. Descartes makes a possible reference to this event in the letter to the Doctors of the Sorbonne that precedes the *Meditations*, AT VII 3. The event raises many interesting question about what Descartes was up to at this point in his career, and about the state of the debate between Aristotelians and their opponents in Paris in the late 1620s. However, it is not even clear exactly when the event occurred, nor is it clear what effect it might have had on Descartes' later work or on his resolve to undertake later work.

60. See Schuster 1980, pp. 55, 88, n. 68; Rodis-Lewis 1971, pp. 84–87.

61. It is possible that the allusion to his reputation is a reference to the Bérulle affair; see Gilson 1967, pp. 276–77. But the evidence for this is scanty.

62. See AT VI 18–19 for the only account of the method that Descartes published during his life.

63. The first mention of the treatise is in a letter to Gibieuf, 18 July 1629, AT I 17, where he refers to "vn petit traité que ie commance." There is a later reference to it in a letter to Mersenne, 8 October 1629, AT I 23, as a project temporarily put aside to work on the problem of the rainbow and other meteorological questions. This seems to have been Descartes' second attempt at such a work. Baillet (1691, 1:153, 170–71) reports an unsuccessful attempt at writing a treatise on divinity in the summer of 1628.

64. For some interesting speculations on what the metaphysics of 1629–30 may have contained, see Rodis-Lewis 1987b.

65. See also Rodis-Lewis 1984, p. 69.

66. For a history of the composition of this work, see Michael Mahoney's introduction in Descartes 1979, pp. vii–xiv.

67. For a discussion of what the original may have contained, see AT XI 698–706.

68. Descartes' idea of beginning an account of the world with a story about

creation seems to date only from the late 1620s, about the time he abandoned the *Rules* and began to work on his system of philosophy. See AT I 561 (K 46–47).

69. It is interesting (though unsurprising) that the Copernicanism clearly present in *The World* is suppressed in the account Descartes gives of his earlier thought in the *Discourse*.

70. See also 281–82, 285–86 (K 25–26); the first two of these letters seem not to have been received.

71. See, e.g., Rodis-Lewis 1971, p. 493, n. 12, for some of the interpretations offered.

72. See *The World*, chap. 8; cf. Aiton 1972, chap. 3, for a standard account of Descartes' vortex theory.

73. See *The World*, chap. 13. For a discussion of the laws of motion that give rise to the centrifugal force, see chap. 7 below.

74. See *The World*, chap. 6, AT XI 34–35. See also Pr III 47.

75. Though the passage is not altogether clear, the implication is that Descartes is at work on his *Dioptrics*.

76. On significance of calling it a "Discourse," see AT I 349 (K 30); AT I 620.

77. Descartes may mean here only that he intends to publish it without *The World*.

78. For example, his work on the rainbow; see AT I 23.

79. A number of undated papers in the manuscript *Cartesius* may date from this period; cf. AT XI 626f with AT VI 298f.

80. Adam and Tannery originally dated this October 1637; the editors of the nouvelle présentation date the letter 22 February 1638 (see AT I 670).

81. On the composition of the *Discourse*, see the references cited above in note 4.

82. See AT I 380–81, 383–84, 457, 558. Descartes originally wanted at least 200 copies for distribution; see AT I 339 (K 28). On the details of the publication, see Cohen 1921, book 3, chap. 13, and Armogathe 1990.

83. See also AT I 563–64 (K 48–49); AT III 39 (K 70–71); Garber 1978, pp. 130–33.

84. See Verbeek 1991; Mouy 1934, pp. 13f; Bouillier 1868, vol. 1, chap. 12.

85. See Mouy 1934, pp. 8f, 14; Bouillier 1868, 1:260.

86. Regius later fell under attack from Descartes too for failing to maintain Cartesian orthodoxy. See AT XI 672–87; Cohen 1921, book 3, chaps. 16–19; Bizer 1965, pp. 21–27; Descartes 1959, pp. 7–17; and Verbeek 1991, chaps. 1 and 3. The texts connected with the Voëtius affair are now collected in French translation with a very helpful introduction in Verbeek 1988.

87. See Mouy 1934, pp. 16, 17; Cohen 1921, book 2, chap. 14; Verbeek 1991, chap. 2.

88. The Latin translation of the *Discourse* didn't actually appear until 1644. The letter cited here, originally dated March 1637 by Adam and Tannery, was redated 30 April 1637 by the editors of the nouvelle présentation; see AT I 668.

89. There is a possible reference to the project in January 1639; see AT II 491–92.

90. See AT III 265 (K 86–87); Rodis-Lewis 1971, pp. 517–18 n. 133.

91. For Descartes' reply, see AT II 682–83; see also AT I 564 (K 49).

92. Note particularly the textual variant given in the notes in AT.

93. Descartes was then contemplating a trip to France; see AT III 228–29.

94. See also AT III 259–60 and a probable reference at AT III 270.

95. It shouldn't be inferred from this that Descartes modeled the *Principles* after the general run of scholastic textbooks; in form, short paragraphs preceded by summaries or descriptions of what the paragraphs contain, the *Principles of Philosophy* is quite unlike any textbook then currently in use, so far as I can determine. However, it does, in some ways, resemble theses posted for disputation, short statements, printed on a placard and posted, which were then defended in an oral disputation. Descartes in fact, refers to the articles of the *Principles* as "theses" when announcing them to Mersenne; see AT III 233 (K 82).

96. There is a brief reference to the work in progress, in both the Paris and Amsterdam editions of the *Fourth Replies:* AT VII 252, note to line 21, and 254. See Garber and Cohen 1982, p. 144 and n. 10, for an account of the rather complex history of that passage.

97. In early January 1643 Descartes reported to Huygens that he was about to start work on the portion of the *Principles* in which the magnet was to be treated; see AT III 801. The full discussion of the magnet isn't completed until Pr IV 144–83. However, the account of the magnet in part IV is grounded in the discussion of grooved particles in Pr III 90–94, 105–9, etc. In January 1644 Descartes reported to Pollot that he had not yet finished the "traité de l'aymant," though he claims to have written the third part of his book, "en contient les principes," and claims to be finishing up the fourth part, in which the account of the magnet is completed (AT IV 72–73). This suggests that in January 1643, Descartes was roughly halfway through part III. This is consistent with what he wrote to Colvius in April 1643 (AT III 646), that he was then working on his account of the heavens, which, of course, appeared in part III.

98. The work in question, *Letter to Vöetius,* appeared in May 1643; see Verbeek 1988.

99. See AT IV 72–73. The book is still in press in May; see AT IV 112–13, 122–23.

100. See Garber and Cohen 1982 for a fuller account of the relations between the *Meditations* and part I of the *Principles.*

101. This and related passages will be discussed below in chapter 6.

102. For a lovely treatment of More's interests in Descartes and the Cartesian philosophy, see Gabbey 1982.

103. For more details on the break between Descartes and Regius, see the references cited above in note 86.

104. The French version also shows some serious misunderstandings; see Costabel 1967, pp. 241f.

105. Also very interesting in this connection is a document found among the papers Leibniz had copied, some remarks entitled "Annotationes quas videtur D. Des Cartes in sua Principia Philosophiae scripsisse," which can be found in AT XI 654–57. A series of isolated remarks, they can quite neatly be coordinated with particular sections of Descartes' *Principles of Philosophy.* My conjecture is that they were marginalia in Descartes' own copy of the *Principles,* and that they may

represent notes he took when rereading it on the occasion of the preparation of the French translation. That is just a conjecture, but whatever their origin, they are almost certainly genuine, and shed important light on certain sections of the *Principles,* as we shall later see.

106. The French version has minor variants.

107. These lines from Pr IV 188L may well have been composed after Pr II 2 L and Pr II 40 L were already being set in type; see AT IV 72–73, where Descartes reports that the printer was already beginning to print the book, even though it was not yet complete.

108. On the dating of the piece, Adam and Tannery note the apparent references in the text to the French *Principles;* see AT XI 220–21.

109. Possible notes for this are found in a piece called "Primae cogitationes circa generationem animalium," AT XI 505–38; the date "Febr. 1648" is found on p. 537.

110. For Descartes' worries in 1632, see AT I 254. For other texts from Descartes' last years that show an interest in the problem of generation, see AT V 112, 170–71, 260–61.

CHAPTER TWO

1. From indications given in the text, Descartes had planned to write three sets of twelve rules each; see AT X 399, 429, 432, 441. The work as it stands contains drafts of eighteen rules, with the titles of three more. Rule 8 is an especially good illustration of the extent to which even the sections we have remain unfinished. As we shall discuss below in this chapter, Descartes introduces there what he characterizes as the noblest example of the method; see AT X 395–98. It is quite clear that there are three successive drafts of the same passage, one after another in the final text; see note 19 below for a fuller discussion. The standard account of the composition of the *Rules* is Weber 1964. Weber claims to find ten distinct strata in the final text. While I think that Weber has the basic chronology correct, I am a bit skeptical that one can make such fine distinctions. The datings I give in the text below are due to Weber unless otherwise attributed. Schuster 1980 and Schuster 1986 provide a useful supplements to Weber on some of the aspects of dating sections of the *Rules.* No autograph of the *Rules* survives, nor was the work published until many years after Descartes' death. There are two somewhat different versions of the text as we have it. One is based on a manuscript copy owned by Leibniz, the so-called Hanover manuscript; the other is based on the first published edition, which appeared in Amsterdam in 1701. For studies that focus on the *Rules,* see, e.g., Beck 1952 and Marion 1981a.

2. "Mind" is the usual translation of the Latin "ingenium." It is probably better translated as "native abilities," though this is somewhat awkward in English.

3. There is a textual problem here; both versions of the text read "induction" instead of "deduction," though in the Hanover manuscript it is crossed out. "Deduction," though, is the generally accepted reading. See the textual note in AT X 694.

4. To avoid confusion, I am breaking with most commentators, who refer to

these as the analytic and synthetic steps, following the distinction Descartes draws in the *Second Replies,* AT VII 155–56 or AT IXA 121–22. See, e.g., Serrus 1933, chap. 1; Beck 1952, chap. 11, etc. This is a distinction that has little direct relevance to the stages of the method of the *Rules.* In the *Rules* we are dealing with a distinction between two parts of a single method; though they are distinct, both are necessary for a true application of the method. But the distinction between analysis and synthesis in the *Second Replies* is completely different. There we are dealing with different ways of setting out a single line of argumentation, and we must choose one or the other. On analysis and synthesis, see also Garber and Cohen 1982.

5. The main exposition of the method in the earliest strata is given in Rules 5 and 6. There, in Rule 6, the principle "secret" of the method is claimed to be the epistemological ordering of "things" from complete *(absolutus)* to relative, from those that contain simple natures to those whose natures are what Descartes calls relative, from those whose natures are comprehensible to us per se to those whose natures must be comprehended through simple natures. While simple natures remain important throughout the composition of the *Rules,* appearing in Rule 12, the last strata of composition, as the objects of knowledge, it is not altogether clear in Rule 6 how the epistemological ordering of "things" and their natures is to advance the method outlined in Rule 5, and how to translate the "secret" of the method into a practical way of performing the reductive, intuitive, or constructive steps of the method Descartes sketches there. The only example Descartes gives in those two rules is itself somewhat obscure, and concerns the ordering of "things" rather than the method itself; the example concerns the theory of proportions on which he was working in 1619, and suggests that a simple proportion is a "thing" simpler than a single mean proportional, which, in turn, is simpler than two mean proportionals. See AT X 384–87 and Costabel 1982, pp. 49–52. On the claim that the method of 1619 as outlined in these rules was more programmatic bravado than practice, see Schuster 1986.

6. Descartes' method is quite closely linked to his work in mathematics in the 1620s. As the discussion in part II of the *Discourse* suggests, the method may well have been a kind generalization of his procedure for solving problems in mathematics; see AT VI 19–20. For recent studies of the method that emphasize its connection to mathematics, see, e.g., Lefèvre 1978; Hintikka 1978; Schuster 1980; Schuster 1986, pp. 47–59; and Grosholz 1990. Important though this observation may be, it is not directly relevant to my concerns in this chapter; the connection between Descartes' method and his mathematics will not be treated in any detail in this study. What interests me most here is understanding how the method of the *Rules* and the *Discourse* does and does not apply outside of mathematics, to Descartes' larger program in metaphysics and natural philosophy, and how the methodological thought of the 1620s is linked with a general view about the large-scale organization of knowledge in those domains.

It is for this reason that I find the version of the method displayed in the anaclastic line example of Rule 8 to be especially attractive; it shows in a clear way precisely how Descartes thought that the method was to apply to problems in physics, problems that while mathematical to a degree, are not problems in pure mathematics. (The anaclastic line problem would have been considered a problem in mixed mathematics for Descartes and his contemporaries.) But it does seem to conflict with expositions of the method that take their starting place from either the discussion of the *mathesis universalis* in the second half of Rule 4, or from the account of method in Rule 14.

A number of commentators have identified Descartes' method in the *Rules* with the *mathesis universalis* in some sense; see, e.g., Milhaud 1921, p. 69; Mouy 1934, pp. 4–5; Van de Pitte 1979; Marion 1981a, §§9–11; Clarke 1982, pp. 166ff. Understood in the most straightforward sense, I find this highly implausible. In the main exposition of the *mathesis universalis* in Rule 4, it is characterized not as method (the term 'method' appears not once in the passage in Rule 4 on the *mathesis universalis* that begins at AT X 374 l. 16), or even as an art (a word that comes up often in that passage), but as a science *(scientia);* it is characterized as the "general science" that deals with order and measure in the most general way (AT X 378). Insofar as Descartes conceives of it as a science, an organized body of certain knowledge, the *mathesis universalis* cannot be the same as the method, it seems to me, which Descartes defines earlier in Rule 4 as a set of rules that lead one to knowledge (AT X 371–72). For this simple reason it seems to me that Descartes' method in the *Rules* simply cannot be identified with *mathesis universalis.* (It is also worth pointing out in this connection that it is not clear just what relation the discussion of the *mathesis universalis* bears to the main text of Rule 4; in the Hanover manuscript, for example, it is not part of the Rule itself, but is found only at the end of the manuscript as a whole, a sort of appendix. On the structure of Rule 4, see Weber 1964, chap. 1, and Weber 1972.)

A better candidate for method is the procedure that Descartes suggests for solving problems in Rule 14. Essential to the procedure recommended in Rule 14 is the idea that all problems about any subject matter whatsoever should be transformed, somehow, into problems that relate to extension and shape alone, so that the mathematics Descartes has been developing, quite possibly the *mathesis universalis* discussed in Rule 4, can be applied to their solution. (This is developed most clearly at AT X 441.) While this method does qualify as a genuine procedure for the solution of problems, it seems quite distinct from the method as explicated in the anaclastic line example of Rule 8; nowhere in the course of that rule does he ever suggest representing the problem posed in geometrical terms.

Which (if either) of these two conceptions of method better represents Descartes' considered view on the question? My own preference is for the method as exhibited in the anaclastic line example. First of all, while the method of Rule 14 may correspond closely to Descartes' mathematical practice, it is never to the best of my knowledge applied anywhere outside of the mathematical writings. That is, nowhere, to the best of my knowledge, does he ever give an example of solving a nonmathematical problem by transforming it into a mathematical problem. Where the method is applied, it seems to me that it is the method as exemplified in Rule 8. (See Garber [forthcoming] for an account of how the method of the anaclastic line example is applied in the case of the rainbow in discourse 8 of the *Meteors.*) Furthermore, it is the method of Rule 8 that reappears in the *Discourse* (AT VI 18–19), I think, and not the more mathematical method suggested in Rule 14. These suggest to me that the conception of method that I discuss in the main text better represents Descartes' considered view on method.

It may well be the case that the two methods are not altogether distinct. It could be that the mathematical method of Rule 14 predates the method of the anaclastic line example, and that the anaclastic line example represents a stage of Descartes' thinking in which he had learned how to abstract the essence of the mathematical method, as it were, and apply it directly to nonmathematical problems, without having to translate such problems into mathematical terms. Or, the method of Rule 14 might represent a later attempt to discipline the rather less mathematical method of the anaclastic line example in terms that more closely correspond to the procedure that worked so well for him in his more purely mathematical work. But if so, it doesn't seem to me that it worked, and Descartes abandoned the idea quickly; the fact that not a single example survives of the transformation of a nonmathematical problem into a mathematical one and solving it in that way suggests to me that he didn't have anything more than a vague and impressionistic idea of the ambitious procedure suggested in Rule 14.

But whatever story is the right one, it seems to me that insofar as we are interested in the conception of method as it applies to problems outside of mathematics, it is that method displayed in the anaclastic line example that should concern us.

7. The example is developed on AT X 394–95.

8. The connection between Rule 8 and Rule 12 will be discussed later in this chapter.

9. Schuster (1980, p. 87, n. 60) dates this passage as post-1626.

10. See Costabel 1982, pp. 53–58, for an account of the historical background to this example. While the example in Rule 8 is illuminating, things can get much more complicated. As I shall note later in this chapter, the account of the rainbow that Descartes gives in discourse 8 of the *Meteors* is meant to be an example of the method. Though it is considerably more complicated than the anaclastic line example, the account Descartes gives of the rainbow does, indeed, exemplify the method, even though it makes explicit appeal to experiment, unlike the anaclastic line example. For an account of the rainbow, together with a discussion of the place of experiment in the method, see Garber (forthcoming).

11. See AT X 371, where Descartes uses this analogy.

12. This is the message of Rule 8; see AT X 392.

13. On the Cartesian critique of Aristotle and Aristotelianism on this question, see Marion 1981a, §2.

14. Descartes' list of simples in Rule 6 includes "cause, simple, universal, single, equal, similar, straight, and others of that sort" (AT X 381), suggesting that statements about those notions would ground statements about effects, complex things, particulars, pluralities, unequals, dissimilars, and nonstraight lines.

15. On the distinction between the intellect and body, see AT X 415–16, 421–22.

16. This does raise the question of sensory and empirical knowledge. When Descartes uses 'science' in these and other passages, like his contemporaries, he is referring to the ordered body of general truths, known through deductive argument. This is opposed to history, which deals with particulars, and need not be deductive or certain. Insofar as we have sensory knowledge, it would seem to fall under history and not science. On the distinction, see, e.g., Goclenius 1613, p. 626, and in the *Second Replies*, AT VII 140. For an account of the role of experiment in Descartes' method, see Garber (forthcoming).

17. On the epistemological nature of order, cf. AT X 381–83, 418. Jean-Luc Marion, in his influential monograph on the *Rules*, *Sur l'ontologie grise de Descartes*, Marion 1981a, especially emphasizes the fact that the notion of order and connection basic to the *Rules* is order with respect to us. While in general terms he is certainly correct, there are important ways in which he pushes the point too far. First of all, Marion argues that the order is, in a strong sense, created by us (§§12–13); it is, Marion claims, "une fiction de la pensée" (p. 77). This is sometimes true, and Descartes acknowledges that when we are dealing with the methodical solution to certain kinds of word games and riddles, we must, indeed, impose an order on the steps of the solution where there is no natural order (see, e.g., AT X 391, 404–5). But while the order in question is epistemological, and thus, in a sense, is dependent on us, it is also a kind of natural and nonarbitrary order in at least the most interesting cases that the method treats. In Rule 6, for example, Descartes notes that we must observe "the mutual connection among things and the natural order" (AT X 382). Furthermore, Marion's emphasis on order as mind-dependent, together with other observations Marion makes about the *Rules*, the unity of science in a unified mind (§§2–3), the transformation of Aristotelian natures and categories into Cartesian simple natures (§§14–15), and so on, motivate, for Marion, the claim that Descartes of the *Rules* had passed from a kind of Aristotelian realism to a kind of idealism; this is the "ontologie grise" of Marion's title (pp. 92–93, 98–99, 131–36, 141–42, 184, 185–90, 191–92). The emphasis of the *Rules* is clearly on knowledge and how we can attain it, and as such is very much concerned with the world from the perspective of the knowing subject. But to suggest, as Marion seems to, that in the *Rules* the world is the creation of the knowing subject seems wrong. While it may perhaps be a philosophical consequence of the position Descartes took, he was certainly not pressing such a view in any explicit way, as Marion seems to recognize in calling it a *gray* ontology. But more than this, it is a view that seems quite distant from any of Descartes' philosophical concerns at this stage in his development. At best this gray ontology seems to be the shadow of Descartes' later concerns, say in the *Meditations*, that Marion is casting back onto the *Rules*.

18. On the appeal to authority, see, e.g., AT X 367. For Descartes' discussion of the logic of the schools, see, e.g., AT X 363–64, 405–6, 430.

19. In what appears to be the first draft, Descartes proposes as an example of his method finding the extent of human knowledge, something that, he says parenthetically, should be done "once in life *[semel in vita]*" by all who seek to develop "good sense *[bona mens]*" (AT X 395). The question remains the same, but the motivation changes subtly in a slightly later passage from Rule 8: "And lest we should remain forever uncertain as to what the mind is capable of, and

lest it should make misguided and haphazard efforts, we must, once in life *[semel in vita]*, before setting ourselves to discover particular things, make careful inquiry into what knowledge human reason is capable of" (AT X 396–97). Here it is not merely an example of the method that we are dealing with, noble though it might be, or with a general desire to cultivate a *bona mens*, but with a project that we must undertake to prevent wasted effort. The reformulation of what appears to be the same basic question, and the repetition of the phrase *semel in vita* strongly suggest to me that we are dealing with a second draft of the same material. The third draft, discussed in the main body of this chapter, changes the question by specifying that we should be interested in the nature as well as the extent of knowledge, and by further clarifying the motivation. It is interesting to note, by the way, that that same phrase, *semel in vita*, that appears three times in Rule 8, once in each draft of the question, appears in the opening sentence of the First Meditation; see AT VII 17.

20. E. M. Curley emphasizes this point as differentiating the concerns of the *Rules* from those of the *Meditations;* see Curley 1978, pp. 44–45.

21. There is also a brief sketch following the first draft of the project; see AT X 395–96.

22. It is interesting to note that Rule 12 itself seems largely unrelated to most of the commentary that follows; it states simply that we should make use of all the help we can get from the intellect, imagination, the senses, and memory in carrying out the method. The first sentence of the commentary then presents this rule as a summary of general advice about the method that went before, and a prelude to the more particular discussions that follow. But Descartes then launches directly into a development of the epistemology sketched out at the end of Rule 8. I suspect that the rule and its first line of commentary were drafted fairly early in the history of the work, and that when, in Rule 8, he came upon the idea of a general epistemological inquiry, he appended the sketch to the preexistent Rule 12, with the intention of coming back and revising the whole rule. The chronology I am suggesting is that he began with the anaclastic line example of Rule 8 sometime during or after 1626, which suggested, in turn the idea of a general epistemological inquiry, again in Rule 8, which gave rise to the sketch as found in Rule 12.

23. Marion compares the assumptions of Rule 12 to those that Descartes uses in the *Dioptrics* and *Meteors,* and claims that the use of assumptions or hypotheses (in all three cases) is not merely provisional, but reflects the fact that to build a science, we don't need to know the real nature of real things: "L'exactitude mathématique, et son universalité, exigent l'abstraction et l'ignorance de l'ousia: . . . hypothèse seule permet la certitude" (Marion 1981a, p. 116; cf. §18 passim). But this seems to ignore Descartes' own testimony in Rule 12 that the hypothetical mode of presention is only provisional, and ought properly to be replaced by a nonhypothetical development of the material (AT X 411–12), not to mention Descartes' comments on the use of hypotheses in connection with the essays of 1637. On this later point, see Garber 1978, pp. 130–33.

24. In an important paper on Descartes' *mathesis universalis,* John Schuster (1980) argues that the account of the mind Descartes gives in Rule 12 is intended mainly to "ground and legitimate universal mathematics" (p. 57). The

claim is that *mathesis universalis,* the study of magnitude in general, is to be legitimated by being construed in terms of the immediate imaginative apprehension of actual physical shapes in the brain, which provide a referent for the extensions and shapes that the *mathesis universalis* deals with, and thus ground mathematical truth, if I understand him properly (pp. 65, 67, 93 n. 129). (Schuster sees this as linking Descartes with Mersenne and his version of mitigated skepticism; see, e.g., pp. 62–63, and 93 n. 129. But that is another issue.) While Schuster's paper is enormously useful for its careful discussion of the background to the *Rules* and the dating of particular portions of the *Rules,* this claim is misguided, I think. First of all, Schuster gives no real argument for his thesis; at best, he appeals to the general intellectual climate in Paris at the time that Rule 12 was being composed, the so-called skeptical crisis, and offers an imaginative, but, so far as I can see, undocumented story as to how in that climate and under the influence of Father Mersenne Descartes could well have hit upon the sort of justificatory program that Schuster attributes to him (see pp. 57, 72). But, so far as I can see, there is no real evidence that he actually did so. (For an alternative point of view that minimizes the influence that Mersenne had on Descartes' intellectual life, see Rodis-Lewis 1971, pp. 82–83, 467 nn. 54, 55.) Furthermore, in emphasizing his own story, and the importance of universal mathematics *(mathesis universalis)* in these texts, Schuster winds up missing some crucial themes in the later strata of the *Rules.* Schuster seems to ignore both in his account of Rule 8 and in his account of Rule 12 the importance Descartes attached to the general problem of the limits of human knowledge. In his account of Rule 8, for example, he quickly dismisses the general epistemological significance of the "noblest example of all," and simply asserts that the project must be understood "in terms of an inquiry into mental function and perception, which—in the first instance—would ground an elaborated version of universal mathematics" (p. 59). And in his account of Rule 12, in concentrating on the physiological psychology Descartes outlines there, what Schuster thinks is directly relevant to the grounding of *mathesis universalis,* he ignores the conclusion that Descartes himself draws from the discussion, that "there are no ways open to human beings for the certain knowledge of truth except evident intuition and necessary deduction" (AT X 425), a conclusion that shows Descartes' more general epistemological concerns here.

25. See Gilson 1967, pp. 486–90, where Gilson presents parallel passages of the *Discourse* and the *Rules,* and the commentary on the four rules of method in the *Discourse,* emphasizing their origin in the *Rules.*

26. A similar point is made in Dijksterhuis 1950b, esp. pp. 43–44.

27. There is an extended discussion of this in Part VI of the *Discourse,* AT VI 76–77. This theme also runs through Descartes' correspondence in the period; see AT I 562–64; AT II 141–43, 199–200 (K 47–49, 55–56, 58–59). These passages suggest the practical utility of a posteriori arguments as means of convincing a reader, though they are imperfect as demonstrations. Elsewhere Descartes suggests that the arguments of the *Essays* can be rendered syllogistically, though the syllogisms are probable; see AT I 422–24, 411 (K 39–40, 41).

28. The example is discussed at length in Garber (forthcoming), where it is shown how the investigation of the rainbow fits the method of the *Rules.* This

example, which makes liberal appeal to experiment, also is used to show how Descartes' method can accommodate experiment and experience in that paper.

29. In a letter of 8 October 1629 Descartes wrote to Mersenne that he has decided to write "a small treatise which will contain the explanation of the colors of the rainbow (to which I have given more care than all the rest) and generally the explanation of all sublunar phenomena" (AT I 23). This small treatise will doubtless become the *Meteors,* and Descartes' words to Mersenne suggest that Descartes probably solved the problem of the rainbow before October 1629.

30. In saying this, I am leaving the purely mathematical writings aside.

31. See Garber 1986 for a reading of the purposes of Meditation I.

32. Other important digressions include the celebrated piece of wax in Meditation II (AT VII 30–33), and the argument for the existence of the external world drawn from the faculty of imagination at the beginning of Meditation VI (AT VII 71–73).

33. See Garber 1986 for an elaboration of some of these themes.

34. See Serrus 1933, chap. 3; Beck 1952, chap. 18; Schouls 1980, chaps. 4–5; etc.

35. For the original title, see AT I 339 (K 28).

36. The commentator is, of course, Gueroult, and the work in question is Gueroult 1984. The order and connectedness of knowledge is one of the central preoccupations of the *Discourse;* see AT VI 11–14. Order also comes up often in connection with the *Meditations;* see AT VII 9–10, 155–56; and in the correspondence, AT III 102–3, 266 (K 87).

37. That section of the *Discourse* is also characterized as metaphysical in AT I 349 (K30); AT I 370. This section of the *Discourse* would seem to correspond to the metaphysics of 1629–30, but it is not entirely clear how closely they correspond in content.

38. Descartes is very careful to avoid anything that would imply that he is a Copernican. And while he does admit that in treating body he avoided appeal to scholastic forms and qualities, he does not give his own account of body as essentially extension and extension alone. See AT VI 42–43.

39. For the parallel passages in *The World,* see AT XI 31–32.

40. On AT IXB 15, the *Discourse* is characterized as containing that logic.

41. One might try to sandwich mathematics in between physics and its metaphysical foundations as the science of extension and number in general, taken apart from their instantiation in countable or extended things. This is suggested in Gaukroger 1980b, p. 126. But it is interesting that in 1647 Descartes does not even mention mathematics, except as a useful preparatory exercise to the real business at hand, the discovery of truth and the construction of the system of knowledge. It is quite possible that in this preface, Descartes is confining himself to natural philosophy, in which mathematics simply does not fit. But it is also not out of the question that Descartes simply lost interest in mathematics in his later years. As early as 1630 he wrote Mersenne: "As for your problems [i.e., mathematical problems Mersenne had proposed to him] . . . I am so tired of mathematics and hold it in such low regard, that I could no longer take the trouble to solve them myself" (AT I 139). Descartes repeated this sentiment a number of times in succeeding years; see Clarke 1982, pp. 119–20.

42. Descartes emphasizes the difference between recognizing a truth and recognizing it as foundational in the preface to the French *Principles*, AT IXB 10–11.

43. See the letter to Mersenne, 15 April 1630, AT I 144 (K 10), where Descartes tells Mersenne that he has found the foundations of his physics in his recent studies of God and himself. In that letter, Descartes goes on to announce his doctrine of the creation of the eternal truths, implying that that will have a central role to play in *The World*. It doesn't, but God certainly does enter that work in a number of crucial ways, particularly in chapter 7, where God grounds the laws of nature. There is no anticipation of this view in any of the earlier writings, so far as I can see.

44. Remember also the definition of intuition in terms of a "pure and attentive mind," AT X 368.

45. Similar ideas come up in the correspondence following the completion of the *Discourse on the Method;* see, e.g., AT I 349–51, 560 (K 30–31, 46).

46. This is not to suggest that there are no important differences between intuitions and deductions on the one hand, and clear and distinct perceptions on the other. But it seems clear that in both cases we are dealing with the immediate apprehension of a truth by a mind appropriately clear of bias and prejudice. Descartes does, on occasion, use his earlier terminology in his later writings, suggesting that it has not been entirely superceded; see, e.g., the *Second Replies*, AT VII 140, where the *Cogito,* the very model of a clear and distinct perception at the beginning of Meditation III (AT VII 35), is said to be known "by a simple intuition of the mind."

47. On the notion of moral certainty, see, e.g., AT VI 37–38; AT VII 141.

48. Descartes' view in the *Rules* of a science as certain as geometry is suggested in the later account he gives of his earlier work in part II of the *Discourse,* AT VI 19–20. For the view that we can have a science more certain than geometry, see, e.g., AT I 144, 350 (K 11, 31); AT VII 141.

49. See AT X 362; Gilson 1979, §408. When possible, I shall cite the scholastic texts collected in Gilson 1979. Gilson's book is no substitute for a fuller acquaintance with the work excerpted. But it is widely available, and it contains a well-chosen assortment of scholastic texts likely to have been known to Descartes.

50. For the distinction in medieval thought, see, e.g., Weisheipl 1978, pp. 467–68; Gilson 1979, pp. 66–67. Descartes makes explicit appeal to a similar distinction in criticizing Galileo; see AT II 433.

51. Among the Aristotelian school texts, the part on logic in Eustachius' *Summa* includes a section on method. For a discussion of the role of method in scholastic textbooks, see Blum 1988, pp. 133ff. Among the anti-Aristotelian texts one finds writings on method by Ramus, Bacon, and Gassendi. For histories of the notion of method in the sixteenth and seventeenth centuries, see Ong 1958; Gilbert 1960; Jardine 1974; and Reif 1969, pp. 28–30.

52. For some discussions of various schemes for the organization of knowledge proposed before Descartes, see, e.g., Weisheipl 1965 and 1978; Clagett 1948, pp. 29–36, etc.

53. Eustachius 1648, part 3, p. 111; cf. Reif 1969, p. 21.

54. Rosenfield 1957 especially emphasizes the criticism of Descartes' hyperbolic doubt among contemporary scholastics.

55. Gilson 1979, p. 182. See also the general references cited above in note 51.

56. The method is identified as a logic in the preface to the French edition of the *Principles,* AT IXB 13–14, though Descartes is quite careful to distinguish the method from the logic of the schools, toward which he was in general unsympathetic, of course.

57. Eustachius 1648, part 4, p. 2; see also Wippel 1982, pp. 385–92.

58. Preface to the French edition of the *Principles,* AT IXB 10. Substance is discussed in Pr I 51f.

59. See Marion 1981a, p. 64 n. 75. The relation between Descartes' conception of metaphysics and that of earlier thinkers is one of the main themes of Marion 1986, chap. 1.

60. See, e.g., Toletus 1599, f. 4v–6r; Eustachius 1648, part 3, pp. 112–13. The discussion derives from Aristotle, *Physics* II.1.

61. See, e.g., Toletus 1599, f. 4v col. a, and Aristotle, *Physics* III.1, 201a10–14. Descartes holds this account of motion up to ridicule in *The World,* AT XI 39–40.

62. See, e.g., Toletus 1599, f. 46r–49r; Eustachius 1648, part 3, p. 112; Aristotle, *Physics* II, 192b 21ff.

63. See the references cited in note 52 above.

64. Some, like the thirteenth-century philosophers Robert Grosseteste and Robert Kilwardby, working within a generally Platonistic and Augustinian tradition, saw physics as dependent in serious ways on mathematics, thus making it essential to study mathematics before undertaking physics. See Weisheipl 1965, pp. 74–78. Others, like Gundissalinus, Albertus Magnus, and Thomas Aquinas, saw physics as an autonomous science, one that can be undertaken without mathematics. See Weisheipl 1965, pp. 70–71, 86, 88–89; Clagett 1948, pp. 34–35; there is a slightly different view on Thomas in Sylla 1979, p. 156. For all that, though, Thomas and Albertus think that mathematics must be studied before physics for apparently pedagogical reasons. This view was widely held. See, e.g., Eustachius 1648, part 1, p. 2; Toletus 1599, f. 3r.

65. This is not to say that the view Descartes took was altogether unheard of. Various writers had suggested that metaphysics might, in certain circumstances, offer proofs of first principles in physics, though one should remember that the first principles they had in mind were different from Descartes' and the proofs the scholastic metaphysician could offer were strictly speaking dialectical proofs, proofs directed against a particular antagonist, using premises he would concede, not the sorts of proofs Descartes would admit. See Sylla 1979, pp. 157–58, 165, 173; Eustachius 1648, part 4, pp. 3–4. Also, the claim that physics must be grounded in metaphysics comes up from time to time as a position one might take, though it was inevitably rejected along with other erroneous positions considered and dismissed. See Weisheipl 1965, 82–83; Toletus 1599, f. 2v. Walter Burley suggests and rejects a view similar to Descartes on this question; see Sylla 1979, p. 175. Though it is not clear which of these particular views and in what precise forms Descartes may have been introduced to the broad idea of grounding physics in metaphysics in some way, it is not unlikely that he was introduced

to the idea in some way or another while in school. But such views were clearly outside the mainstream of scholastic thought, and it is unlikely that such views would have been treated sympathetically, or even very seriously.

66. This novel feature of Descartes' position is emphasized in Roth 1937, p. 4; Hatfield 1985.

67. See, e.g., Rosenfield 1957; Gilson 1975, pp. 316–33.

68. See Regius 1646, chap. 12. Regius' order here is quite close to the standard order for a scholastic physics textbook; cf., e.g., the order in Regius' *Fundamenta* with that of part III of Eustachius' *Summa*.

69. See Regius 1646, p. 1, on excluding God. One can get a sense of how different the two philosophers are by comparing Regius' chap. 1 with part II of Descartes' *Principles*.

70. The preface is unpaged. On hypotheses, see Rohault 1969, 1:13–14.

71. See, e.g., the systematic treatment of their views in Hobbes' *De Corpore* and Gassendi's *Syntagma Philosophicum*.

CHAPTER THREE

1. Gäbe 1983 emphasizes the affinities between the discussions of body in Rule 14 and the mature doctrines of body. There is reason to be cautious, though. The point of Rule 14 is to get us to represent all problems in mathematical terms, and to get us to use the imagination as a help to representing and solving problems so conceived in mathematical terms. The point of the passages about extension, subject, and body is to show that to represent extension *in imagination* is to represent it as a *body,* and that *in the imagination,* extension is inseparable from its subject, body. Certain comments Descartes makes in this connection can be construed as conceding that as regards the intellect, and, perhaps, as regards the truth, extension is separable from body. See, e.g., AT X 443 ll. 8–9; 445 ll. 9–11. (In the last passage, the phrase, "ut sint verae" or "ut sint licet verae" may be interpreted several ways). But, on the other hand, Descartes does say in Rule 14 that if anyone conceived extension without a subject, "utetur . . . solo intellectu malè judicante" (AT X 443). Thus Schuster, correctly, I think, argues that in the *Rules* Descartes makes no claims about the essence of body. See Schuster 1980, pp. 62–63, 93 n. 129.

2. For Descartes' uses of the term 'substance', see AT X 424 ll. 11–12, and 449 l. 4. For the other terms he uses, see the citations (or lack thereof) in Armogathe and Marion 1976.

3. The use of 'form' in this context is not to be understood as form in the physical sense, the sense in which the form is appealed to to explain the characteristic behavior of a body. In this case it is simply synonymous with 'nature'. See, e.g., Eustachius 1648, part 4, p. 18; Suarez, *Disputationes Metaphysicae* 15.11.3–4 in Gilson 1979, §311; Gracia 1982, pp. 216–17. Gracia 1982, pp. 175–279, is an extremely useful glossary of late scholastic technical terminology, together with numerous references from primary sources.

4. Cf., e.g., Eustachius in Gilson 1979, §177.

5. 'Mode' was used that way as early as 1631; see AT I 216 (K 20–21).

6. There is also a list of definitions in the appendix to the *Second Replies,* where

Descartes attempts to set out some of the arguments of the *Meditations* in geometrical fashion; see AT VII 160–62. Though probably written at roughly the time that Descartes was beginning work on part I of the *Principles,* there are real differences between the definitions given there and those given in the *Principles,* and differences in metaphysical vocabulary. In none of these definitions does he use the terms 'mode' or 'attribute' the way he does in the *Principles,* using instead the terms 'property', 'quality', 'attribute', and 'accident' synonymously (see especially definitions V and VII). Another exposition of Descartes' metaphysical framework is given in a letter to an unidentified correspondent dated 1645 or 1646; see AT IV 348–50 (K 186–88).

7. Of course, as Descartes recognized, God is the only thing that satisfies this definition, strictly speaking. Finite substances are things that can exist independently of everything except for God. This definition of substance is also found in the *Fourth Replies;* see AT VII 226. But the formal definition of substance given in the geometrical appendix to the *Second Replies* defines substance differently, as a substratum; see definition V on AT VII 161. Furthermore, the definition of real distinctness (definition X on AT VII 162) implies that it would be possible for there to be two substances that could not exist apart from one another, insofar as it allows for the possibility that there may be substances that are not really distinct from one another.

8. In the French version, the claim is that all properties of a substance "depend on" the special attribute that constitutes its nature or essence.

9. In Pr I 56 Descartes sanctions the use of 'quality' in connection with substance.

10. See also the account Descartes gives in the letter to Elisabeth, 21 May 1643, AT III 665 (K 138).

11. In the preceding section, Pr I 55, though, Descartes calls duration a mode. The notion of duration, like the notion of substance, though it applies to all substances, minds, bodies, and God, is not univocal; just as 'substance' as applied to God means something somewhat different than 'substance' as applied to finite creatures (see Pr I 51), the duration of God is different from the duration of his creatures. For finite creatures, both minds and bodies, duration is successive, Descartes says, one event temporally following another, while God's duration is what Descartes calls simultaneous. See Pr I 23; AT V 193 (K 232); AT V 223 (K 236); AT V 165–66 (Descartes 1976, pp. 31–32).

12. Descartes does recognize that we have a notion of time, its measurement and its passage that is, in a sense, independent of this or that enduring body, a sense in which one can say that two different events in finite substances, say a thought in one and a motion in the other, happen during the same interval of time. But, Descartes claims, "when we distinguish time from duration taken generally, and say that it is the measure *[numerus]* of motion, it is only a mode of thought" (Pr I 57).

13. At least, there would be no duration in the creaturely sense, successive duration.

14. The apparent parallelism between extension and duration leads to some trouble insofar as the arguments for indefinite extension would seem to trans-

late into arguments against a time of creation, in obvious contradiction to Christian doctrine. Cf. AT V 52–53 (K 221–22); AT XI 656.

15. Later, in chapter 9, we shall have to look into the cause of duration. Though duration is an attribute of every finite substance, Descartes holds that there must be something that maintains any finite substance in existence, and that something cannot be the body or the mind by itself; see, e.g., AT VII 110, 369–70, etc. And so, as with the notion of motion to which it is closely linked, Descartes carefully separates out the notion of duration, which pertains to bodies and minds, finite substances, and distinguishes it from our notion of its cause, God, who, as we shall see, sustains the world through his continual recreation.

16. Motion (which essentially involves duration) also fits into Descartes' geometrical conception of body in this way, as we shall discuss below in chapter 6.

17. It is important to distinguish this notion of form, synonymous with nature or essence, from physical conceptions of form; see note 3 above.

18. See, e.g., Aquinas 1950, p. 81 (Aquinas 1965, pp. 8–9); and Aquinas 1948, pp. 3–6 (Aquinas 1965, pp. 34–36). See also the discussion in Thomas, *Summa Theologiae* Iq76a4c.

19. Aquinas 1948, p. 43 (in Aquinas 1965, p. 63).

20. Eustachius 1648, part 4, p. 53. "Accidental form" is used here synonymously with nonessential accident.

21. See Aquinas 1950, p. 83 (in Aquinas 1965, pp. 10–11); Gracia 1982, pp. 253–54; D. P. Henry 1982, p. 128.

22. See Aristotle, *Topics* I.5, 102a 17ff; Goclenius 1613, p. 28; Gracia 1982, p. 178.

23. Previous writers seem to used the term either for divine attributes, or more generally for any predicate, accident, etc. See Goclenius 1613, p. 131.

24. See the *Notes against a Program,* AT VIIIB 350, 355, in which Descartes chides Regius for departing from the standard philosophical use of the term 'mode'. For an account of the wide variety of uses the term 'mode' had, see Goclenius 1613, pp. 694ff, and Suarez, *Disputationes Metaphysicae* 7.1.17 (Suarez 1947, p. 28).

25. For a summary account of the late scholastic theory of distinctions, see disputation VII of the *Disputationes Metaphysicae,* translated in Suarez 1947.

26. For a discussion of Descartes' view on the modal distinction in historical context, see Wells 1965.

27. *Disputationes Metaphysicae* 7.1.17 (Suarez 1947, p. 28); Suarez makes clear here that this is just one among many uses of the term. See also Gracia 1982, pp. 231–32.

28. For Descartes' answer to Caterus, see AT VII 120–21. In Pr I 62, which could not have been written too long after he wrote the *First Replies,* Descartes admits to his readers that in that answer, he had run together the modal distinction and the distinction of reason.

29. It is interesting to observe that almost all of the uses of the term 'mode' cited on p. 65 appear to postdate the reply to Caterus.

30. The quotations below all come from AT VII 79–80; for fuller treatment of the argument, see Gueroult 1984, vol. 2, chap. 14; Garber 1986, pp. 104–7.

31. See AT VII 52, where this claim is first announced.

32. See AT VII 80–90, where Descartes acknowledges the general reliability of certain "teachings of nature" and reintroduces the senses in a limited way, under the control of reason, leading ultimately to the final argument of the *Meditations,* the answer to the dream argument of Meditation I.

33. On the apparent weakening of the claim ("we *seem* to ourselves clearly to see"), see Descartes' remarks to Burman, AT V 167 (Descartes 1976, p. 34).

34. See, e.g., Pr I 46, 66–70. The senses are also treated briefly in Pr IV 189–98, but Descartes' main interest here is to give a brief foretaste of the physiology he had intended to present in the unwritten fifth and sixth parts of the *Principles.*

35. On the relations between the *Meditations* and part I of the *Principles,* see Garber and Cohen 1982.

36. Prost 1907, chap. 2, argues for reading Descartes' (infrequent) use of the term 'occasion' in contexts like this as an indication that he accepted occasionalism in something like the sense that the later Cartesians did; Prost's claims are rejected in Gouhier 1926, pp. 83–88; Laporte 1950, pp. 225–26. For general discussions of the term, see Battail 1973, pp. 141–46; Cordemoy 1968, pp. 322 n. 10; for a general discussion of the language of indirect causality in Descartes and the later scholastics, see Specht 1966, chaps. 2 and 3.

37. See also the review of the state of mind of the meditator at the start of the project that Descartes gives in the beginning of Meditation VI, AT VII 74–75.

38. See also AT VII 8. Elsewhere Descartes talks of things existing objectively in an idea; see AT VII 161. For accounts of the history of this notion, and where Descartes may be borrowing it from, see, e.g., Dalbiez 1929, Cronin 1966, Nadler 1989, pp. 147–65, and Normore 1986. While Descartes clearly seems to be borrowing the term from the schoolmen, Gilson argues that Descartes understands objective existence in the mind in a sense much stronger than at least his Thomist contemporaries would. According to Gilson, for a Thomist, the object of an idea is an entity of reason, and needs no cause, unlike in Descartes, where the necessity of a cause for an objective reality generates the Meditation III argument for the existence of God, and, indeed, the argument for the existence of body. See Gilson 1967, pp. 321–23. Hence Caterus' objections to the Meditation III argument, AT VII 92 ff.

39. See also Gassendi's dismissive characterization at AT VII 288–89. For an account of the notion, see O'Neill 1983, chaps. 1 and 2; O'Neill 1987, pp. 235–40.

40. See also the accounts of the wax in Garber 1986, pp. 99–101, and Carriero 1986, pp. 214–17.

41. The French translation adds the following clause, paraphrasing the first sentence of the quotation: "that is, that one can conceive everything necessarily found in the wax, without having to think of them [i.e. color, hardness, and shape]" (AT IXA 136).

42. Pr II 4F adds a positive statement about their nature: "and that its nature consists in this alone, that it is a substance which has extension."

43. One can find very similar arguments in both Galileo and Locke, for example; see Drake 1957, p. 274; Locke, *Essay Concerning Human Understanding,* bk. II, chap. VIII, sects 8ff.

44. See Aristotle, *Physics* 209a 5–6; Toletus 1599, ff. 103r, 106r; but see also Eustachius 1648, part 3, p. 149.

45. It is interesting to note, though, that Descartes does make appeal to a distinction very similar to that the schoolmen drew between essence and proper accidents in writing to Henry More, 5 February 1649, AT V 269 (K 238). This passage will be discussed below in chapter 5.

46. Note that insofar as it is a thing, it has certain properties, like duration, that are not geometrical in the narrowest sense, i.e., not modes of extension taken narrowly, as discussed earlier in this chapter.

47. On the care with which Descartes chose titles for the individual Meditations, see AT III 297 (K 94).

48. See Descartes' remarks to Gassendi, AT VII 380–81.

49. Descartes here ignores the possibility that these objects might be not in our minds, but in God's. Later in the century, Malebranche will argue for exactly such a position. See, e.g., *De la recherche de la vérité*, bk. 3, part 2, chap. 3, RM I 422–28 (Malebranche 1980a, pp. 222–25). There Malebranche appeals to an argument similar to the one we have been discussing in Descartes to establish that we cannot be the authors of such ideas.

50. Or, perhaps, to union of mind and body; see the discussion later in this chapter.

51. In Meditation III such qualities are called materially false, getting Descartes into some trouble; see Wilson 1978, pp. 101–19.

52. Malebranche was much struck by this contrast; for him it entailed that we have no idea of mind. See *De la recherche de la vérité*, bk. 3, part 2, chap. 7, §4, RM II 451–53 (Malebranche 1980a, pp. 237–39).

53. At the time he drafted Meditation VI, it is not clear that Descartes had settled on his formal definition of substance; note the discussion of Descartes' concept of substance earlier in this chapter. Descartes never uses the word *substantia* in the Latin text of the proof for the distinction between mind and body. But the term does appear in the French translation; compare AT VII 78 with AT IX A 62. Furthermore, the term appears earlier, in Meditation III, in connection with examples of minds and bodies (AT VII 44) and in the paragraph immediately following the argument for the distinction between mind and body (AT VII 78). I shall assume, then, that the argument for the distinction between mind and body is intended to establish that they are separate substances.

54. See Donagan 1978, pp. 192–93.

55. 'Act *[actus]*' is not to be understood as an action, but in the scholastic sense, as an actuality, the realization of a potentiality and thus something real. See, e.g., Goclenius 1613, pp. 46ff.

56. Again, 'act' is to be understood as a technical term. The French translation of this passage has an interesting variant; instead of saying that all corporeal acts "agree in the common concept of extension," the French says that "they agree with one another insofar as they presuppose extension" (AT IXA 137). See also AT VII 78, 121, 423–24.

57. See also Pr I 63. Descartes seems to take a somewhat different point of view in his conversation with Berman; see AT V 156 (Descartes 1976, p. 17).

58. At the end of the century, Locke will challenge Descartes' confident

claims to have seen into the natures of mental and material substance, and propose exactly this, that for all we know, mind and body may share a common substratum; see *Essay Concerning Human Understanding,* bk. 4, chap. 3, §6. In his comments on the *Meditations,* Gassendi is similarly skeptical that Descartes has shown us the natures of mind and body; see AT VII 266, 338; but he does not seem to draw the radical conclusion that Locke does.

59. Also relevant is Descartes' use of the verbal form, 'inform' in both the *Rules* and the *Principles* in connection with the mind and body; see AT X 411; Pr IV 189.

60. See, e.g., the letters from Elisabeth to Descartes, 6/16 May 1643 and 10/20 June 1643 (AT III 661–62, 684–85), and Gassendi's reaction in the *Fifth Objections* (AT VII 343–45).

61. Though his main point is that sensations and passions more generally are modes of both mind and body at the same time, "stradling modes" in his terminology, Paul Hoffman has recently suggested that sensations are modes of the mind-body union; see Hoffman 1990, p. 318. Some have suggested further that Descartes saw the human being as a third kind of substance over and above mind and body; see, e.g., Laporte 1950, p. 183; Hoffman 1986; Broughton and Mattern 1978; Rodis-Lewis 1971, pp. 353, 543 n. 29. Cottingham 1986, pp. 127–32, suggests a kind of intermediate view. On that view, there are three kinds of attributes, mental, material, and sensory. Though there are substances associated with the first two kinds of attributes, he argues that Descartes did not associate the third with a distinct substance.

CHAPTER FOUR

1. Descartes' general ignorance of the philosophy of the schools did not prevent Mersenne from telling those who questioned Descartes' knowledge of the tradition that he "knows it as well as the masters who teach it" (AT II 287).

2. See, e.g., Eustachius 1648, part 3, pp. 118–19.

3. Eustachius 1648, part 3, p. 119. See also the excerpts from the Coimbrian commentary on Aristotle's *Physics* I.9, in Gilson 1979, §§272–73; St. Thomas' account in chaps. 1 and 2 of his *De principiis naturae,* Aquinas 1950, pp. 79–86 (in Aquinas 1965, pp. 7–14); Gracia 1982, pp. 229–30; McMullin 1963. Though quite central to medieval Aristotelianism, it is not clear that this notion is actually found in Aristotle himself; see, e.g., Charleton 1970, pp. 129–45.

4. See, e.g., Eustachius 1648, part 3, pp. 123–24; Gracia 1982, pp. 216–17.

5. Gilson 1979, §209. The passage in question is from the Coimbrian Commentary on *Physics* I.9–10. In this context, natural things are simply things that have their own natures. Thus substantial forms are sometimes identified with natures; see, e.g., Eustachius 1648, part 3, p. 140. The substantial form was not the only kind of form the schoolmen recognized. In addition to the substantial form, which distinguished one kind of substance from another, there were accidental forms, like being white or black, which distinguished things that may be the same kind of substance. On the distinction between substantial and accidental forms, see the excerpt from Toletus' commentary on *Physics* V.1 in Gilson 1979, §210.

6. Descartes mentions real qualities in a number of places: AT III 648 (K 135); AT III 667 (K 139); AT V 222 (K 235); AT VII 441–42; etc. The notion of a real quality was not a common one in scholastic philosophy, so far as I can tell. Goclenius mentions it only briefly in his lexicon entry on 'qualitas', saying that "real [qualities] are the principle qualities," in contrast with "intentional qualities," which derive from real qualities, and that "the alteration of a real quality is the way for generation," where generation presumably means the change of a thing from one sort of substance to another. See Goclenius 1613, p. 912. This is consistent with Descartes' use of the term. His most frequent example of a real quality is heaviness, a quality that, on standard Aristotelian doctrine, pertains to the elements earth and water by their nature (form), as lightness pertains to elemental air and fire. Such qualities are often called motive qualities, and are usually treated together with the closely related primary qualities, moistness, dryness, heat, and cold, which are also used to characterize the elements. See, e.g., Eustachius 1648, part 3, pp. 207–11, and Goclenius 1613, pp. 912–16. Descartes probably meant to include the scholastics' primary qualities among his real qualities. Unlike the scholastics Descartes does not carefully distinguish what he calls real qualities from substantial forms, and insofar as I am here attempting to give a sense of what the scholastic opponent looked like to Descartes, neither shall I.

7. Gilson 1979, §355; the passage is from the commentary on *Physics* VIII.4. See also Eustachius 1648, part 3, pp. 210–11.

8. For examples of some of the discussion of substantial forms in connection with the Voëtius affair, see, e.g., AT III 488, 512ff (Verbeek 1988, pp. 98–99, 105ff); AT VIIIB 62, etc.

9. In this passage Descartes is actually talking about real qualities, which in general he doesn't distinguish carefully from substantial forms.

10. Descartes uses a similar critique in attacking Roberval, who had proposed a kind of theory of universal gravitation; see AT IV 401. See also AT III 424 (K 112); AT III 667 (K 139); AT V 222–23 (K 236).

11. On the connection between will and intellect in Descartes, see, e.g., AT III 372 (K 102–3).

12. See Garber 1986, §I.

13. Color is a favorite example in this context.

14. See AT III 420 (K 109); AT III 667 (K 139); AT VII 441–42. Other prejudices still are noted in Pr I 71.

15. The question of the vacuum among the atomists is treated below in chapter 5. For a summary of Kepler's views on color and vision, see Schuster 1980, pp. 61–62.

16. Scholastic philosophy and common sense are also linked in AT VII 441–42, AT II 213. See Gilson 1975, pp. 168–73.

17. For a more explicit development of what is basically the same idea, see Malebranche's remarks in his *De la Recherche de la Vérité*, bk. 1, chap. 16 (RM I 165–70) (Malebranche 1980a, pp. 73–75) and bk. 6, part 2, chap. 2 (RM II 300–308) (Malebranche 1980a, pp. 440–45).

The case I have been making is most obvious for notions like heaviness, which Descartes' remarks allow us readily to see as an imposition of the mental onto the world. A similar case can also be made for other notions. Heat and cold, two of the four primary qualities which, for the scholastics, derive from the form (cf., e.g., Eustachius 1648, part 3, pp. 207–10) and thus are taken really to be in bodies, can be seen as impositions of the (mental) sensations of warmth and cold onto the world. Color, which was sometimes also taken to be in things (see, e.g., the excerpts from the Coimbrian commentary on *De anima* II.7 in Gilson 1979, §105), could be similarly regarded.

18. This question is taken up in Gilson's essay, "De la critique des formes substantielles au doute méthodique," in Gilson 1975, pp. 141–90, still the best discussion of Descartes' refutation of substantial forms in the literature. Gilson's account of the later scholastic conception of the substantial form emphasizes the sense in which it was considered an active cause and its connection with the idea of the soul (pp. 154–62); see also Reif 1969, p. 27. But Gilson argues (pp. 162–63) that Descartes departs seriously from scholastic doctrine in seeing matter and form as separate substances. Here, though, Gilson seems to be too much the Thomist, and seems to ignore other currents in scholastic thought that allow the form and its matter greater autonomy. See, e.g., Fontialis 1740, pp. 73f; Reif 1969, p. 26; Weisheipl 1963; Knight 1962. On the relation between Descartes' conception of the soul as form and the scholastic conception, see Gilson 1975, pp. 245–55; Gilson 1967, pp. 431–32; Gouhier 1978a, chap. 13. For a different view, see Adams 1987, 2:690–95. Adams argues that it is wrong to think of Ockham's view as an anticipation of Descartes'. But whether it is a good interpretation of Ockham or not, this possible misreading of the schoolmen may have been important in suggesting the position to Descartes and his contemporaries.

19. Later in *The World,* the extension of a body is described as "its true form and its essence" (AT XI 36), and in a contemporary letter, from 1632, Descartes tells Mersenne that his goal is "to know a priori all of the different forms and essences of terrestrial bodies" (AT I 250). This use of form seems to persist into the 1640s; see the letter to Regius, January 1642, where Descartes contrasts the scholastic use of the notion of form with his own more intelligible conception (AT III 506). The word 'form' is, of course, also used to mean 'shape'; see, e.g., AT II 485, Pr III 47.

20. This dependence on the scholastic terminology of matter and form is also obvious in later Cartesians. Henricus Regius, for example, giving what he intends to be a representation of Descartes' views in his *Fundamenta Physices* of 1646, begins the account by dividing the study of natural things into two parts, the study of matter, i.e., material substance, and the study of form, i.e., shape, size, motion, and the human soul; see Regius 1646, pp. 1–29. This use of scholastic terminology to characterize Cartesian doctrine was also common among other followers of Descartes. It is important, though, to distinguish this tendency from the view, popular among many thinkers in the seventeenth century, that Descartes was genuinely consistent with Aristotelian teachings. For a good treatment of this movement, see especially Mercer 1989, chap. 2; see also Weier 1970; John Henry 1982, pp. 213–15, 233; Thijsson-Schoute 1950, pp. 233–35; Mouy 1934, pp. 172–74.

21. It is alluded to in Pr IV 201L, an allusion that is made explicit in Pr IV 201F. There is another possible allusion in Pr IV 203F.

22. Note also Descartes' remarks to J.-B. Morin in 1638, AT II 201–2. Descartes' dealings with Morin are treated at greater length in Garber 1988a or Garber 1988b.

23. See also AT IV 157; Pr IV 200. These passages pursue the theme that Descartes' philosophy is not novel, but, in fact, quite ancient. In the letter to Dinet, AT VII 580, this theme is repeated, but with the twist that it is really Aristotle, not Descartes, who is the innovator!

24. This is also evident in Regius' exposition of Cartesian physics. After his exposition of matter (i.e., material substance), and form, both general form (i.e., size, shape, motion), and special form (i.e., the human mind), Regius makes clear that he rejects "the primary matter of the vulgar" and "all substantial forms"; see Regius 1646, pp. 30ff.

25. As I noted earlier, in its earliest conception, in late 1640, the *Principles* was to contain a direct refutation of portions of Eustachius' textbook, a direct confrontation between the old philosophy and the new. Part of the reason Descartes abandoned this project was because of Eustachius' death on 26 December 1640, which prevented him from getting Eustachius' permission to use his book in that way; see AT III 260, 286. But, as Descartes told Mersenne in December 1641, he also came to see that such a direct confrontation was unnecessary:

> It is certain that I had chosen the compendium of Father Eustachius as the best, had I wanted to refute anyone. But it is also true that I have completely given up the plan of refuting this philosophy, since I see that it is so completely and clearly destroyed by the establishment of mine alone, that no other refutation is needed. (AT III 470)

This echoes comments Descartes made earlier that year to Mersenne, in January 1641, before the *Meditations* appeared. There Descartes suggested that direct confrontation is not only unnecessary, but also unwise, since it might cause otherwise receptive readers to reject his new idea; see AT III 297–98 (K 94). This attitude is also reflected in Descartes' advice to Regius in January 1642, where he explains why the scholastic view is only barely mentioned in the *Discourse* and *Essays:*

> I should also greatly prefer that you never put forward any new opinions, but retaining all of the old ones in name, only bring forward new arguments [*rationes*], something no one could criticize. Those who properly grasped your arguments would spontaneously conclude from them that which you wanted to be understood. For example, why did you openly reject substantial forms and real qualities? Don't you remember that I quite explicitly remarked in the *Meteors,* page 164, that I in no way rejected or denied them, but only did not need them for explaining my position [*rationes*]? If you had done the same, then everyone in your audience would have rejected them when they saw that they were useless, but you would not have fallen into such unpopularity with your colleagues. (AT III 491–92 [K 126–27])

The passage from the *Meteors* is AT VI 239 (Ols. 268), discussed above. Descartes' rather accommodating remarks to Jesuits like Charlet can also be explained in this way.

26. The reference is, of course to the affair with Regius and Voëtius.

27. On the dispute, see Bizer 1965, pp. 21-27; Cohen 1920, pp. 535-78; Verbeek 1988; and Verbeek 1991, chap. 1. The supposed denial of substantial forms was a main theme in Voëtius' attacks on the new philosophy in December 1641, attacks that started the dispute off; see the excerpts from the theses debated: AT III 488, 511-19 (Verbeek 1988, pp. 98-99, 104ff). For Descartes' remarks on the theses, see AT III 494f, AT VII 587; AT VIIIB 32; AT VIIIB 210. While Descartes was not named in the theses Voëtius published on that occasion, he made it clear in the public disputation that it was Descartes who was his target; see AT VII 587. The question of substantial forms and why they must be rejected is thus prominent in Descartes' advice to Regius in January 1642 as to how he should respond, advice that Regius seems to have taken in his reply on 16 February 1642. For Descartes' advice, see AT III 500, 501-2, 503, 505, 507; on Regius' reply to Voëtius, see Descartes' remarks in AT III 528ff and Verbeek 1988, pp. 116ff, as well as the excerpts given in its condemnation by the Academic Senate at the University of Utrecht, AT III 552-53, and AT VII 591-92 (Verbeek 1988, pp. 121-22).

28. See also Schoock's remarks in the context of a proceeding against him initiated by Descartes in Groningen in 1645; AT IV 180.

29. The theses are now lost. Descartes refers to them in a letter to Pollot, 8 January 1644, particularly emphasizing the issue of substantial forms; see AT IV 77-78. Excerpts from Revius' answer to Heereboord are given in AT IV 655-56. At that disputation Heereboord claimed to be for, but was really against, forms, causing some confusion, as Descartes reports in AT IV 78. For an account of the incident, see Verbeek 1991, chap. 2.

30. Gervais de Montpellier 1696. One suspects, though, that by this time hylomorphism had largely been defeated. The satire was published anonymously; for the attribution to Gervais de Montpellier, see Bouillier 1868, 1:577 note 1.

31. On logic, see AT X 363-64, 405-6, 430, 439-40, and on the general uncertainty of scholastic philosophy, see AT X 363, 367-68.

32. It is interesting in this connection to note also the treatment of the vacuum in chapter IV of *The World,* where the strong a priori arguments against the void, so prominent in later works are overlooked in favor of rather weaker arguments; we shall discuss this more fully below in chapter 5.

33. See AT I 402, where Fromondus gives a dismissive characterization of Descartes' mechanistic program. Descartes' correspondence with Fromondus is discussed in Garber 1988a and 1988b. This does resemble Bacon's critique of the Aristotelian philosophy as all talk and no works. In the preface to the *Great Instauration,* for example, Bacon writes that "that wisdom which we have derived principally from the Greeks is but like the boyhood of knowledge, and has the characteristic property of boys: it can talk, but it cannot generate, for it is fruitful of controversies but barren of works" (Bacon 1960, pp. 7-8). However, unlike

Bacon, Descartes is not thinking of technological success but explanatory success.

34. For some of the serious problems with mechanical explanation in seventeenth-century thought, see Gabbey 1985, pp. 9–28, and Gabbey 1990. Gabbey argues that "many mechanistic explanation of natural phenomena were in their own way just as circular or tautologous as the corresponding explanations from the Peripatetic tradition" (1990, p. 274).

35. The French version is a little different: "even if I have, perhaps, imagined causes which could produce effects similar to those that we see, we ought not to conclude from that that those we see are produced by them [i.e., by these causes]."

36. Many years later, in 1699, Leibniz used such an argument to try to convince the Cartesian de Volder to accept substantial forms; see G II 194 (Leibniz 1969, p. 522).

37. See the development of this argument in "De la critique des formes substantielles au doute méthodique" (Gilson 1975, pp. 141–90). Gilson sees this as Descartes' principal argument.

38. See the references cited above in note 18.

39. See Eustachius 1648, part 3, p. 247; AT I 154 (K 16).

40. See Rosenfield 1968 for the standard account.

41. Descartes' view is that while sensation and imagination require a body attached to the soul, in pure reasoning the soul acts alone. Margaret Wilson emphasizes this feature of Descartes' dualism; see Wilson 1978, pp. 177–85.

42. See also AT IV 574 (K 206); AT V 278 (K 244–45). Descartes makes it clear in these passages that he excludes language purely expressive of passions.

43. Something is morally certain if it has certainty sufficient for daily life, even if it isn't sufficiently certain for philosophical purposes. See Pr IV 205F–206F, and Gilson 1967, pp. 358–59. When Descartes refers here to a moral impossibility, his point is, presumably, that we can be morally certain that it is impossible.

44. See AT I 413 (K 36); AT III 121; for the sense in which they may have sensation, see *Treatise on Man*, passim, and AT VII 436–37. In the latter passage, Descartes distinguishes three grades of sensation. Animals can partake of the first grade, but not the second or third.

45. Descartes sometimes implies that they have passions in an extended sense; see, e.g. AT IV 574 (K 206–7).

46. It may be this that Descartes had in mind as early as 1619 when, reflecting on the unusual ability animals have for performing certain very specific tasks, Descartes noted: "From certain very perfect actions of animals we suspect that they lack a free will" (AT X 219).

CHAPTER FIVE

1. See esp. Schmitt 1983; Brockliss 1987.

2. See Aristotle, *De Generatione et Corruptione* I.2, 8; *De Caelo* III.4. See also the account of atomism in Maimonides, *Guide of the Perplexed* I.73.

3. See Grant 1974, pp. 312–24; Murdoch 1982, pp. 573ff. See Lasswitz 1890, bk. 1, for a general account of atomism before the Renaissance.

4. See Boas 1952, pp. 422f; Jones 1981, pp. 214–15; Boas (1952, pp. 424, 431f) also cites Galen, Cicero, and Hero of Alexandria as sources for Renaissance atomism.

5. On discussions of atomism in the sixteenth century, see Boas 1952, pp. 425–26; Jones 1981, pp. 215–25; Lasswitz 1890, 1:263–401.

6. See Boas 1952, pp. 426–28, 435–37, 439–42; Jones 1981, p. 299, n. 24; Le Grand 1978.

7. See Boas 1952, pp. 437–39. Beeckman seems also to have been an influence on Gassendi; see Jones 1981, p. 282; Joy 1987, pp. 39, 241.

8. On early seventeenth-century atomism, see Gregory 1964, Gregory 1966, John Henry 1982, and Kargon 1966. On Gassendi, see especially Joy 1987. It should be noted that while these atomists shared a belief in the existence of atoms in *some* sense, they differed quite radically from one another and often strayed from the sort of classical Epicurean atomism of the sort that Gassendi, for example, espoused.

9. See, e.g., Spink 1960, part 1; Kargon 1966; John Henry 1982.

10. See Diogenes Laertius, *Lives* X 54–55; Lucretius, *De rerum natura* II 333f, 730f; Descartes' views on these questions are discussed above in chap. 3.

11. On the boundlessness of the universe, see Descartes, Pr II 21; Lucretius, *De rerum natura* I 958f; Diogenes Laertius, *Lives* X 41–42. On the multiplicity of stars and planetary systems, see Descartes Pr III 53, 65ff; Lucretius, *De rerum natura* II 1048f; Diogenes Laertius, *Lives* X 45. It should be noted that in these sections, Descartes prefers to call the world indefinitely extended rather than infinite. Furthermore, he denies that there is a plurality of worlds. This is not meant to deny that there is a plurality of suns or planetary systems, but a plurality of worlds in the Aristotelian sense, a plurality of distinct physical systems, collections of suns, planets, etc., each with its own center, lunar sphere, etc., that stand outside our universe; see Duhem 1985, part 5.

12. See Descartes, Pr III 47, AT II 485; Lucretius, *De rerum natura* V 416f.

13. See Descartes, Pr III 47; Lucretius, *De rerum natura* V 416ff. For Lucretius, the world starts with uniformity, and chaos arises from a swerve; see Lucretius, *De rerum natura* II 216ff. This is not in Epicurean texts that survive; see Rist 1972, p. 48. It is important to note that Descartes, almost certainly worried about the biblical account of creation in *Genesis,* presents his view as a myth in *The World* and in the *Principles,* argues that his creation story is a false hypothesis, like those astronomers use in their systems; see *The World,* chap. 6, and Pr III 43–47. On Descartes and the book of *Genesis,* see also AT III 296 (K 93), AT IV 698, and the charming story about Descartes and Anna Maria van Schurman, AT IV 700–701.

14. See Descartes, Pr I 28; Diogenes Laertius, *Lives . . .* X 81; Lucretius, *De rerum natura* IV 823f. It should be noted that Descartes does not seem altogether consistent in his rejection of final causes. E.g., he seems to allow them in Meditation VI, where he argues that the senses were given us to preserve the mind-body union. They even seem to enter into his physics from time to time, as we shall see in connection with the discussion of the laws of impact below in chapter 8.

15. A notable exception is the careful and scholarly Henry More, who cites Lucretius *against* Descartes; see AT V 239, 241.

16. See also AT II 51, 396; AT III 166. For an account of Descartes' correspondence with Fromondus, see Garber 1988a or 1988b.

17. This is evident, e.g., in Descartes' rather ungenerous response to Beeckman in his letter to Beeckman, 17 October 1630, AT I 157–67 (excerpted in K 16–17). Beeckman had referred to Descartes as "son écolier et son serviteur," calling down Descartes' wrath. See Baillet 1691, 1:206.

18. See Diogenes Laertius, *Lives* X 63–67; Lucretius, *De rerum natura* III 161f.

19. For an account of some of the so-called Christian mortalists in the early seventeenth century, see Burns (1972).

20. On the ancient atomists, see cf. Rist 1972, chap. 8; on Descartes' attempts to disassociate his mechanism from these views, see, e.g., AT III 215 (K 80); AT III 428 (K 114–15). Gassendi also worked very hard to separate the aspects of the Epicurean system he accepted from these and other non-Christian doctrines; see, e.g., Osler 1985. Despite his caution, contemporaries accused Descartes of both materialism and atheism; see Caton 1973, pp. 10ff.

21. On the atomists, see Diogenes Laertius, *Lives* X 60–61, and Rist 1972, p. 46f; on Descartes, see Pr IV 202. Ultimately, of course, whatever tendencies bodies have derive from God, for Descartes.

22. See Rist 1972, chap. 3.

23. Diogenes Laertius, *Lives* X 41–42; the translation is by R. D. Hicks in the Loeb Classical Library edition. See also X 54 and Lucretius, *De rerum natura* I 483ff.

24. JB and AT differ slightly here, JB (properly) treating 'atomus' as feminine, and AT treating it as masculine.

25. One cannot infer that Descartes believed in atoms simply from his use of the term. In *The World,* AT XI 14–15, he uses the term, though, as we shall see, correspondence shows that he had rejected atoms in the strict Epicurean sense by 1630, and earlier in that same chapter he had claimed that all matter is indefinitely divisible. Later in this chapter I shall discuss the possibility that in his youth Descartes also accepted the void.

26. See Sabra 1967, pp. 112ff.

27. It is interesting to note that Descartes seems unaware of or uninterested in a tradition of arguments trying to establish that the geometrical continuum requires indivisibles of a sort. See, e.g., Fromondus 1631, chaps. 32f. Descartes refers to this book on one occasion, in a letter intended for its author; see AT I 422 (K 39). But there is no real evidence that he actually read it with any care.

28. Descartes is often quite happy to talk about the infinite divisibility of body; see, e.g., AT I 422 (K 39), AT III 36, AT VI 239 (Ols. 268), AT VII 106, etc. But he is sometimes more cautious with the notion of infinity. Reserving the notion of infinity for God alone, he sometimes argues that we should use the weaker term 'indefinite' when discussing things in the created world. See, e.g., Pr I 26, AT V 274 (K 242). For a general account of the distinction, see Ariew 1987a. It is not entirely clear what he means by the term, either in general or in its specific application to the doctrine of the divisibility of matter. (Nor was it clear to his

contemporaries; Gassendi, for one, found it quite puzzling. See Gassendi 1658, 1:263A.) Sometimes the point seems to be that however far we may divide a body, we can always divide it further. See, e.g., Pr I 26; Pr II 34. Sometimes the point seems to be more epistemological, that while we can see no end to the division of body, God may be able to. See, e.g., AT V 167 (Descartes 1976, pp. 33–34), AT XI 12, 656, AT V 273–74 (K 241–42). Since a finite magnitude can be made up of an infinity of parts, each of which can be divided still, as Descartes recognized (see AT IV 445–47 [K 198–99]; AT IV 499–500), both of these conceptions of indefinitude are consistent with the division of a body into an actual infinity of extended parts, something Descartes thinks arises on occasion, as we shall see later in this chapter.

29. See Sorabji 1982, p. 65f; Gassendi 1658, 1:263ff; Charleton 1654, pp. 23, 85, 95; Jones 1981, pp. 286–87; Joy 1987, chap. 7.

30. Gassendi 1658, 1:264A.

31. On Aristotle, see Sorabji 1982, p. 55; but cf. Miller 1982, p. 100. Toletus 1599, f. 170r, lists mathematical atomism and physical atomism as competing positions. Fromondus 1631 is directed explicitly against Epicurus and Epicurean atomism (see *Ad Lectorum* **), though the emphasis is on mathematical atomism. See, e.g., p. 56, where Fromondus distinguishes mathematical arguments from physical arguments. It is interesting to note that the physical arguments include, e.g., the tortoise paradox, where the position in question is still mathematical atomism. On Gassendi's efforts to correct the historical mistake of confusing the two atomisms, see Joy 1987, chap. 7.

32. See Rist 1972, chap. 8.

33. See, e.g., Gassendi's discussion of causes in the *Syntagma Philosophicum*, part II, book IV (Gassendi 1658, 1:283ff). There he asserts that God is the first cause (chap. 2), that he is the author of the world (chap. 5), that God is the "governing cause" of the world (chap. 6), and that he takes a special interest in humankind (chap. 7). Furthermore, in his *Syntagma Epicuri*, his exposition of the views of Epicurus and his school, Gassendi takes special care to let his readers know that he himself does not subscribe to the Epicurean views on the gods; see Gassendi 1658, 3:14.

34. Gassendi 1658, 1:308 A.

35. It is important to note that for Descartes, not *all* bodies are in this state. In addition to this fluid matter, actually divided into parts indefinitely small, what Descartes calls the *first element,* there are two other elements, the *second element,* composed of tiny balls, and the *third element,* composed of larger, largely irregular chunks. See, e.g., *The World,* chap. 5, and Pr III 52. But though the particles of the second and third elements are in actuality undivided, they are made up of the same extended substance as the first element, and so they are in principle (and in practice) divisible.

36. See, e.g., AT I 139–40 (K 9–10); AT XI 24; AT II 483–84. In this claim, Descartes would seem to depart from the strict Aristotelian position, in accordance with which the continuum, while divisible to infinity, can never be divided "through and through," that is, the infinite divisibility is a *potential* infinity, and a body cannot actually be divided into an actual infinity of parts at the same time. See *De Generatione et Corruptione,* I.2, 316a 15ff; Miller 1982, pp. 90ff; Toletus

1589, f. 171r; Eustachius 1648, part 3, pp. 151–53. Some later scholastics went so far as to say that not even God could understand or perform such an actually infinite division as Descartes imagines. See Toletus 1589, ff. 171v–172r, 176r–176v; Eustachius 1648, part 3, pp. 153–54. But despite this departure from Aristotelian orthodoxy, much of Descartes' position against indivisibles appears very Aristotelian in inspiration. Like Aristotle and his followers, Descartes sees the world of body as continuous and indefinitely divisible, and like the Aristotelian tradition, Descartes sees the mathematical divisibility of any extended quantity as being central to the refutation of the atomist claim that there are genuine indivisibles in nature; cf. Miller 1982, pp. 100–102. But by the early seventeenth century there were numerous twists in the Aristotelian position, numerous arguments for atomism to be responded to and numerous arguments against indivisibles to be considered. Descartes, though, seems utterly uninterested in entering into the full debate. For an account of the arguments surrounding indivisibles as viewed from an Aristotelian point of view, see, e.g., Toletus 1589, ff. 169v–178v; Fromondus 1631. For an atomist answer, see Gassendi's discussion in his *Syntagma,* Gassendi 1658, 1:256B–266A.

37. Lucretius, *De rerum natura* I 421; see also Diogenes Laertius, *Lives* X 39–40.

38. See Diogenes Laertius, *Lives* X 41–42; Lucretius, *De rerum natura* I 368–69, 1008–1020; Rist 1972, pp. 56–58.

39. See Aristotle, *Physics* IV.6–9.

40. See Grant 1974, pp. 45–50, for an account of the affair and a selection of articles from the condemnation most relevant to natural philosophy.

41. See Grant 1979; Duhem 1985.

42. Grant 1974, p. 48 §49.

43. See Grant 1981, chaps. 5–6; Duhem 1985, chaps. 9–10.

44. See Grant 1981; Grant 1976; Duhem 1985.

45. Quoted in Grant 1981, p. 186.

46. Ibid., pp. 183–99.

47. See Grant 1981, pp. 199–206; Henry 1979; Schmitt 1967.

48. See Grant 1981, pp. 213–15; Henry 1979.

49. See Grant 1981, pp. 155–56, 158–59, 162–63, 165–74.

50. *Disputationes Metaphysicae* 51.1.13. The passage is about the notion of '*ubi*' strictly speaking, but in 51.2.2, Suarez identifies '*ubi*' and '*locus*'.

51. See ibid., 30.7.35.

52. See JB IV 44; de Waard 1936, pp. 77–79.

53. For the discussions in 1618, see JB I 260–61 (AT X 58); JB I 263 (AT X 60); JB IV 49–52 (AT X 75–78). For discussions of motion through a vacuum in Descartes' later writings, see AT I 71–72; AT I 228, AT II 442, AT XI 629. For discussions of motion in a vacuum in the medievals, see Grant 1981, chap. 3; Duhem 1985, chap. 9; Weisheipl 1985, chap. 6. It is interesting to note that in at least one discussion of motion in a vacuum, a letter to Mersenne from 15 November 1638, the vacuum is paraphrased as "a space filled only with matter that neither augments nor diminished the motion [of a body]" (AT II 442).

54. See the discussion of Rule 14 in chapter 3, note 1.

55. For the earlier discussion (which occurred two years earlier, in 1629), see AT I 71f.

56. The chapter titles were probably not in the original manuscript; see AT XI iv–v.

57. Later in this chapter we shall discuss Descartes' celebrated doctrine of the creation of the eternal truths as a possible response to this theological problem.

58. See also, e.g., AT V 223 (K 236); AT V 271 (K 239–40); AT V 403 (K 257).

59. See also AT V 194 (K 233); AT V 272 (K 241). Descartes' first reference to the example occurs in a letter to Mersenne, 9 January 1639, in apparent answer to a question that Mersenne had raised; see AT II 482. Descartes' view is that were all body within the vessel annihilated, the sides would touch *"immediatement,"* as he noted in the French edition of the *Principles* (Pr II 18F). Similarly, Samuel Sorbière reported to his friend Pierre Gassendi about a conversation he held with Descartes in early May 1644 where he asked Descartes what would happen if God destroyed the body within a cube, and if the space were thereby destroyed, whether the sides of the cube would have to move in order to meet one another. Descartes, "that most acute gentleman," is reported to have replied that "having eliminated the space by which the walls were separated while the body remained along with the matter, no motion would be necessary" (AT IV 109). This is a conclusion one must accept, Descartes told Mersenne, "or else there would be a contradiction in your thought" (AT II 482). But it is by no means easy to picture what exactly the vessel would look like the moment after God did the deed. Jammer suggests that what Descartes imagines is that the vessel would simply implode due to the pressure of the external atmosphere, though he claims, wrongly as we shall later show, that Descartes had no conception of atmospheric pressure. See Jammer 1969, pp. 43–44. But surely this is *not* what Descartes imagined.

60. For later reactions to this argument, see, e.g., Gassendi 1658, 1:184A; Cordemoy 1968, pp. 103–4, 309–10; Rohault 1969, 1:28.

61. Descartes' use of this distinction is one of the clearest evidences of his recent reacquaintance with scholastic literature, and his desire to write in a way familiar to the schools. The distinction makes its first serious appearance in Descartes' writings only in the *Principles,* and was probably picked up from Eustachius; cf. Pr II 15 with Eustachius 1648, part 3, pp. 155–56, in Gilson 1979, §254. On the history of the distinction, see Grant 1981, pp. 14–19.

62. It is interesting to note that Descartes seemed to have rejected such a notion earlier in the *Rules;* see AT X 426, 433.

63. For an account of the circumstances of the visit and the report, see Pascal 1964–, 2:346–48.

64. See de Waard 1936, chaps. 8–9.

65. For Petit's account of the meeting, see Pascal 1964–, 2:349–59, esp. 351–55.

66. Pascal 1964–, 2:498–508, 677–90.

67. Ibid., 2:507.

68. Ibid.; cf. p. 562 on the caution. For Roberval's version of the experiments and what they show, see Pascal 1964–, 2:459–77. See also Dijksterhuis 1961, pp. 444–48.

69. Pascal 1964–, 2:507.

70. See Adam 1887–88, pp. 65–66. Adam's conjecture is based on a remark

in Jacqueline Pascal's letter to her sister, Madam Périer, relating Pascal's meetings with Descartes in late September 1647; see Pascal 1964–, 2:482.

71. The only copy of that letter is in Pascal's published *Récit*. Bernard Rochot has suggested that the letter was not sent to Périer that early, but that the decision to do the experiment dates only from midsummer 1648. He suggests that the dating of the letter in the published *Récit* was intended to safeguard Pascal's priority in having the idea to do the experiment against claims Descartes had made to have thought of the experiment first; see Rochot 1963, pp. 75–76. Descartes' claims will be discussed below in this section.

72. Pascal 1964–, 2:688. For a general account of Pascal's scientific thought that pays special attention to his work on the vacuum, see Guenancia 1976. See also Fanton d'Andon 1978, where Pascal's epistemology is given special attention in connection with the experiments on the vacuum.

73. See also Adam 1887–88, pp. 66–69. Descartes also considered antiperistasis as a possible explanation, the view that what prevents the liquid from falling in the tube is the fact that the space outside is full, and there is no way in which an appropriate circular flow can be set up. See AT V 116 and de Waard 1936, pp. 88–93.

74. See de Waard 1936, chap. 10.

75. For a general account of the relations between Descartes' thought and Pascal's, both scientific and philosophical, see Le Guern 1971.

76. See Pascal 1964– , 2:481; see also Descartes' remarks in a letter to Mersenne, 31 January 1648, AT V 116.

77. See Pascal 1964–, 2:482; see also Baillet 1691, 2:345.

78. Carcavy had taken Mersenne's place as Descartes' contact in Paris after the latter's death on 1 September 1648; see Baillet 1691, 2:377.

79. On the history of Descartes' acquaintance with the results, see AT V 365–66, 370.

80. Pascal 1964–, 2:689.

81. Ibid., 2:507; AT V 98, 653.

82. See Jacqueline Pascal's remarks on the meeting, Pascal 1964–, 2:481, and Le Tenneur to Mersenne, 16 January 1648, CM XVI 55ff.

83. Carcavy seems to have related to Descartes the results of the experiments, without ever sending him a copy of Pascal's actual pamphlet. See AT V 365–66, 370, 391, 412.

84. Le Tenneur goes on to ask Mersenne: "do you think that carrying a glass tube and 20 pounds of mercury up a mountain that high is easy?"

85. See the discussion in Dijksterhuis 1961, pp. 452–54.

86. Adam (1887–88, passim, esp. pp. 620–21), emphasizes this point. But I think that he is overly generous to Pascal in assuming that he saw it too.

87. Baillet 1691, 2:345; see also the excerpts from Roberval's account of the discussions, AT XI 687–90. The suggestion in Baillet is that Roberval had by then accepted the weight of air as explaining the phenomena.

88. Pascal 1964–, 2:513f, 571–72. For another systematic attempt to reconcile the experiments with Cartesian principles, see Picot to Carcavy, AT V 603–6; note Costabel's comments on AT V 608–9.

89. Pascal 1964–, 2:518–27, 559–76.

90. Ibid., 2:508. Baillet (1691, 2:332) and, following him, Adam (1887–88, p. 621) state that Pascal appended additional arguments against the subtle matter to the printed version of the *Expériences nouvelles* that Huygens sent to Descartes. But the letter on which Baillet bases his claim, Descartes to Mersenne, 13 December 1647, AT V 98, apparently unknown to Adam in 1888, doesn't seem to support such a claim.

91. See Carcavy's last surviving letter to Descartes, 24 September 1649, AT V 412.

92. Baillet 1691, 2:346.

93. See Pascal 1964–, 2:518, 529.

94. Ibid., 2:513.

95. Ibid., 2:514–15.

96. Ibid., 2:520–21.

97. See Gassendi 1658, 1:205A; Charleton 1654, pp. 42–44.

98. On More's later views and their influence, see Grant 1981, pp. 221–28, 244–45; on the general relations between More's thought and Descartes', see Gabbey 1982.

99. See also AT V 301, 379. In these latter passages More adds that the human soul is extended, as well as God and the angels.

100. It is interesting to compare More's position here with that of his arch-enemy, Thomas Hobbes, who also holds that everything that really exists occupies space. See Hobbes 1951, pp. 428–29, 689–90. Hobbes, of course, concludes from this that everything is body, More, that there must be something to body other than just extension.

101. See also AT V 270 (K 239); AT V 343 (K 250); AT V 403 (K 257).

102. Here I am indebted to the brilliant discussion in Gabbey 1980, pp. 299–300 n. 27.

103. See also Pr II 4, where Descartes explicitly excludes the sensory notion of hardness *(durities, dureté)* from the nature of body.

104. See also Gabbey's textual note, AT V 676.

105. See, e.g., G VI 580 (Leibniz 1989, p. 258); Locke *Essay* 4.7.12–13; Gabbey 1980, p. 299, claims Newton did not make such a mistake.

106. Grant 1976, pp. 139–40, 152, 156–59. Not all medievals denied vacua within nature; see, e.g., the discussion of Nicholas of Autrecourt in Grant 1981, pp. 74–77.

107. See Grant 1981, pp. 109, 324 n. 30, 390–91.

108. See Grant 1979, pp. 225–26; Suarez, *Disputationes Metaphysicae.*

109. See Grant 1981, pp. 84–85, 122–23, 335–36; Grant 1976, pp. 151–52.

110. See Grant 1981, pp. 127–34, 157–63.

111. Ibid., pp. 135–44.

112. Ibid., chaps. 6, 7 passim.

113. See Toletus 1589, ff. 108r–109r, 112v–113r, 116v–122v.

114. Ibid., f. 122v.

115. Ibid. Toletus pays little attention to the distinction between internal and external as it pertains to imaginary place.

116. For a more general account of the distinction among the schoolmen, see

Grant 1981, pp. 14–19. The distinction goes back at least as far as Buridan in the fourteenth century.

117. Toletus 1589, f. 123r col. 1.

118. Ibid., f. 123v col. 1.

119. Ibid., f. 123r col. 1.

120. Ibid., f. 123r col. 2.

121. Ibid., f. 123v col. 1. Proper accidents were discussed briefly in chapter 3. Toletus attributes this view to Philoponus, the sixth-century commentator whose views were very influential in the sixteenth century; see Grant 1981, p. 15.

122. Toletus 1589, f. 123r col. 2.

123. Ibid., f. 122v col. 2.

124. Ibid.

125. Ibid., f. 123r col. 1.

126. Ibid., f. 124r col. 1.

127. Ibid., f. 122r col. 1.

128. Ibid, f. 122v col. 2.

129. See Grant 1981, chaps. 2, 5–7.

130. Toletus 1589, f. 123v col. 1.

131. Toletus 1614, f. 34r col. 1.

132. See also Toletus' characterization of the notion of a property, ibid., f. 33r col. 1; f. 33v col. 1. See also the discussion in chapter 3 above.

133. Toletus 1589, f. 18r col. 2.

134. Ibid., f. 53r col. 1. Leibniz argues similarly that the notion of extension presupposes some (nonmathematical) quality, the quality which is said to be extended; see, e.g., G IV 393–94 (Leibniz 1989, p. 251).

135. See Grant 1979, p. 242.

136. An earlier note in the same series suggests a different position. Descartes wrote: "It is a great argument for truth that something cannot fail to be conceived, and for falsity that it cannot be conceived: for example the vacuum, an indivisible, a finite world, etc. For the one implies being, the other implies non-being" (AT XI 654). The note given in the text, though, seems to represent Descartes' considered view on the matter.

137. The issue of the creation of the eternal truths has received a great deal of attention in recent years, and the literature on this issue is quite enormous. For some recent treatments of Descartes' views, see, e.g., Frankfurt 1977 and Curley 1984. For the view of a neglected contemporary in many respects strikingly similar to the view I am attributing to Descartes here, see Jacobus Fontialis, *De idea mirabilis matheseos de Ente* (1660?), in Fontialis 1740, pp. 437ff. This view is also reputed to be in his *Nodi perplexi, sive Ontologia Labyrinthea* (1655?, pub. 1722); see Gabbey 1970, p. 48. But I have not been able to locate a copy.

138. See AT I 145 (K 11). It is important to remember that Descartes saw the doctrine of the creation of the eternal truths as central to his new physics; so far as I can see, almost none of the many readings of this doctrine in Descartes take account of that fact.

139. See, e.g., Funkenstein 1986, chap. 3; Grant 1979, p. 215.

140. See Gerauld de Cordemoy, *Six discours sur la distinction & l'union du corps*

& de l'âme, in Cordemoy 1968, pp. 85–189. For an account of his thought, see Battail 1973, chaps. 3–4.

141. See, e.g., Boyle's "Author's Discourse to the Reader" in *The Origin of Forms and Qualities according to the Corpuscular Philosophy,* in Boyle, *Works* (1772), 3:7. See also Kargon 1966, pp. 99–100.

142. For an account of some of the battles over atoms and the void in the later seventeenth century, see Prost 1907.

CHAPTER SIX

1. Descartes never indicates whom he might have in mind as those who fail to make the necessary distinction.

2. The notion of motion is discussed in Pr II 24–35, and the causes of motion are discussed in Pr II 36 ff.

3. See JB I 260–63 (AT X 58–61); JB IV 49–52 (AT X 75–78); AT X 219–20.

4. The use of a geometrical notion of motion suggests motion without time. This will be discussed below when we take up the relation between motion and time.

5. This account of motion as change of place is also suggested in the *Rules,* where in Rule 12 Descartes points out that the ambient surface of a body can "be moved *[moveri]* with me in such a way that although the same [surface] surrounds me, yet I am no longer in the same place" (AT X 426).

6. It is not clear to me whom specifically Descartes has in mind with this definition.

7. See the letter to Morin, quoted above, AT II 364.

8. In his letter to Descartes on 23 July 1649, More had asked Descartes "whether . . . that very force or action of motion is in the thing moved" (AT V 384). Unfortunately, Descartes did not respond.

9. For a more standard account of Descartes as relativist, see, e.g., Prendergast 1972; Dugas 1958, pp. 172–73; Westfall 1971, pp. 57–58. For an account of the question of absolute and relative notions of local motion in scholastic thought, see Masi 1947 and Claggett 1959, pp. 585ff, 681.

10. I say 'genuine' rather than 'real' distinction because in the technical sense of the term, the distinction is not real but modal, the sort of distinction that holds between distinct modes of a given substance. My main point is that the distinction between motion and rest is not a mere distinction of reason. See Pr I 60–62; AT XI 656–57.

11. When Descartes defines motion in the proper sense, it is not just in terms of the inner surface, but in terms of the neighboring bodies themselves. But since Descartes holds that the inner surface is itself a mode of those bodies, properly speaking, the difference is not significant; see AT VII 433–34.

12. Gueroult 1980 also attempts to provide such a distinction, though in a way different from the way in which I will try to draw it in what follows. Gueroult, unlike most of Descartes' readers, realizes that it is important for Descartes that motion and rest be two distinct and different modes of body (pp. 204–8). However, he sees the doctrine of the reciprocity of transference as undermining the distinction, and calls for supplementary criteria to enable Descartes to make the

distinction (p. 208). Gueroult finds two such criteria implicit in Descartes' discussion of motion, what he calls *single displacement* and *displacement as a whole*. The criterion of single displacement is the principle that "motion cannot be attributed to a body if, in attributing motion, one must attribute not *one* but several motions to it" (p. 208). This principle can be seen at work in an example Descartes gives in Pr II 30. There Descartes considers two bodies on the surface of the Earth, one AB, being transferred from east to west, and the other, CD, being transferred from west to east at the same time. In this circumstance, Descartes says, "so that we depart least from common ways of speaking, we should say here not that the Earth moves, but only bodies AB and CD" (Pr II 30), for otherwise we would have to attribute two opposing motions to the same body, the Earth. On Gueroult's second criterion, displacement as a whole, "motion cannot be attributed to a given body if, in virtue of this attribution, only part of it and not the *whole body* moves" (p. 208). This is illustrated by Descartes' discussion of the Tychonic theory in part III of the *Principles*. There, as we shall see later in this chapter, Descartes argues that on the Tychonic theory, it is really the Earth and not the Heavens that move diurnally since the whole surface of the Earth is separating from its neighborhood, while only the very small inner concave surface of the Heavens are changing; see Pr III 38. Though ultimately unsuccessful, in Gueroult's opinion, he argues that Descartes thought that these two criteria are sufficient for distinguishing between motion and rest. Motion taken by itself is inevitably relativistic, Gueroult argues, and we can only break this relativity and introduce a genuine distinction between motion and rest by going to the level of cause and introducing forces (pp. 203, 211–12). But whether adequate or inadequate to the task, the supplementary criteria he attributes to Descartes for distinguishing between motion and rest are simply not *intended* to do the work that Gueroult thinks they are. What Gueroult calls the criterion of single displacement is, indeed, suggested in the course of Descartes' discussion of the proper definition of motion. But its function is not to introduce a new supplementary criterion, but only to explain why, despite the reciprocity of transference, we all tend to think, for example, that when we walk on the Earth, it is *us* that is moving and not the Earth. The opening sentence of Pr II 30, where Descartes considers the example of two bodies moving on the surface of the Earth, clearly indicates that what he intends to discuss is the fact that the doctrine of the reciprocity of transference discussed in the preceding section "clashes . . . with the common way of speaking" (Pr II 29) and why, when we see a body in transference with respect to the Earth, "we do not on that account think that [the Earth] itself is moving" (Pr II 29). And immediately after explaining why we *think* that the Earth is at rest, Descartes reminds the reader that "meanwhile we must remember that everything that is real and positive in bodies that move . . . is also found in the others contiguous with them" (Pr II 30). Whatever the so-called criterion of single displacement is supposed to do, it *isn't* supposed to undermine the reciprocity of transference in any way, and it seems not to be intended as a way of providing a genuine distinction between motion and rest. The case is similar with respect to the criterion of displacement as a whole. Descartes begins his evaluation of the Tychonic hypothesis with a statement of the reciprocity of transference, pointing out that the mutual transference of the Earth and the Heavens gives us "no reason . . . why we attribute motion to the Heavens rather than to the Earth" (Pr III 38). The claim that he goes on to make, that we, indeed, have more reason to attribute motion to the Earth than to the Heavens on Tycho's view seems, at best, to be an extra flourish in the argument. Furthermore, so far as I can see, this supposed criterion does not appear at all in the basic exposition of the notion of motion in part II of the *Principles.* How strange it would be if a criterion as central to Descartes' account of motion and rest as Gueroult thinks it is would have no place in Descartes' careful exposition of the nature of motion, and would enter only much later in the course of a discussion of quite something else.

13. So far as I can see, though, it is not a distinction of reason in the strict sense of Pr I 62, where a distinction of reason seems to be limited to the distinction between a substance and its attribute(s), or between two attributes of a single substance.

14. The notion of rest as a cessation of action, also a feature of the vulgar notion of rest, is not at issue here.

15. Newton 1972, pp. 46–47 (Newton 1966, pp. 6–7).

16. Though it was not missed by Newton himself. See his very perceptive discussion of Descartes' account of motion in the early "De gravitatione," in Newton 1962, pp. 82ff (Latin), or pp. 123ff (English translation).

17. These passages will be discussed in detail below in chapter 8.

18. See Descartes' remarks to Hobbes in the *Third Replies*: "nor can we imagine *[fingi]* that it is one substance which is the subject of shape, and another which is the subject of local motion, etc., since all of those acts agree in the common concept *[communis ratio]* of extension" (AT VII 176).

19. Koyré 1978, pp. 259–60.

20. A stone is called a substance in AT VII 44; Pr II 11F; for clothing as substance see AT VII 435, and for a hand or arm, see AT VII 222, 228.

21. See, e.g., Spinoza, *Ethics,* part I, prop. 15, scholium, and the exchange between Leibniz and de Volder, G II 166, 183–84 (Leibniz 1969, p. 519). See also Lewis 1950, pp. 45–51.

22. See Lewis 1950, pp. 39–41, 43–45; Rodis-Lewis 1971, p. 548, n. 60; Marshall 1979, pp. 54–57.

23. Of course, since space is full, there will always be something at every point; but even though there will be body at the center of gravity, the body at the center of gravity may well not be a part of the body whose center of gravity it occupies.

24. Cordemoy 1968, p. 99.

25. See Cordemoy 1968, pp. 95–96; for the Cartesian response, see Prost 1907, p. 167.

26. Gueroult 1980, pp. 211–12.

27. The full letter is translated in the appendix to chapter 8.

28. "De Ipsa Natura" §13, G IV 513 (Leibniz 1989, p. 163); all further quotations are from that section.

29. More 1662, Preface General, p. xi.

30. On the objection to the Ptolemaic system from the observations of Venus, see Ariew 1987b.

31. In a letter to an unknown correspondent that Adam and Tannery date to

1644, Descartes refers to his own system as "celuy de Copernic, expliqué en la façon que ie l'explique" (AT V 550). For other thoughts on the dating, see AT V 665.

32. Though Kepler's work was certainly available to Descartes, there is no hint of elliptical orbits in his cosmology.

33. For an account of Descartes vortex theory of planetary motion, as well as its later history, see Aiton 1972.

34. It is not entirely clear here whether Descartes is talking about the annual motion or the diurnal motion. Though, of course, the stellar parallax due to the annual motion was too small to be observed in Descartes' day, suggesting that the argument is concerned with the diurnal motion, Pr III 29 would seem to be a continuation of the discussion in Pr III 28, which quite clearly does seem to deal with the annual motion. Quite possibly he means to be speaking in general terms about both.

35. By the Copernican view in this passage, Descartes would seem to mean his own version of the theory.

36. At least, one would assume so. Descartes is not at all careful to separate the annual and diurnal motions in the discussion of his own views.

37. Koyré 1978, pp. 261, 265.

38. Blackwell 1966, p. 227.

39. For other similar accounts, see, e.g., Dugas 1958, pp. 172–73; Taliaferro 1964, p. 22; Aiton 1972, pp. 33, 41–42.

40. It is worth pointing out that even the doctrine of the reciprocity of transference, which Descartes does subscribe to, is not applicable in the case of two bodies in collision, strictly speaking. Descartes' principle applies only when we are dealing with bodies in mutual contact, a body and its surrounding neighborhood. But for two bodies in impact, this never arises. For this observation I am indebted to an anonymous reviewer.

41. See, e.g., Koyré 1965, pp. 81–82; Westfall 1971, pp. 57–58. Neither Koyré or Westfall think that the proper definition eliminates as much of the relativity as I think it does.

42. I agree with many commentators that Descartes' discussion of the motion of the Earth in the opening sections of part III of the *Principles* is an attempt (rather transparent at that) to convince the Church of his orthodoxy. But at least one commentator has denied this. Milhaud (1921, pp. 17–22), argues that Descartes saw his account of an immobile Earth as necessary to answer certain standard objections against Copernicanism.

43. For discussions of the history of the notion of determination, see, e.g., Costabel 1967, pp. 236–40; Gabbey 1980, pp. 248–50. In this section I am particularly indebted to Gabbey's insightful discussion of determination.

44. See also AT III 250–51; AT IV 185. The distinction between speed and determination is also at issue in Descartes' discussion of the laws of composition both quantities follow. Descartes claims that determinations are composed by the parallelogram rule, while speeds are composed through simple addition; see AT II 19; AT III 288; AT III 325–26. For a helpful discussion of the Descartes/Hobbes exchange, where this issue is especially important, see Knudsen and Pedersen 1968–69.

45. See also AT III 112–13, 163. Though somewhat untidy, no real difficulties are raised by the variety of contrasts Descartes uses in connection with the distinction. Even though they are distinct notions, speed, motion, and force are closely connected in the Cartesian scheme. Motion is, of course, a mode of body, and, as we shall shortly note, it is to this mode that both speed and determination pertain, strictly speaking. But 'motion' is often used by Descartes as an abbreviated way of referring to *quantity* of motion, which is measured by size times speed for a given body, as we shall see when discussing the conservation law in later chapters. And, as we shall see in connection with our discussion of collision, the force a moving body has to persist in its motion is identical to its quantity of motion. Thus Descartes wrote to Mersenne on 23 March 1643 that "the impression and the [quantity of] motion and the speed, considered in one body, are the same thing" (AT III 636; see also 180–81; by 'impression' here I take Descartes to mean what he calls 'force of going on' in the laws of impact, to be discussed below). From his point of view, the contrasts he draws between determination on the one hand, and speed, [quantity of] motion, and force on the other come to much the same thing.

46. See also AT III 355. Gabbey 1980, pp. 311–12 n. 125, takes note of the use here of the unusual word *'motion,'* which appears here as a synonym for speed, and attempts to link it to earlier uses of the French term and its Latin cognate, *motio,* a reasonably uncommon word closely connected with the more commonly used *'motus.'* I think that there is little mystery here. Descartes wants to distinguish between motion considered as a mode in a body, characterized by both speed and direction (a vector quantity in modern usage) with the (scalar) notion of motion, considered as the bare (nondirectional) speed or quantity of motion, speed times size. To do this he called upon a near synonym for the more commonly used *mouvement,* motion, and pressed it into service.

47. *'Varietez'* might be *'proprietez';* see Adam and Tannery's note on line 7, and Gabbey 1980, p. 258.

48. Descartes writes to Mersenne on 3 December 1640 that determination without motion is *"chymerique & impossible"* (AT III 251). Determination is also linked in substantive ways with speed, its coordinate mode of the mode of motion. Since every motion must have some speed or other, a given "determination cannot be without some speed" (AT II 18), nor can there be any speed (force) that lacks determination (AT III 113). Indeed, as we shall later see in more detail in connection with refraction, determination is sometimes "forced to change in different ways insofar as it is required to adjust itself to speed" (AT II 20).

49. The misunderstanding began as early as Fermat; see AT I 465–66. See also Milhaud 1921, p. 110; Mouy 1934, p. 55; Clarke 1982, pp. 219–20; etc.

50. For the identification of the writing which Descartes is discussing, see AT V 661, note to p. 208 l. 26.

51. It is important to remember here that what is halved is the *full* speed of the ball, and not just its vertical component.

52. Despite the implication of this passage, F is not on the circle.

53. It should be noted that just as Descartes distinguishes carefully between the actual division of a body into parts, and imaginary divisions that are not actually made (see Pr II 23), he distinguishes between actual and imaginary

decompositions of determination. This comes out clearly in his responses to Fermat's objections; for a very helpful account of the exchange, see Sabra 1967, pp. 116–35. Fermat's argument turns on the fact that there seem to be alternative ways of decomposing the determination the tennis ball has. Fermat's claim is that "of all the divisions of the determination to motion, which are infinite, the author [i.e., Descartes] has chosen only the one which suffices for his conclusion" (AT I 359). Descartes responds thus:

> From the fact that I wrote that 'the determination to move can be divided' (I understand here really divided, and not only by the imagination) 'into all of the parts one can imagine it to be composed of,' he has no reason to conclude that the division which is made by the surface CBE, which is a real surface, namely a polished body CBE, is only imaginary. (AT I 452; the case under discussion here is that of reflection, where the ball is bounced off of a hard surface; see also AT II 17, AT III 112–13)

Just as motion can *really* divide a body into its parts, so "the direction *[sens]* in which the surface CBE is opposed to the motion of the ball" (AT I 452) can *really* divide a given determination into its components, one of which persists unchanged (the horizontal) and one of which changes (the vertical). It is by no means clear what exactly Descartes means by a real division of a given determination, and how a real division is to be distinguished from an imaginary division. But one thing Descartes seems to have in mind is that the real parts of a determination can change independently of one another, and have a life of their own, in a sense, just as the real parts of a body do.

54. This point is emphasized in Sabra 1967, pp. 110–11; see also AT III 291.

55. While in general Descartes regards determination as a quantitative notion, there are other passages that suggest that he sometimes thought of it as if it were not a quantity at all. For example, in writing to Mersenne on 3 December 1640 Descartes considers Bourdin's objections to his *Dioptrics,* in particular, the refractive case in which the tennis ball enters water after passing through surface CBE (see fig. 6.2). He notes:

> If the ball which is pushed from A to B, being at point B is determined by the surface CBE to go toward I, whether it is air beneath this surface or water, this would not change its determination at all, but only its speed, which will decrease much more in the water than in the air. (AT III 250–51)

Descartes' point here seems to be that even though the water may slow the body down, it does not change its direction, the fact that it is moving toward I; he appears to suggest that the speed of a body in a given direction can change without changing its determination *in that direction.* This appears inconsistent with the more strictly quantitative notion of determination used in the *Dioptrics,* where a change in the speed by which a body moves in a particular direction is, per se, a change in its determination. This conception is also suggested in an earlier reply to Bourdin from 29 July 1640, where Descartes compares the relation of speed and determination to that which holds between shape and size:

Although there cannot be force *[vis]* without determination, nevertheless one can have the same determination joined to a greater or lesser force, and the same force can remain, although the determination is changed as much as you like. As, for example, although shape cannot exist without quantity, the one can be changed without, however, changing the other. (AT III 113)

Here the suggestion is that just as the same shape can be instantiated in a larger or smaller body, the same determination can be found in greater or lesser magnitudes, which in turn suggests that in at least this one passage, determination is a nonquantitative notion, simply identified with direction.

56. See Sabra 1967, pp. 110–11; Gabbey 1980, p. 256. One can construe the rules of impact, Pr II 46–52, as providing some guidance on the change in determination. However, as we shall see below in chapter 8, the laws deal only with one very special case, that of direct collision (the change in determination due to a body moving with a determination in the exact opposite direction) and cannot readily be generalized to a full account of the change of determination.

57. Gabbey 1980, p. 258; Sabra 1967, p. 121.

58. For overviews of scholastic conceptions of motion, see Maier 1982, pp. 21–39; Murdoch and Sylla 1978; Weisheipl 1982, pp. 526–36.

59. Eustachius 1648, part 3, p. 161; see also Toletus 1589, ff. 79v–80r; the Coimbrian commentary on *Physics* III.3, excerpted in Gilson 1979, §290. It is important here to remember the fact that there were accidental forms as well as substantial forms; Eustachius means to include both in this quotation.

60. Eustachius 1648, part 3, p. 161; Toletus 1589, f. 80r; Gilson 1979, §290. The original source of the definition is Aristotle, *Physics* 201a 10f.

61. See Eustachius 1648, part 3, pp. 161, 165; Toletus 1589, ff. 79v–80r; the Coimbrian commentary on *Physics* V.2, excerpted in Gilson 1979, §291. The question as to whether substantial change counts as a kind of motion was disputed; the point is unclear in Aristotle; compare *Physics* 201a 4f with 225b 10f; see also Toletus 1589, ff. 152v–153v. For Descartes' discussion of this feature of the definition, see AT XI 39.

62. See, e.g., Eustachius 1648, part 3, pp. 162–63; Toletus 1589, ff. 85r, 87v, 163r.

63. In this doctrine, Descartes may not have been quite as distant from the schools as he may have thought; see Maier 1982, chap. 1.

CHAPTER SEVEN

1. See also JB IV 51 (AT X 78), the document Beeckman is referring to, where Descartes makes explicit reference to Beeckman's principle.

2. See AT X 224; Gouhier 1958, p. 24.

3. For other pre-*World* discussions of laws of motion, see AT I 71–72, 117.

4. The phrase 'eademque ratione' in the Latin version is somewhat obscure. In translating it as 'the same laws' I am following the French version.

5. The French version is slightly different, and will be discussed below. For a discussion of the curious phrase, 'quantum in se est', see Cohen 1964. Note that

this formulation would seem to apply to mind as well as to body. Regius, Descartes' one-time disciple, makes appeal to this principle in his account of mind/body unity: "Vinculum, quo mens corpori unita manet, videtur esse lex naturae universalis, qua unumquodque manet in eo statu, in quo est, donec inde ab alio deturbetur" (Regius 1646, p. 250). See also AT VIIIB 344. For Descartes' reaction, see the *Notes against a Program,* AT VIIIB 357.

6. It should be pointed out here that there is a scholastic distinction between universal and particular causes. Eustachius gives it as follows:

As far as what pertains to the most important sorts of efficient causes, various distinctions *[divisiones]* among them are to be treated, of which the most important are these: I. the distinction between universal and particular [causes]. The former are those [causes] which have the power *[vis]* of producing many different sorts of effects by coming together indiscriminately with various particular causes, as, for example, are the heavens and the stars. God is also a cause of this sort, on whose general concourse depends all efficacy of the causes of created things. The latter are those [causes] which are limited *[determinatus]* to the production of a certain effect, as, for example, fire is limited to the production of fire similar to itself. (Eustachius 1648, part 3, p. 141; also in Gilson 1979, §64)

The distinction between universal and particular causes is, thus, a distinction between efficient causes capable of bringing about more or fewer different sorts of things. And so, Goclenius notes, "a cause is more universal if it is the cause of more effects, and less universal if it is [the cause] of fewer" (Goclenius 1613, p. 356B). Similarly, the scholastic distinction seems also to distinguish causes indeterminate in the sense that they do not, by themselves, suffice for producing a specific effect, from other causes that either produce specific effects by themselves, or direct a universal cause to produce one particular effect rather than another. Descartes was certainly aware of the scholastic meaning of this distinction at the time of the publication of the *Principles;* see, e.g., his letter to Mesland, 2 May 1644, AT IV 111–12 (K 146–47). But I don't see how this distinction is directly relevant to the distinction Descartes draws between God as primary cause of motion, and the three laws of nature. Furthermore, writing to Elisabeth on 6 October 1645, he suggests that "the distinction made in the schools between universal and particular causes" is out of place when talking of God, who is "the universal cause of everything" (AT IV 314 [K 180]).

7. On this point Stephen Menn suggested that we must consider the three laws as genuine agents, distinct from God, in a sense, though grounded in him. The idea, as I understand it, is that for Descartes, the laws are intermediate between God and the motion he causes, genuine causal agents of sorts. Peter Dear suggested connecting this up to certain of Descartes' remarks in the 1630 letters about the eternal truths. There Descartes writes to Mersenne:

The mathematical truths, which you call eternal were established by God and depend on him entirely, just as all the rest of his creatures do. . . . God has established these laws in nature, just as a king establishes the laws in his kingdom. . . . They are all inborn in our minds, just as a king would imprint his laws on the hearts of all of his subjects, if he had the power to do so. (AT I 145 [K 11])

Dear's suggestion, as I understand it, is that just as the eternal truths are laws of thought that God imposes on minds, the three laws of nature are laws that, in the same way, God imposes on bodies. The idea is an interesting one, but I hesitate to ascribe it to Descartes in the *Principles*. As we shall later see, in the *Principles* the laws are derived *directly* from the way God acts on bodies. Furthermore, there is no hint of such a view of law anywhere else but in these few lines of Pr II 37.

8. I am leaving aside certain exceptions Descartes flags in Pr II 36; these will be discussed below in chapter 9, when we discuss occasionalism and the question of whether mind can cause motion in the world.

9. This, in any case, is Descartes' intention, despite the fact that he never works out how exactly the collision law is to apply in the general case of oblique collisions. See chapter 8 below.

10. It might be argued here that law 2 isn't conditional in any way insofar as it expresses the existence of a rectilinear tendency that is *always* found in body. But it is important here to note the gloss Descartes gives on 'conatus' in Pr III 56, where the attribution of a conatus to a body is said to mean that the body will in fact travel in a rectilinear direction, unless it is prevented by doing so by another cause. If 'conatus' is synonymous with 'tendency' here (as I think it must be, given the context and the apparent appeal to law 2 in Pr III 56), this suggests that the appeal to the notion of 'tendency' in law 2 makes it an implicitly conditional statement. The notion of tendency will be discussed further in this chapter, when law 2 is discussed in more detail.

11. See, e.g., Gabbey 1980, pp. 236–37 for such a reading.

12. See, e.g., the controversy surrounding Leibniz's attempt to replace the Cartesian conservation principle with a law of the conservation of what he called living force, *vis viva*. A nice account of this is given in Iltis 1971.

13. The entire paragraph containing this passage appears almost word for word in a letter dated March or April 1648; see AT V 135–36 (K 228–29). It is possible that Descartes repeated himself here, but the manuscripts for both of these letters are missing, and it is also possible that Clerselier introduced a corruption.

14. See Costabel 1967, pp. 250–51. Costabel sees the use of the plural again a few lines later as a positive indication that the phrase is not just a typographical error.

15. See also AT V 135–36 (K 228–29). The relatively crude analysis of impact here, more like that in *The World* than in the *Principles,* suggests strongly to me that this passage belongs to the 1639 letter, and that its appearance in the 1648 letter (as remarked in note 13 above) is mistaken. See chapter 8 below for an account of the development of Descartes' ideas about impact from *The World* to the *Principles*.

16. This, in essence, is suggested in Costabel 1967, p. 251.

17. However, see AT XI 629–30, a note in which Descartes seems to suggest that a uniformly applied force *(vis)* will result in a uniform acceleration. Given

its context, it probably dates from the mid 1630s, and it would not be surprising to find that Descartes had a different conception of force here than he did in his later writings. For a discussion of the contents of the note in question, as well as a discussion of its dating, see Garber 1983b, pp. 129–30, n. 34.

18. See Iltis 1971 and Gueroult 1967, chaps. 3–5, for accounts of the various arguments Leibniz used.

19. Descartes granted such a premise in his little treatise on mechanics, sent to Constantijn Huygens in October 1637; see AT I 435–36.

20. Descartes considers and rejects such a principle in a letter to Mersenne, June or July 1648; see AT V 205. It is interesting to note, though, that this letter did not appear in Clerselier's edition of Descartes' correspondence, and so was probably unknown to Leibniz.

21. See the Preliminary Specimen to Leibniz's *Dynamica*, GM VI 288–90 (Leibniz 1989, pp. 107–9). The same basic argument is offered in many varients in the Leibniz corpus; it seems to have been one of Leibniz's favorite arguments. See Iltis 1971.

22. See JB IV 51 (AT X 78); AT I 88f, esp. p. 90 note a; AT I 71–72. The treatment of free-fall as an abstract mathematical problem was later set aside, as Descartes developed his own style of a priori physics. See, e.g., Descartes' important letter to Mersenne concerning his differences with Galileo, 11 October 1638, AT II 380ff. There (AT II 385) he argues that all of Galileo's studies of free-fall are "basti sans fondement" since one must determine what exactly gravity is before one can calculate the speed of a body in free-fall. Since Descartes' own theory of gravity involves the interaction between a heavy body and the subtle matter flowing around the Earth, a simple law of free-fall is very difficult, indeed, impossible to derive. See, e.g., Descartes' remarks to Debeaune, 30 April 1639, AT II 544. For an account of Descartes' dealings with the problem of free-fall, see Koyré 1978, pp. 79–94.

23. Descartes makes much the same use of the principle ten years later, in a letter to Mersenne, 28 October 1640, AT III 208.

24. See, e.g., the account of light given in chapter XIII of *The World*, AT XI 84ff, and Pr III 55.

25. Indeed, as stated in the *Principles* it would also seem to include the states of minds within its scope.

26. See, e.g., the treatment of heat in Pr IV 29.

27. Descartes' use of his analysis of circular motion in his theory of light is a matter of enormous complexity. For a lucid account, see Shapiro 1974.

28. See also Pr III 55f and AT XI 44, 84f. Though the law is stated in terms of circular motion, it is straightforwardly extensible to the general case of curvilinear motion.

29. Note that in Pr II 39 the term used is "tendere." It seems to me, though, that Descartes is using 'conatus' and 'tendency' synonymously in these passages.

30. See also Descartes' use of 'determinare' in Pr III 57.

31. Of course, the "circles" in question are not real geometrical circles, but simply closed curvilinear paths. But as I remarked above, law 2 can be extended to this case too.

32. We do know one argument that Descartes definitely rejected. Roberval

proposed a sort of Leibnizian argument from the principle that "la nature ne souffre rien d'indeterminé," arguing that the tangent to a circle is the only path that is "determiné," an argument that Descartes ridiculed; see AT V 204.

33. See also Pr III 58–59. Strictly speaking, I should think that the two "motions" or "inclinations" into which Descartes divides the rectilinear motion (tendency) are determinations.

34. Descartes is often given credit for having been the first to publish a complete statement of the law. See, e.g., Drake 1970, p. 241; Weisheipl 1985, pp. 50–51. But the issue is quite complex, and turns on important differences between principles of persistence proposed by Descartes, Gassendi, and Galileo in the 1630s and early 1640s; on this, see Rochot 1952.

35. See, e.g., Aristotle, *Physics* IV.8, 215a 15; *Physics* VIII.10, 267a 15–21; *De Caelo* III.2 301b 25ff. See also Murdoch and Sylla 1978, p. 210; Weisheipl 1985, pp. 27ff.

36. Gilson 1979, §300; see also Eustachius 1648, part 3, pp. 167–68; Maier 1982, chaps. 4–5; Clagett 1959, pp. 505–25; Wallace 1981, chap. 15; Wallace 1985, pp. 191–202; Weisheipl 1985, chaps. 2–3.

37. The impetus theory is sometimes presented as an ancestor of views like Descartes', in which bodies in motion remain in motion, a kind of middle position between the strict Aristotelian view and the Cartesian, Galilean, or Newtonian views. But recent scholars have emphasized the differences between the impetus theorists and the moderns, and the Aristotelian credentials of the impetus theory. On this, see esp. Maier 1982, chap. 4.

38. See Aristotle, *De Caelo* I.2.

39. For an account of the vortex theory in Descartes and his followers, see Aiton 1972.

40. See, e.g., Drake 1970, chap. 12; Weisheipl 1985, pp. 49ff.

41. Galileo 1890–1909, 8:268ff (Galileo 1974, pp. 217ff).

42. See Galileo 1890–1909, 7:52–53, 171–74 (Galileo 1967, pp. 28, 145–48); Galileo 1890–1909, 8:242–45 (Galileo 1974, pp. 196–99).

43. This is a point Galileo particularly stresses in the *Two Great World Systems;* see Galileo 1890–1909, 7:53, 174 (Galileo 1967, pp. 28, 148).

44. These arguments are rehearsed in the second day of the *Two Great World Systems* (Galileo 1890–1909, vol. 7, and Galileo 1967). See Grant 1984 for an account of the matter from the point of view of the anti-Copernicans.

45. Galileo 1890–1909, 7:212–13 (Galileo 1967, pp. 186–87).

46. This is one of the main themes of Gabbey 1980.

CHAPTER EIGHT

1. I should acknowledge from the start that my own reading of Descartes' account of impact is greatly indebted to the careful analysis found in Gabbey 1980.

2. See, e.g., JB I 265f. See also Taliaferro 1964, pp. 17–19; Gabbey 1980, p. 244; CM II 632–44.

3. I am indebted to John Schuster for this observation.

4. See the discussion of these passages in Gabbey 1980, pp. 246–47.

5. Ibid., pp. 243ff.

6. See, e.g., AT I 246–47; AT II 467, 543, 627, etc. For some discussions of impact in the early 1640s but before the publication of the *Principles*, see, e.g., AT III 208, 210–11, 450–51, 452, 634–36, 651–53. Some of these discussions are in apparent contradiction with the views Descartes defends in the *Principles*.

7. It is important to translate the first part of the law as indicating that the body will be deflected in 'another direction', and not 'the other direction', also allowable by the Latin. The examples discussed in the sections following involve only direct collisions, and thus all rebound is in the opposite direction of the original motion. However, the law is also meant to cover oblique collisions as well. The French translation leaves the phrase out altogether.

8. The French translation contains what can only be a mistake in translation: "one motion is in no way contrary to another motion of greater speed" (Pr II 44F). See Costabel 1967, p. 241. On the distinction between motion and determination in the law of impact, see Dubarle 1964, pp. 124–25. Dubarle makes an interesting comparison between that distinction and the distinction between form and matter in scholastic thought.

9. By a hard body here Descartes presumably means one which is not deformed in impact, that is, one whose parts do not change position with respect to one another in impact. The reference back to Pr II 37 in this passage is a bit curious. As noted in chapter 7, the law is taken to apply to "things" that are "simple and undivided." This is most naturally interpreted as bodies (and, perhaps, minds) whose "states," properties like being square or being in motion, are taken to persist. But in this Descartes seems to imply that the things in question, now "not composite but simple" rather than "simple and undivided" are the states, like being in motion. I think that it is reasonably clear that this is simply inconsistent with the passage in Pr II 37. I think that we cannot interpret the "simple and undivided" things of Pr II 37 as states like motion; what sense could one make of such a state being "undivided"? Furthermore, in the French version of the section, Pr II 41F, Descartes (or Picot) deletes the reference back to Pr II 37 where it is found in the Latin, making reference only to the general principle that motion persists.

10. B may be said to lose some of its determination insofar as it may lose some of its speed in that direction, but it will maintain some determination in the original direction, unlike the case with respect to the first part of the law.

11. By a perfectly hard body, Descartes seems to mean one which is not deformed in impact. Descartes does offer some hints on how he might handle collisions between deformable bodies in a letter to Mersenne, 26 April 1643, AT III 652–53. The influence of a fluid medium on the impact of bodies is discussed in Pr II 53ff. A fluid medium can change the way bodies in impact behave considerably. As Descartes notes in Pr II 49F and Pr II 56, a hard body emersed in a fluid medium can be moved quite easily, even if it is at rest, in apparent (but only apparent) contradiction with rule 4 in Pr II 49. While oblique collisions come up in the analysis of reflection and refraction in the *Dioptrics*, and while Descartes does take up specific cases of oblique collision in his correspondence from time to time (see, e.g., AT III 651–52), to the best of my knowledge,

Descartes nowhere gives a systematic account of how his collision law might apply in the case of oblique collision.

12. See, e.g., Leibniz's detailed examination of Descartes' seven rules in his "Animadversiones" on Descartes' *Principles,* G IV 375–84 (Leibniz 1969, pp. 397–403). Leibniz's strategy is to show radical discontinuities in Descartes' rules. As the cases approach one another (as the speed of one body approaches rest, for example), the outcomes should approach one another, Leibniz argues. But, he shows, they don't. Leibniz elsewhere argues that the rules are also defective insofar as they assume the conservation of quantity of motion, and not the conservation of mv^2; see chap. 7 above, G III 45–46, and GM VI 123–24. Descartes himself did realize that the seven rules were in contradiction with ordinary experience; see Pr II 53F. But Descartes' point is just that the idealized conditions assumed in the rules are rarely, if ever, found in nature. Leibniz's point is that the rules are *in principle* defective.

13. All of the rules of impact, both in the Latin and French editions, are translated in the appendix.

14. See Costabel 1967 p. 251.

15. If this is, indeed, what Descartes is saying here, then there is an interesting problem for the view: how does C know what speed B is moving at? Certainly, there is no force of resisting in C before the actual moment at which B collides with C. But at that moment, B exerts a force for proceeding, proportional to its size times its speed. But how can C distinguish at that moment between a small body moving very fast, and a larger body moving proportionally slower?

16. See the penetrating discussion of the measure of the force of resisting in a resting body in Gabbey 1980, pp. 269–70. Gabbey reads rules 4 and 5 a bit differently than I do, seeing the force of rest in C as proportional not to the speed of B, but to the quantity of motion C *would* receive were B to be able to impose it on C; similarly, he reads the force for proceeding in B as the quantity of motion B *would* have if it were to impose enough motion on C for them both to be able to move at the same speed. Though, as we shall see, this reflects some later texts where, I think, Descartes is rethinking his account of impact, I don't see any hints of this kind of counterfactual reading of force of rest in the Latin *Principles.*

17. See, e.g., Blackwell 1966, pp. 226–27. See also the discussion in chap. 6 above.

18. One of the standard questions in the literature is whether or not the seven rules of impact are strictly deducible from the law of impact, law 3. See, e.g., Dubarle 1937; Blackwell 1966; and Clarke 1982, pp. 211ff. Given the fact that law 3 does not cover the cases in which the force for proceeding is equal to the force of resisting, it seems obvious that the rules cannot be deduced in any strict sense from the law.

19. See also the discussion in Clarke 1982, pp. 222–23, and Gabbey 1980, pp. 307–8 n. 90.

20. See Gabbey 1980, pp. 262–63.

21. The letter to Clerselier is found in AT IV 183–87. See also AT IV 180–83, written to Picot on the same day, 17 February 1645, discussing some details of the translation in progress.

22. The letter is given in translation in the appendix. All quotations in the text of this chapter are from that translation, with the references given by paragraph number.

23. This, in essence, is the account Gabbey gives of these cases; see Gabbey 1980, pp. 269–70. However, Gabbey seems to hold that this account is behind the one Descartes gives in the Latin *Principles* as well.

24. See Gabbey 1980, pp. 276ff. As I pointed out earlier, Gabbey takes it for granted that the account of the force of resisting for a body at rest that Descartes gives in this paragraph of the Clerselier letter is what he had in mind all along in the Latin *Principles*. While it is possible, I see no reason to think so.

25. Assuming that Descartes would not alter his other rules, i.e., R1–R3 and R7, this, of course, complicates his account of impact considerably, forcing him to hold that the force for proceeding a moving body has is measured differently in colliding with a resting body than in colliding with another body in motion.

26. Indeed, in earlier discussions Descartes seems to have asserted quite explicitly that a body can communicate more than half of its motion to another at rest in a letter to Mersenne, 25 December 1639, AT II 627.

27. Gabbey 1980, pp. 263–64, seems to assume that the discussion here is an explication of what Descartes was doing in the Latin *Principles;* again, I see no reason to assume that he had this in mind when drafting the Latin *Principles*.

28. This principle is what Alan Gabbey has called the Principle of Least Modal Mutation (PLMM). More generally, Gabby translates what I call 'change' ('changement' in the French) as 'mutation'. He argues that 'mutatio' was a scholastic term of art, a precise notion different from the more colloquial 'change', and that this is what Descartes intended here; indeed, he notes, the Latin translation of this letter renders 'changement' as 'mutatio'. See Gabbey 1980, pp. 263–64; 306–7 n. 77; 313 n. 147. I am unconvinced. If Descartes wanted to make reference to the scholastic term, he could easily have used the French 'mutation' instead of 'changement'. The appeal to the Latin version of Descartes' letters seems inconclusive at best, since Descartes neither made the translations nor reviewed them. Furthermore, given that there is no Latin cognate of 'changement', I'm not sure what Latin word the translator could have used to signal a less technical use of the term.

29. For a discussion of the distinction between the two terms Descartes uses in this passage, 'mouuement' and 'motion', see chapter 6 above.

30. Here it is crucial to remember that determination is a quantitative notion, as discussed in chapter 6 above.

31. See Gabbey 1980, p. 264.

32. E.g., let $m(B)=2$, $m(C)=4$, $v(B)=3$, $v(C)=2$. Then if B pushed C, $v(B)'=v(C)'=2\ 1/3$, a change that one would think would be less than the total change of determination in B.

33. See Leibniz's comment on this in his "Specimen Dynamicum," GM VI 239 (Leibniz 1989, pp. 122–23).

34. See, e.g., Gabbey 1980, pp. 264–65.

35. See Costabel 1967, pp. 240f.

36. Costabel (1967, p. 241) points out a difference between the French and Latin versions of Pr II 44 that can only be a mistake.

37. In the appendix to this chapter, the Latin and French versions of Pr II 46–52 are translated separately and printed in parallel columns to emphasize the differences between the two.

38. I suspect that this is the sort of argument Descartes has in mind in the equality case added to R7 (Pr II 52F) in the French edition, though he offers no justification there.

39. In the French version, Descartes offers no new justification of the outcome of R5, as such, strangely enough. In Pr II 50F Descartes only justifies the claim that the ever so slightest difference between the sizes of B and C should produce such a different result.

40. The only direct references I can find to law 3 in the later parts of the *Principles* are in Pr III 65 and 88, where the law is used qualitatively rather than quantitatively. This isn't to say that there aren't connections between the problem of impact and Descartes' scientific program. Descartes' treatment of reflection and refraction in Discourse 2 of the *Dioptrics* does involve the behavior of tennis balls colliding with surfaces of different sorts. However, the way the questions are treated there involves only a principle of persistence of determination, and reasoning about what happens when the speed of the ball is changed; see the discussion in chapter 6 above. But as John Schuster has emphasized to me, even though the law of impact might not be applied in Descartes' scientific writings, his analysis of reflection may have exerted a major influence on the case treated in R4. In reflection, a tennis ball bounces off of a hard surface without losing any of its motion, and without moving the surface at all. Schuster quite plausibly sees this as the inspiration behind R4, the case in which the moving B rebounds after collision with the larger C at rest. Schuster sees the continued commitment to this rule as a reflection of his continued commitment to his analysis of reflection in the *Dioptrics*.

41. See the discussion of this in Gabbey 1980, p. 265.

42. Kepler 1937–, 8:94; see also 3:25 and 7:301. For an account of the role that inertia plays in Kepler's theory of the motion of the planets, see Stephenson 1987, pp. 141–45.

43. Debeaune mentioned this natural sluggishness of bodies in a letter to Mersenne, 13 November 1638; see CM VIII 172. Gabbey (1980, p. 288) plausibly conjectures that Mersenne mentioned this in a now lost letter to Descartes, citing Mydorge as one who denied such a view. I know of no extant text where Mydorge discusses the issue.

44. This passage is identical with AT V 136 (K 229). See also AT II 627. The latter kind of inertia, which Descartes relates to the extension of the surface of a body, is connected with the fact that a body immersed in a fluid medium will be more or less easy to set in motion, depending upon the details of the situation. See Pr II 53f and AT V 551–52.

CHAPTER NINE

1. 'Hyperaspistes' is the name Descartes gave to an unknown correspondent who wrote him with objections to the *Meditations*. Literally translated it means

defender or champion of the cause, in this case, champion of views opposed to those of Descartes.

2. The example of the sun's light turned out to be an unhappy one, as Descartes was forced to acknowledge, insofar as there are certain minerals which once illuminated by the sun will continue to glow in the dark. See AT III 405, 429 (K 115–16).

3. For Gassendi's objection, see AT VII 300.

4. It was just a little more than a year before he drafted the reply to Gassendi, on 25 December 1639, that Descartes wrote to Mersenne saying that he had brought with him from France a copy of the Bible and "vne Somme de S. Thomas" (AT II 630 [K 69]).

5. See *Summa Theologiae* I q104 a1 c.

6. Ibid. I q104 a1 ad4.

7. Amesius, *Medula theologica* (1628), quoted in Heppe 1935, p. 208 (Heppe 1950, p. 257); see also Heppe 1935 and 1950, chap. 12. Heppe is a handy compendium of theological views of the Reformed Church, drawn from numerous theological tracts, many published in the seventeenth century. Heppe 1935 is the German edition, with quotations mostly in Latin, and Heppe 1950 is an English translation of the same. See also Eustachius 1648, part III p. 177, also in Gilson 1979, §112.

8. Quoted in Heppe 1935, p. 208 (Heppe 1950, p. 258).

9. See, e.g., Descartes' remarks to Clerselier in the letter that replaces Gassendi's *Fifth Objections* in the French edition of the *Meditations*, AT IXA 213.

10. See Gueroult 1984, 1:193–202; see also Vigier 1920; Smith 1902, pp. 131–32; Wahl 1920; Alquié 1950, pp. 125–28; Gilson 1967, pp. 340–42; Rodis–Lewis 1971, p. 533 n. 23.

11. Gueroult 1984, 1:195. Subsequent references are in the text.

12. See Beyssade 1979. The position is developed earlier in Laporte 1950, pp. 158–60. For a more recent development of the position, see Arthur 1988.

13. See Beyssade 1979, pp. 130–31.

14. See ibid., pp. 134–40.

15. Ibid., pp. 140–42.

16. Ibid., p. 348; see also p. 353.

17. Ibid., p. 349.

18. See Sorabji 1982, and Sorabji 1984, pp. 297, 375–81, 394.

19. *Summa Theologiae* I q104 a2.

20. Ibid. I q104 a1 c.

21. Ibid. I q104 a2 c.

22. Ibid. I q104 a2 ad 1.

23. For later statements of this same doctrine in the sixteenth and seventeenth century, see Heppe 1935, pp. 200–201, 209–11 (Heppe 1950, pp. 258–61).

24. This difference is emphasized in Gilson 1967, pp. 341–42.

25. See, e.g., Gueroult 1980, pp. 218–19; Gueroult 1984, 1:196; Hatfield 1979, p. 127.

26. For a discussion of the view in de la Forge and Malebranche, see Garber 1987.

27. The passage raises an obvious question about the relation between an attribute and a substance, a question the young Burman raised to Descartes in conversation. See AT V 154–55 (Descartes 1976, p. 15). See also John Cottingham's comments in Descartes 1976, pp. 77–80.

28. Dubarle 1964, p. 121, e.g., suggests that the essential move in the Cartesian theory of motion is the rejection of the Aristotelian principle that everything in motion requires an external mover. This exchange with More suggests that things are much more complex than that.

29. There is a passage from the last letter to More from August 1649 that appears to go against the divine impulse view I have been developing. Descartes writes:

> I think that what gives you difficulty in this matter is the fact that you conceive a certain force in a resting body, through which it resists motion, as if that force were a positive something, indeed a certain action, that is different from the rest itself, although it [i.e., the positive something] is plainly nothing but a modal entity. (AT V 404 [K 258])

This would appear to count against the view that resistance, for Descartes, is an impulse that God imposes on bodies in order to oppose the motion of another body. But I don't think so. In order to understand the remark, we must read it in the context of the view More had attributed to Descartes. In the letter to which Descartes is replying here More had interpreted Descartes as holding that rest is to be identified with action, the action of resisting (AT V 380), and that it is a "positive something," and indeed *must* be if a body at rest is to have the ability to resist (AT V 382). In a comment on Pr II 26 appended to that letter, he asked whether what Descartes had in mind by rest was "a certain perpetual force of preservation, or an action sustaining itself, and girding itself against all impetus, . . . a certain force or internal action of body" (AT V 384). It is against this that Descartes is reacting, I think. He sees More as attributing *to the resting body itself* "a positive something" that is "different from the rest itself." What Descartes wants More to do is reject this "positive something" that he attributes to bodies and return to his conception of rest, not an action, but the simple nontransference of a body with respect to its neighborhood. As he had told More earlier in the letter (AT V 403), resistance *arises* from rest, but it is not the same *thing* as rest; rest is in bodies as a mode, while the resistance that arises when another body attempts to set a resting body into motion comes from outside the body, from God. The "positive something" that More imagines to be in resting bodies themselves is only the state of rest, and *that* is only a mode. Although the writing in this passage is obscure and compressed, this is how we must interpret the last phrase of this passage, I think.

30. See chap. 7 above for a discussion of the phrase 'eademque ratione' in this passage.

31. See *Summa Theologiae* I q104 a3 and a4. See also Dubarle 1964, pp. 121–22.

32. Gueroult points out what amounts to the same problem, I think. He notes that in the timeless instant, there is no difference between a body in a state of motion and that same body in a state of rest. And so, on the cinematic view of how God causes motion, there can be no difference between the two, or, at least,

Descartes cannot ground the difference that there obviously is. Since Gueroult takes Descartes to hold the cinematic view of how God causes motion, he takes this to be a very basic criticism of the Cartesian program. See Gueroult 1980, pp. 217–22.

33. Beeckman seems to have held a similar measure of quantity of motion; see JB III 129. The note is from 1629, but may represent views Beeckman held much earlier, when he was in regular contact with Descartes. Gabbey 1973, p. 383, suggests that this measure of motive force may well have been quite widespread in the early seventeenth century.

34. It is interesting to note that this kind of proof for law 1 has a consequence for an old problem in physics. Aristotle held (*Physics* VIII.7, 261b 1ff) that at the moment when a body changes its direction (bouncing off of a surface, falling after having been projected upwards) it is at rest. But for Descartes it cannot be, "since, if its motion were once interrupted . . . one could find no cause that could afterwards start it up again" (AT VI 94 [Ols. 76]). See also AT I 172; AT III 288–89. And so Descartes holds that in collision, a body undergoes an instantaneous change in its direction and speed, without coming to rest.

35. See the discussion in chap. 7 above.

36. See also the discussion of nature on AT XI 37.

37. There is an interesting change in the French edition. As noted earlier, in Pr II 37, the "external cause" that prevents the persistence of motion in the Latin edition is replaced by the more specific "collision with others" in the French version of that section. There is a corresponding change in the French version of this section.

38. See Gabbey 1980, pp. 243ff.

39. The claim that the impact-contest forces derive from law 1, though ingenious, is not unproblematic. Leibniz, who wants to deny the Cartesian ontology of geometrical bodies and explicitly add force as something over and above extension, makes the following remark on this claim to the Cartesian De Volder:

> You deduce inertia from the force any given thing has for remaining in its state, something that doesn't differ from its very nature. So you judge that the simple concept of extension suffices even for this phenomenon. . . . But even if there is a force in matter for preserving its state, that force certainly cannot in any way be derived from extension alone. I admit that each and every thing remains in its state until there is a reason for change; this is a principle of metaphysical necessity. But it is one thing to retain a state until something changes it, which even something intrinsically indifferent to both states does, and quite another thing, much more significant, for a thing not to be indifferent, but to have a force and, as it were, an inclination to retain its state, and so resist changing. (G II 170 [Leibniz 1989, p. 172])

40. This also seems to be a point that Leibniz is making in the passage from the letter to De Volder quoted above.

41. See, e.g., Hatfield 1979. Though his title ("Force [God] in Descartes' Physics") indicates that he holds such a view, and though he is often cited in this

connection, in reality I think that the paper is not really concerned with this issue of force and its status at all. Though Hatfield claims to be concerned with the question of how merely extended bodies can have forces as properties (e.g., pp. 113–14, 129, 134–35), the real question he takes up is that of causal agency, whether bodies can be genuine causal agents in Descartes' world. Hatfield writes in conclusion:

> In the end, then, the view that assigns to Descartes a wholly geometrical treatment of matter and motion . . . stands; Descartes removed causal agency from the material world and placed it in the hands of God and created minds, that is, in the hands of immaterial substances endowed with the power to act upon matter. (pp. 134–35)

I think that Hatfield is entirely right in his conclusion, as I shall further discuss later in this chapter when we turn to Descartes and occasionalism. But while not unconnected, the problem of agency in Descartes' world is somewhat different than the problem of what sense can be made of the apparent attributions of force to bodies in Descartes' laws of motion, particularly law 3; even after we settle the question of what the real causal agents are, we must still ask where the forces that enter into the laws fit into Descartes' world.

42. Gueroult 1980, p. 198.

43. Ibid., p. 197.

44. See Gabbey 1980, p. 235: "We may supplement Gueroult's account. . . . "

45. Ibid., pp. 236, 237, 238.

46. Alan Gabbey (1980, p. 239) offers a different explanation for why Descartes doesn't discuss the ontological questions raised by his notion of force. Gabbey refers to the passage in the August 1649 letter to More, where Descartes discusses the causes of motion in the world. There Descartes distinguishes motion, the mode, from its cause, which can be either God, or a created substance, such as our mind, to which he gives the force *[vis]* to move a body. The passage then continues (in Gabbey's translation):

> And that force is also in created substance as its mode, but it is not so in God. Because this point is not so easily understood by everyone, I did not want to deal with it in my writings, lest I should seem to favour the opinion of those who view God as a world-soul united to matter. (AT V 404 [K 257])

Gabbey then continues with his explanation:

> Descartes' strategy was well chosen. Animistic talk of force as an attribute [mode?] of body, in a systematic public presentation, would have squared ill with a philosophical quest for the nature of things in terms of what was most simple, most intelligible and most in accordance with the attributes of Descartes' God.

I think that Gabbey is misreading the passage he quotes from the letter to More. As I read it, the passage has nothing directly to do with the forces that enter into laws 1 or 3. The forces he is talking about are not forces in any physical sense, but the force or power that God and created minds have to move bodies.

His point is that while such power is a mode of mind, it has a more difficult status in God; the caution he expresses about discussing the issue in public seems to derive from certain theological worries he has, rather than from any worry about unnecessarily complicating his mechanistic physics.

47. For general accounts of occasionalism among the members of the Cartesian school, see, e.g., Prost 1907; Gouhier 1926, chap. 3; Battail 1973, pp. 141–46; Specht 1966, chaps. 2, 3.

48. Indeed, when it first appears, it is closely associated with Descartes himself. It is an integral part of de la Forge's commentary on Descartes' *Treatise on Man,* and it is one of the central points of a letter Clerselier, Descartes' literary executor, wrote to de la Forge in December 1660, a letter which appeals to the authority of "nostre Maistre" on a number of occasions, and which Clerselier published along side Descartes' own letters in one of his volumes of Descartes' collected correspondence. On de la Forge, see Gouhier 1926, pp. 93–94; for the Clerselier letter, see Clerselier 1667, pp. 640–46. I am indebted to Alan Gabbey for calling the Clerselier letter to my attention.

49. This, I think, is the main point of Hatfield 1979.

50. This letter appeared in the first volume of Clerselier's edition of Descartes' correspondence in 1657. For the notion of a universal cause, see chap. 7 above. A total cause (in contrast with a partial cause) is simply a cause that by itself produces the whole effect. Eustachius 1648, part III, p. 141, uses the following example. A single horse that pulls a chariot is the *total cause* of its motion, while in a chariot pulled by a team of horses, each horse in the team is a *partial cause.*

51. Forge 1974, p. 240. A similar argument can also be found in Dialogue VII of Malebranche's *Dialogues on Metaphysics;* see RM XII–XIII 161–62 (Malebranche 1980b, p. 159).

52. To these arguments one might also call attention to the several passages in which Descartes uses the word 'occasion' to characterize particular causal relations. But as argued in Gouhier 1926, pp. 83–88, this is hardly worth taking seriously as an argument.

53. For a recent discussion of some of this larger theological debate, see Freddoso 1988.

54. This argument is developed at greater length in Garber 1987.

55. In Garber 1983b I argue that, in fact, the laws of motion that Descartes posits for inanimate nature do not hold for motion caused by minds, and in this way, animate bodies, bodies attached to minds, stand outside of the world of physics. I argue that the position widely attributed to Descartes, that the mind can change the direction in which a body is moving but not add or subtract speed (thus apparently violating the conservation principle) is not actually Descartes' view.

56. Hoenen (1967, p. 359) claims that he did include bodies here.

57. There is another place that is sometimes thought to support the view that bodies cannot be the genuine cause of sensations in the mind, and thus support the view that Descartes is an occasionalist. The passage in question is the celebrated passage from the *Notes against a Program* (1648). He writes:

Nothing reaches our mind from external objects through the sense organs except certain corporeal motions. . . . But neither the motions themselves nor the shapes arising from them are conceived by us exactly as they occur in the sense organs, as I have explained at length in my *Dioptrics*. Hence it follows that the very ideas of the motions themselves and of the shapes are innate in us. The ideas of pain, colors, sounds, and the like must be all the more innate if, on the occasion of certain corporeal motions, our mind is to be capable of representing them to itself, for there is no similarity between these ideas and the corporeal motions. (AT VIIIB 359)

The use of the word 'occasion' in this context (as well as in a previous sentence on the same page) does lend some support to the claim that the use of the corresponding French word in the French translation of Pr II 1, published in the same year, is no accident, and may be significant for the way in which Descartes is thinking about body/mind causality. But it is important to recognize that the claim that the sensory idea is innate in the mind is, I think, irrelevant to the issue of Descartes' occasionalism, the question about whether God is the real cause of mind-body interaction. Descartes' worry here is not (primarily) the *causal connection* between the sensory stimulation and the resulting sensory idea; what worries him is their utter *dissimilarity,* the fact that the sensory idea is nothing like the motions that cause it. To make an analogy, consider, for example, a computer with a color monitor capable of displaying complicated graphics and pictures. Suppose that if I tap in a certain sequence of keystrokes, a picture of the Notre Dame in Paris appears on the screen. One might perhaps want to point out that the actual sequence of motions (that is the keystrokes) that causally produce the picture in no way "resembles" the picture, and one might infer from that fact to the claim that the picture must be innate in the machine, that is, stored in its memory. But one probably wouldn't want to infer from that that the keystrokes are not in some sense the direct cause of the picture's appearing, that the keystrokes did not really elicit the picture, and one *certainly* wouldn't want to infer that it was God who somehow connected the keyboard with the screen of the monitor. I think that the situation is similar with respect to Descartes' point in the passage quoted from the *Notes against a Program;* in this case, as in the computer case, Descartes' main point is simply that sensory ideas cannot come directly from the motions that cause them, but must, at best, be innate ideas that are elicited by the motions communicated to the brain by the sense organs.

58. Clerselier 1667, p. 642. Clerselier argues here that while a finite incorporeal substance, like our mind, cannot add (or destroy) motion in the world, it can change its direction, because unlike motion itself, "the determination of motion . . . adds nothing real in nature . . . and says no more than the motion itself does, which cannot be without determination." This, though, would seem to conflict with what Descartes himself told Clerselier in the letter of 17 February 1645, that motion and determination are two modes of body which "change with equal difficulty" (AT IV 185, translated in the appendix to chap. 8 above.)

59. Cordemoy 1968, p. 143.

60. RM II 309 (Malebranche 1980a, p. 446).
61. See RM II 309–12 (Malebranche 1980a, pp. 446–48).
62. RM II 316 (Malebranche 1980a, p. 450).

Afterword

1. See Descartes' remarks on Galileo, criticizing him for his approach to natural philosophy, AT II 380.

BIBLIOGRAPHY

Adam, Charles. 1887–88. "Pascal et Descartes: les expériences du vide (1646–1651)." *Revue Philosophique* 24:612–24 and 25:65–90.

Adams, Marilyn McCord. 1982. "Universals in the Early Fourteenth Century." In Kretzmann, Kenny, and Pinborg 1982, pp. 411–39.

———. 1987. *William Ockham*. 2 vols. Notre Dame, Ind.: University of Notre Dame Press.

Aiton, E.J. 1972. *The Vortex Theory of Planetary Motions*. New York: Neale Watson Academic Publications.

Alquié, Ferdinand. 1950. *La découverte métaphysique de l'homme chez Descartes*. Paris: Presses Universitaires de France.

Aquinas, St. Thomas. 1948. *Le "De Ente et Essentia" de S. Thomas d'Aquin*. Ed. by M.-D. Roland-Gosselin, O.P. Paris: J. Vrin.

———. 1950. *De Principiis Naturae*. Ed. by J. J. Pauson. Fribourg: Société Philosophique, and Louvain: E. Nauwelaerts.

———. 1965. *Selected Writings of St. Thomas Aquinas*. Ed. and trans. by R. P. Goodwin. Indianapolis: Bobbs-Merrill.

Ariew, Roger. 1979. Review of Descartes 1976. *Studia Cartesiana* 1:182–87.

———. 1987a. "The Infinite in Descartes' Conversation with Burman." *Archiv für Geschichte der Philosophie* 69:140–63.

———. 1987b. "The Phases of Venus before 1610." *Studies in History and Philosophy of Science* 18:81–92.

Armogathe, J.-R. 1990. "La publication du *Discours* et des *Essais*." In G. Belgioioso, G. Cimino, et al., eds., *Descartes: Il Metodo e i Saggi*, 1:17–25. Roma: Instituto della Enciclopedia Italiana.

Armogathe, J.-R., and Carraud, V. 1987. "Texte original et traduction française d'un inédit de Descartes." *Bulletin Cartésien* 15:1–4, bound with *Archives de philosophie* 50, cahier 1.

Armogathe, J.-R., Carraud, V., and Feenstra, R. 1988. "La Licence en Droit de Descartes: Un placard inédit de 1616." *Nouvelles de la République des Lettres* 1988–II, pp. 123–45.

Armogathe, J.-R., and Marion, J.-L. 1976. *Index des Regulae ad Directionem Ingenii de René Descartes*. Rome: Edizioni dell'Ateneo.

Arthur, Richard. 1988. "Continuous Creation, Continuous Time: A Refutation of the Alleged Discontinuity of Cartesian Time." *Journal of the History of Philosophy* 26:349–75.

Bacon, Francis. 1960. *The New Organon and Related Writings*. Ed. by F. H. Anderson. Indianapolis: Bobbs-Merrill.

Baillet, A. 1691. *La vie de Monsieur Des-Cartes*. Paris: Daniel Horthemels.

Balz, A. G. A. 1951. *Cartesian Studies*. New York: Columbia University Press.

Battail, Jean-François. 1973. *L'avocat philosophe: Géraud de Cordemoy, 1626–1684*. The Hague: Martinus Nijhoff.

Beck, L. J. 1952. *The Method of Descartes: A Study of the Regulae*. Oxford: Oxford University Press.

Berkel, Klaas van. 1983a. "Beeckman, Descartes, et la philosophie physico-mathématique." *Archives de Philosophie* 46:620–26.

———. 1983b. *Isaac Beeckman (1588–1637) en de mechanisering van het wereldbeeld*. Amsterdam: Rodopi.

Beyssade, Jean-Marie. 1979. *La philosophie première de Descartes*. Paris: Flammarion.

Bizer, Ernst. 1965. "Reformed Orthodoxy and Cartesianism." In Robert Funk et al., eds., *Translating Theology into the Modern Age*, pp. 20–82. New York: Harper & Row.

Blackwell, Richard J. 1966. "Descartes' Laws of Motion." *Isis* 57:220–34.

Blum, P. R. 1988. "Der Standardkursus der katholischen Schulphilosophie im 17. Jahrhundert." In E. Kessler, C.H. Lohr, and W. Sparn, eds., *Aristotelismus und Renaissance: In memoriam Charles B. Schmitt*, pp. 127–48. Wiesbaden: Otto Harrassowitz.

Boas, Marie. 1952. "The Establishment of the Mechanical Philosophy." *Osiris* 10:412–541.

Bouillier, Francisque. 1868. *Histoire de la philosophie cartésienne*, 3rd ed., 2 vols. Paris: Delagrave.

Broughton, Janet, and Mattern, Ruth. 1978. "Reinterpreting Descartes on the Notion of the Union of Mind and Body." *Journal of the History of Philosophy* 16:23–32.

Brockliss, L. W. B. 1987. *French Higher Education in the Seventeenth and Eighteenth Centuries*. Oxford: Oxford University Press.

Burns, Norman T. 1972. *Christian Mortalism from Tyndale to Milton*. Cambridge, Mass.: Harvard University Press.

Buzon, F. de. 1987. *Descartes: Abrégé de musique, présentation, traduction, et notes*. Paris: Presses Universitaires de France.

Carriero, John P. 1986. "The Second Meditation and the Essence of the Mind." In Rorty 1986, pp. 199–221.

Caton, Hiram. 1973. *The Origin of Subjectivity: An Essay on Descartes*. New Haven: Yale University Press.

Charleton, W., trans. and ed. 1970. *Aristotle's Physics I, II, Translated with Introduction and Notes*. Oxford: Oxford University Press.

Charleton, Walter. 1654. *Physiologia Epicuro-Gassendo-Charltoniana*. London: Printed by Tho. Newcomb, for Thomas Heath.

Clagett, Marshall. 1948. "Some General Aspects of Physics in the Middle Ages." *Isis* 39:29–44.

——. 1959. *The Science of Mechanics in the Middle Ages.* Madison: University of Wisconsin Press.

Clarke, Desmond M. 1982. *Descartes' Philosophy of Science.* University Park: Pennsylvania State University Press.

Clerselier, Claude. 1667. *Lettres de Mr Descartes . . . [tome III].* Paris.

Cohen, I. Bernard. 1964. "'Quantum in se est': Newton's Concept of Inertia in Relation to Descartes and Lucretius." *Notes and Records of the Royal Society of London* 19:131–55.

Cohen, Gustave. 1921. *Écrivains français en Hollande dans la première moitié du XVIIe siècle.* Paris: Champion.

Cordemoy, Gerauld de. 1968. *Oeuvres philosophiques.* Ed. by P. Clair and F. Girbal. Paris: Presses Universitaires de France.

Cosentino, G. 1970. "Le matematiche nella *Ratio Studiorum* della Compagnia di Gesù." *Miscellanea Storica Ligure* 2:171–213.

——. 1971. "L'insegnamento delle Matematiche nei collegi Gesuitici nell'Italia settentrionale." *Physis* 13:205–17.

Costabel, Pierre. 1967. "Essai critique sur quelques concepts de la mécanique cartésienne." *Archives internationales d'histoire des sciences* 20:235–52. Reprinted in Costabel 1982, pp. 141–58. (Pagination in the text follows that of the original publication.)

——. 1982. *Démarches originales de Descartes savant.* Paris: J. Vrin.

——. 1983. "L'intuition mathématique de Descartes." *Archives de Philosophie* 46:637–46.

Cottingham, John. 1979. "Response to Ariew." *Studia Cartesiana* 1:187–89.

——. 1986. *Descartes.* Oxford: Basil Blackwell.

Cronin, T. J. 1966. *Objective Being in Descartes and Suarez.* Rome: Gregorian University Press.

Curley, E. M. 1978. *Descartes against the Skeptics.* Cambridge, Mass.: Harvard University Press.

——. 1984. "Descartes on the Creation of the Eternal Truths." *Philosophical Review* 93:569–97.

——. 1987. "Cohérence ou incohérence du *Discours*?" In Grimaldi and Marion 1987, pp. 41–64.

Dainville, François de. 1940. *La Naissance de l'humanisme moderne.* Paris: Beauchesne.

——. 1954. "L'enseignement des mathématiques dans les Collèges Jésuites de France du XVIe au XVIIIe siècle." *Revue d'histoire des sciences et de leur application* 8:6–21, 109–23.

Dalbiez, R. 1929. "Les sources scholastiques de la théorie Cartésienne de l'être objectif." *Revue d'histoire de la philosophie* 3:464–72.

Dear, Peter. 1988. *Mersenne and the Learning of the Schools.* Ithaca: Cornell University Press.

Descartes. 1936–63. *Descartes: Correspondence publiée avec une introduction et des notes.* Ed. by Charles Adam and Gérard Milhaud. Paris: Alcan and Presses Universitaires de France.

——. 1959. *Lettres à Regius et remarques sur l'explication de l'esprit humain.* Ed. and trans. by G. Rodis-Lewis. Paris: J. Vrin.

————. 1964–74. *Oeuvres de Descartes*. Ed. by Charles Adam and Paul Tannery, nouvelle présentation. Paris: J. Vrin.

————. 1965. *Discourse on Method, Optics, Geometry, and Meteorology*. Trans. by Paul J. Olscamp. Indianapolis: Bobbs-Merrill.

————. 1970. *Philosophical Letters*. Ed. and trans. by Anthony Kenny. Minneapolis: University of Minnesota Press.

————. 1976. *Descartes' Conversation with Burman*. Ed. and trans. by J. Cottingham. Oxford: Oxford University Press.

————. 1979. *Le Monde, ou Traité de la lumière*. Ed. and trans. by Michael Sean Mahoney. New York: Abaris Books.

————. 1983. *Principles of Philosophy*. Trans. by V. R. Miller and R. P. Miller. Dordrecht: D. Reidel.

————. 1985. *The Philosophical Writings of Descartes*. Ed. and trans. by J. Cottingham, R. Stoothoff, and D. Murdoch, 2 vols. Cambridge: Cambridge University Press.

Dijksterhuis, E. J., et al. 1950a. *Descartes et le cartésianisme hollandais*. Paris and Amsterdam: Presses Universitaires de France and Éditions Français d'Amsterdam.

Dijksterhuis, E. J. 1950b. "La méthode et les essais de Descartes." In Dijksterhuis 1950a, pp. 21–44.

————. 1961. *The Mechanization of the World Picture*. Oxford: Oxford University Press.

Donagan, Alan. 1978. "Descartes's 'Synthetic' Treatment of the Real Distinction between Mind and Body." In Hooker 1978, pp. 186–96.

Doney, Willis, ed. 1967. *Descartes*. Garden City: Doubleday.

Drake, Stillman. 1957. *Discoveries and Opinions of Galileo*. Garden City: Doubleday.

————. 1970. *Galileo Studies: Personality, Tradition, and Revolution*. Ann Arbor: University of Michigan Press.

Dubarle, Dominique. 1937. "Remarques sur les régles du choc chez Descartes." *Cartesio nel terzo centenario del "Discorso del metodo."* Milan: Vita e pensiero.

————. 1964. "Sur la notion cartésienne de quantité de mouvement." In *Mélanges Alexandre Koyré*, 2:118–28. Paris: Hermann.

Dugas, René. 1958. *Mechanics in the Seventeenth Century*. Neuchatel: Éditions de Griffon.

Duhem, Pierre. 1985. *Medieval Cosmology*. Ed. and trans. by Roger Ariew. Chicago: University of Chicago Press.

Eustachius a Sancto Paulo. 1648. *Summa Philosophae Quadripartita* [1609]. Cambridge: Roger Daniel.

Fanton d'Andon, Jean-Pierre. 1978. *Horreur du vide: Expérience et raison dans la physique pascalienne*. Paris: Éditions du Centre National de la Recherche Scientifique.

Fitzpatrick, Edward A., ed. 1933. *St. Ignatius and the Ratio Studiorum*. New York: McGraw-Hill.

Fontialis, Jacobus. 1740. *Opera Posthuma*. Namur.

Forge, Louis de la. 1974. *Oeuvres Philosophiques*. Ed. by P. Clair. Paris: Presses Universitaires de France.

Frankfurt, Harry. 1977. "Descartes on the Creation of the Eternal Truths." *Philosophical Review* 86:36–57.

Freddoso, Alfred. 1988. "Medieval Aristotelianism and the Case against Secondary Causation in Nature." In Thomas V. Morris, ed., *Divine and Human Action: Essays in the Metaphysics of Theism*, pp. 74–118. Ithaca: Cornell University Press.

Fromondus, Libertus. 1631. *Labyrinthus sive de Compositione Continui Liber Unus.* Antwerp.

Funkenstein, Amos. 1986. *Theology and the Scientific Imagination from the Middle Ages to the Seventeenth Century.* Princeton: Princeton University Press.

Gabbey, Alan. 1970. "Les trois genres de découverte selon Descartes." *Actes du XIIe Congrès international d'histoire des sciences*, pp. 45–49. Paris: Albert Blanchard.

———. 1973. "Essay Review of W. L. Scott, *The Conflict between Atomism and Conservation Theory: 1644–1860.*" *Studies in History and Philosophy of Science* 3:373–85.

———. 1980. "Force and Inertia in the Seventeenth Century: Descartes and Newton." In Gaukroger 1980a, pp. 230–320.

———. 1982. "Philosophia Cartesiana Triumphata: Henry More, 1646–1671." In T. M. Lennon, J. M. Nicholas, and J. W. Davis, eds., *Problems of Cartesianism*, pp. 171–250. Kingston and Montreal: McGill-Queen's University Press.

———. 1985. "The Mechanical Philosophy and its Problems: Mechanical Explanations, Impenetrability, and Perpetual Motion." In J. C. Pitt, ed., *Change and Progress in Modern Science*, pp. 9–84. Dordrecht: D. Reidel.

———. 1990. "Explanitory Structures and Models in Descartes' Physics." In G. Belgioioso, G. Cimino, et al., eds., *Descartes: Il Metodo e i Saggi*, 1:273–86. Roma: Instituto della Enciclopedia Italiana.

Gäbe, L. 1983. "La Règle 14. Lien entre géométrie et algèbre." *Archives de philosophie* 46:654–60.

Galileo. 1890–1909. *Opere di Galileo Galilei.* Ed. by A. Favaro, 20 vols. Florence: G. Barbèra.

———. 1967. *Dialogue Concerning the Two Chief World Systems.* Trans. by Stillman Drake. Los Angeles and Berkeley: University of California Press.

———. 1974. *Two New Sciences.* Trans. by Stillman Drake. Madison: University of Wisconsin Press.

Gadoffre, Gilbert. 1987. "La chronologie des six parties." In Grimaldi and Marion 1987, pp. 19–40.

Garber, Daniel. 1978. "Science and Certainty in Descartes." In Hooker 1978, pp. 114–51.

———. 1983a. "Understanding Interaction: What Descartes Should Have Told Elisabeth." *Southern Journal of Philosophy* 21 (supplement):15–32.

———. 1983b. "Mind, Body, and the Laws of Nature in Descartes and Leibniz." *Midwest Studies in Philosophy* 8:105–33.

———. 1986. "*Semel in Vita:* the Scientific Background to Descartes' *Meditations.*" In Rorty 1986, pp. 81–116.

———. 1987. "How God Causes Motion: Descartes, Divine Sustenance, and Occasionalism." *Journal of Philosophy* 84:567–80.

———. 1988a. "Descartes, les aristotéliciens et la révolution qui n'eut pas lieu en 1637." In Méchoulan 1988, pp. 199–212.

———. 1988b. "Descartes, the Aristotelians, and the Revolution That Did Not Happen in 1637." *Monist* 71:471–86. [This is an English version of Garber 1988a.]

———. 1988c. "Does History Have a Future?" In P. H. Hare, ed., *Doing Philosophy Historically,* pp. 27–43. Buffalo: Prometheus Books.

———. Forthcoming. "Descartes and Experiment." Proceedings of the San Jose Descartes Conference, April 1988.

Garber, Daniel, and Cohen, Lesley. 1982. "A Point of Order: Analysis, Synthesis, and Descartes's *Principles.*" *Archiv für Geschichte der Philosophie* 64:136–47.

Gassendi, Pierre. 1658. *Opera Omnia.* 6 vols. Lyon.

Gaukroger, Stephen, ed. 1980a. *Descartes: Philosophy, Mathematics and Physics.* Sussex: Harvester Press.

———. 1980b. "Descartes' Project for a Mathematical Physics." In Gaukroger 1980a, pp. 97–140.

Gervais de Montpellier. 1696. *Histoire de la conjuration faite à Stokolm contre Mr Descartes.* In P. Gabriel Daniel, *Suite du voyage du monde de Descartes . . . ,* pp. 219–47. Amsterdam: Chez Pierre Mortier.

Gilbert, Neal. 1960. *Renaissance Concepts of Method.* New York: Columbia University Press.

Gilson, É. 1967. *René Descartes: Discours de la méthode, texte et commentaire,* 4th ed. Paris: J. Vrin.

———. 1975. *Études sur le rôle de la pensée médiévale dans la formation du système cartésien,* 4th ed. Paris: J. Vrin.

———. 1979. *Index scolastico-cartésien,* 2nd ed. Paris: J. Vrin.

Goclenius, Rudolph. 1613. *Lexicon Philosophicum.* Frankfurt.

Gouhier, H. 1926. *La vocation de Malebranche.* Paris: J. Vrin.

———. 1958. *Les premières pensées de Descartes.* Paris: J. Vrin.

———. 1973. *Descartes: Essais sur le "Discours de la méthode," la métaphysique et la morale,* 3rd ed. Paris: J. Vrin.

———. 1978a. *La pensée métaphysique de Descartes.* Paris: J. Vrin.

———. 1978b. *Cartésianisme et Augustinisme au XVIIe siècle.* Paris: J. Vrin.

Gracia, J. J. E. 1982. *Suarez on Individuation: Metaphysical Disputation V: Individual Unity and Its Principle.* Milwaukee: Marquette University Press.

Grant, Edward. 1971. *Physical Science in the Middle Ages.* New York: John Wiley.

———, ed. 1974. *A Source Book in Medieval Science.* Cambridge, Mass.: Harvard University Press.

———. 1976. "Place and Space in Medieval Physical Thought." In Machamer and Turnbull 1976, pp. 137–67.

———. 1979. "The Condemnation of 1277, God's Absolute Power, and Physical Thought in the Late Middle Ages." *Viator* 10:211–44.

———. 1981. *Much Ado about Nothing: Theories of Space and Vacuum from the Middle Ages to the Scientific Revolution.* Cambridge: Cambridge University Press.

———. 1984. "In Defense of the Earth's Centrality and Immobility: Scholastic Reaction to Copernicanism in the Seventeenth Century." *Transactions of the American Philosophical Society* 74 (part 4).

Gregory, Tulio. 1964. "Studi sull'atomismo del seicento. I. Sebastiano Basson." *Giornale critico della Filosofia Italiana* 18:38–65.

———. 1966. "Studi sull'atomismo del seicento. II. David van Goorle e Daniel Sennert." *Giornale critico della Filosofia Italiana* 20:44–63.

Grene, Marjorie. 1986. "Die Einheit des Menschen: Descartes unter den Scholastikern." *Dialectica* 40:309–22.

Grimaldi, Nicolas, and Marion, Jean-Luc, eds. 1987. *Le Discours et sa méthode.* Paris: Presses Universitaires de France.

Grosholz, Emily. 1990. *Cartesian Method and the Problem of Reduction.* Oxford: Oxford University Press.

Guenancia, Pierre. 1976. *Du vide à Dieu.* Paris: François Maspero.

Gueroult, Martial. 1967. *Leibniz: Dynamique et Métaphysique,* 2nd ed. Paris: Aubier-Montaigne.

———. 1980. "The Metaphysics and Physics of Force in Descartes." In Gaukroger 1980a, pp. 196–229.

———. 1984. *Descartes' Philosophy Interpreted According to the Order of Reasons.* Ed. and trans. by Roger Ariew, 2 vols. Minneapolis: University of Minnesota Press.

Hatfield, Gary. 1979. "Force (God) in Descartes' Physics." *Studies in History and Philosophy of Science* 10:113–40.

———. 1985. "First Philosophy and Natural Philosophy in Descartes." In A. J. Holland, ed., *Philosophy, Its History and Historiography,* pp. 149–64. Dordrecht and Boston: D. Reidel.

Henry, D. P. 1982. "Predicables and categories." In Kretzmann, Kenny, and Pinborg 1982, pp. 128–42.

Henry, John. 1979. "Francesco Patrizi da Cherso's Concept of Space and its Later Influence." *Annals of Science* 36:549–73.

Henry, John. 1982. "Atomism and Eschatology: Catholicism and Natural Philosophy in the Interregnum." *British Journal for the History of Science* 15:211–39.

Heppe, Heinrich. 1935. *Die Dogmatik der evangelische-reformierten Kirche.* Ed. by Ernst Bizer. Neukirchen, Kreis Moers: Buchhandlung des Erziehungsvereins.

Heppe, Heinrich. 1950. *Reformed Dogmatics.* Ed. by Ernst Bizer and trans. by G. T. Thomson. London: George Allen & Unwin.

Hintikka, Jaakko. 1978. "A Discourse on Descartes's Method." In Hooker 1978, pp. 74–88.

Hobbes, Thomas. 1951. *Leviathan.* Ed. by C. B. Macpherson. New York: Penguin Books.

Hoenen, P. H. J. 1967. "Descartes's Mechanicism." In Doney 1967, pp. 353–68.

Hoffman, Paul. 1986. "The Unity of Descartes' Man." *Philosophical Review* 95:339–70.

———. 1990. "Cartesian Passions and Cartesian Dualism." *Pacific Philosophical Quarterly* 71:310–32.

Hooker, Michael. 1978. *Descartes: Critical and Interpretive Essays.* Baltimore: Johns Hopkins University Press.

Iltis, Carolyn. 1971. "Leibniz and the *vis viva* Controversy." *Isis* 62:21–35.

Jammer, Max. 1969. *Concepts of Space: The History of Theories of Space in Physics,* 2nd ed. Cambridge, Mass.: Harvard University Press.

Jardine, Lisa. 1974. *Francis Bacon: Discovery and the Art of Discourse.* Cambridge: Cambridge University Press.

Jones, Howard. 1981. *Pierre Gassendi, 1592–1655: An Intellectual Biography.* Nieuwkoop: B. De Graaf.

Joy, Lynn Sumida. 1987. *Gassendi the Atomist: Advocate of History in an Age of Science.* Cambridge: Cambridge University Press.

Kargon, Robert H. 1966. *Atomism in England from Hariot to Newton.* Oxford: Oxford University Press.

Kepler, Johannes. 1937–. *Gesammelte Werke.* Ed. by W. Van Dyck, M. Caspar, F. Hammer, 20 vols. Munich: Beck.

Knight, David. 1962. "Suarez's Approach to Substantial Form." *Modern Schoolman* 39 (1961–62):219–39.

Knudsen, Ole, and Pedersen, Kurt M. 1968–69. "The Link between 'Determination' and Conservation of Motion in Descartes' Dynamics." *Centaurus* 13:183–86.

Koyré, Alexandre. 1965. *Newtonian Studies.* Cambridge, Mass.: Harvard University Press.

———. 1978. *Galileo Studies.* Trans. by John Mepham. Atlantic Highlands, N.J.: Humanities Press.

Kretzmann, Norman, ed. 1982. *Infinity and Continuity in Ancient and Medieval Thought.* Ithaca: Cornell University Press.

Kretzmann, N., Kenny, A., and Pinborg, J., eds. 1982. *The Cambridge History of Later Medieval Philosophy.* Cambridge: Cambridge University Press.

Laporte, Jean. 1950. *Le rationalisme de Descartes.* Paris: Presses Universitaires de France.

Larmore, Charles. 1980. "Descartes' Empirical Epistemology." In Gaukroger 1980a, pp. 6–22.

Lasswitz, Kurd. 1890. *Geschichte der Atomistik vom Mittelalter bis Newton.* 2 vols. Hamburg and Leipzig: Leopold Voss.

Lefèvre, Roger. 1978. "Méthode cartésienne et modèle mathématique." *Modèles et interprétation,* pp. 89–116. Villeneuve-d'Ascq: Presses Universitares de Lille.

Le Guern, Michel. 1971. *Pascal et Descartes.* Paris: A. G. Nizet.

Le Grand, H. E. 1978. "Galileo's Matter Theory." In R. E. Butts and J. C. Pitt, eds., *New Perspectives on Galileo,* pp. 197–208. Dordrecht: D. Reidel.

Leibniz. 1849–63. *G. W. Leibniz: Mathematische Schriften.* Ed. by C. I. Gerhardt, 7 vols. Berlin: A. Asher, and Halle: H. W. Schmitt.

———. 1875–90. *G. W. Leibniz: Die philosophischen Schriften.* Ed. by C. I. Gerhardt, 7 vols. Berlin: Weidmannsche Buchhandlung.

———. 1969. *Philosophical Papers and Letters,* 2nd ed. Ed. and trans. by L. Loemker. Dordrecht: D. Reidel.

———. 1989. *Philosophical Essays.* Ed. and trans. by R. Ariew and D. Garber. Indianapolis: Hackett.

Lennon, T. M. 1974. "Occasionalism and the Cartesian Metaphysic of Motion." *Canadian Journal of Philosophy* 1 (Suppl., pt. 1): 29–40.

Lenoble, R. 1943. *Mersenne ou la naissance du mécanisme.* Paris: J. Vrin.

Lewis, G. 1950. *L'individualité selon Descartes.* Paris: J. Vrin.

Lindberg, David C., ed. 1978. *Science in the Middle Ages.* Chicago: University of Chicago Press.

Lukács, Ladislaus, S.J., ed. 1986. *Ratio atque Institutio Studiorum Societatis Iesu (1586, 1591, 1599).* Monumenta Paedagogica Societatis Iesu, vol. 5. Monumenta Historica Societatis Iesu, vol. 129. Rome: Institutum Historicum Societatis Iesu.

Machamer, Peter, and Turnbull, Robert, eds. 1976. *Motion and Time, Space and Matter.* Columbus: Ohio State University Press.

Maier, A. 1982. *On the Threshold of Exact Science.* Trans. by S. D. Sargent. Philadelphia: University of Pennsylvania Press.

Malebranche, Nicholas. 1958–84. *Oeuvres Complètes de Malebranche.* Ed. by André Robinet, 20 vols. Paris: J. Vrin.

———. 1980a. *The Search after Truth.* Trans. by T. M. Lennon and P. J. Olscamp. And *Elucidations of the Search after Truth.* Trans. by T. M. Lennon. Columbus: Ohio State University Press.

———. 1980b. *Entretiens sur la métaphysique/Dialogues on Metaphysics.* Trans. by Willis Doney. New York: Abaris Books.

Marion, Jean-Luc. 1981a. *Sur l'ontologie grise de Descartes,* 2nd ed. Paris: J. Vrin.

———. 1981b. *Sur la théologie blanche de Descartes.* Paris: Presses Universitaires de France.

———. 1986. *Sur le prisme métaphysique de Descartes.* Paris: Presses Universitaires de France.

Marshall, David J., Jr. 1979. *Prinzipien der Descartes: Exegese.* Freiburg: Verlag Karl Alber.

Masi, Roberto. 1947. *Il movimento assoluto e la posizione assoluta secondo il Suarez.* Rome: Facoltà di Filosofia del Pont. Ateneo Lateranense.

McMullin, Ernan, ed. 1963. *The Concept of Matter in Greek and Medieval Philosophy.* Notre Dame: University of Notre Dame Press.

Méchoulan, Henry, ed. 1988. *Problématique et réception du Discours de la méthode et des essais.* Paris: J. Vrin.

Menn, Stephen. 1989. *Descartes and Augustinianism.* Ph.D. diss., University of Chicago.

Mercer, Christia. 1989. *The Origins of Leibniz's Metaphysics and the Development of his Conception of Substance.* Ph.D. diss., Princeton University.

Mersenne, Marin. 1932–88. *Correspondance du P. Marin Mersenne, religieux minime.* Ed. by C. de Waard et al., 17 vols. Paris: Beau-Chesne (vol. 1), Presses Universitaires de France (vols. 2–4), CNRS (vols. 5–17).

Mesnard, Pierre. 1956. "La pédagogie des Jésuites." In J. Chateau, ed., *Les grands pédagogues,* pp. 45–107. Paris: Presses Universitaires de France.

Milhaud, Gaston. 1921. *Descartes savant.* Paris: Félix Alcan.

Miller, Fred D., Jr. 1982. "Aristotle against the Atomists." In Kretzmann 1982, pp. 87–111.

More, Henry. 1662. *A Collection of Several Philosophical Writings of Dr Henry More . . .* London: James Flesher, for William Morden.

Mouy, Paul. 1934. *Le développement de la physique cartésienne: 1646–1712.* Paris: J. Vrin.

Murdoch, John. 1982. "Infinity and Continuity." In Kretzmann, Kenny, and Pinborg 1982, pp. 564–91.

Murdoch, John E., and Sylla, Edith D. 1978. "The Science of Motion." In Lindberg 1978, pp. 206–64.

Nadler, Steven. 1989. *Arnauld and the Cartesian Philosophy of Ideas*. Princeton: Princeton University Press.

Newton, Sir Isaac. 1962. *Unpublished Scientific Papers of Isaac Newton*. Ed. by A. R. Hall and M. B. Hall. Cambridge: Cambridge University Press.

———. 1966. *Mathematical Principles of Natural Philosophy*. Trans. by A. Mott and F. Cajori, 2 vols. Los Angeles and Berkeley: University of California Press.

———. 1972. *Philosophiae Naturalis Principia Mathematica*. Ed. by A. Koyré and I. B. Cohen, 2 vols. Cambridge, Mass.: Harvard University Press.

Normore, Calvin. 1986. "Meaning and Objective Being: Descartes and His Sources." In Rorty 1986, pp. 223–41.

O'Neill, Eileen. 1983. *Mind and Mechanism: An Examination of Some Mind-Body Problems in Descartes' Philosophy*. Ph.D. diss., Princeton University.

———. 1987. "Mind-Body Interaction and Metaphysical Consistency: A Defense of Descartes." *Journal of the History of Philosophy* 25:227–45.

Ong, Walter J. 1958. *Ramus, Method, and the Decay of Dialogue*. Cambridge, Mass.: Harvard University Press.

Osler, Margaret J. 1985. "Baptizing Epicurean Atomism: Pierre Gassendi on the Immortality of the Soul." In M. J. Osler and P. L. Farber, eds., *Religion, Science, and Worldview: Essays in Honor of Richard S. Westfall*, pp. 163–83. Cambridge: Cambridge University Press.

Pachtler, G.-M. 1887. *Ratio Studiorum et Institutiones Scholasticae Societatis Jesu*. Monumenta Germaniae Paedagogica, vol. 5. Berlin: A. Hoffmann.

Pascal, Blaise. 1963. *Oeuvres complètes*. Ed. by Louis Lafuma. Paris: Éditions du Seuil.

———. 1964–. *Oeuvres complètes*. Ed. by Jean Mesnard. 2 vols. to date. Brussels: Desclée de Brouwer.

Popkin, Richard. 1979. *The History of Scepticism from Erasmus to Spinoza*, 4th ed. Los Angeles and Berkeley: University of California Press.

Prendergast, Thomas L. 1972. "Descartes and the Relativity of Motion." *Modern Schoolman* 49:64–72.

Prost, Joseph. 1907. *Essai sur l'atomisme et l'occasionalisme dans l'école cartésienne*. Paris: Paulin.

Regius, Henricus. 1646. *Fundamenta Physices*. Amsterdam: Ludovicus Elzevirius.

Reif, Patricia. 1969. "The Textbook Tradition in Natural Philosophy, 1600–1650." *Journal of the History of Ideas* 30:17–32.

Rist, J. M. 1972. *Epicurus: An Introduction*. Cambridge: Cambridge University Press.

Rochemonteix, C. de. 1899. *Un collège des jésuites aux XVIIe et XVIIIe siècles: Le Collège Henri IV de la Flèche*, 4 vols. Le Mans: Leguicheux.

Rochot, Bernard. 1952. "Beeckman, Gassendi et le principe d'inertie." *Archives Internationales d'Histoire des Sciences* n.s. 31:282–89.

———. 1963. "Comment Gassendi interprétait l'expérience du Puy de Dôme." *Revue d'histoire des sciences* 16:53–76.

(Rodis-)Lewis, G. 1950. See Lewis 1950.

Rodis-Lewis, G. 1971. *L'oeuvre de Descartes*. Paris: J. Vrin.

———. 1983. "Quelques questions disputées sur la jeunesse de Descartes." *Archives de Philosophie* 46:613–19.

———. 1984. *Descartes: Textes et débats*. Paris: Le livre de poche.

———. 1987a. "Descartes et les mathématiques au collège." In Grimaldi and Marion 1987, pp. 187–211.

———. 1987b. "Hypothèses sur l'élaboration progressive des *Méditations* de Descartes." *Archives de philosophie* 50:109–23.

Rohault, Jacques. 1969. *A System of Natural Philosophy: A Facsimile of the Edition and Translation by John and Samuel Clarke Published in 1723*, 2 vols. New York: Johnson Reprint.

Rorty, Amélie, ed. 1986. *Essays on Descartes' Meditations*. Los Angeles and Berkeley: University of California Press.

Rosenfield, Leonora Cohen. 1957. "Peripatetic Adversaries of Cartesianism in 17th century France." *Review of Religion* 22:14–40.

———. 1968. *From Beast-Machine to Man-Machine: Animal Soul in French Letters from Descartes to La Mettrie*, 2nd ed. New York: Octagon Books.

Roth, Leon. 1937. *Descartes' Discourse on Method*. Oxford: Oxford University Press.

Sabra, A. I. 1967. *Theories of Light from Descartes to Newton*. London: Oldbourne.

Schmitt, Charles B. 1967. "Experimental Evidence For and Against a Void: The Sixteenth-Century Arguments." *Isis* 58:352–66.

Schmitt, Charles B. 1983. *Aristotle and the Renaissance*. Cambridge, Mass.: Harvard University Press.

Schouls, Peter. 1980. *The Imposition of Method*. Oxford: Oxford University Press

Schuster, John A. 1977. *Descartes and the Scientific Revolution, 1618–1634: An Interpretation*. Ph.D. diss., Princeton University.

———. 1980. "Descartes' *Mathesis Universalis*, 1619–28." In Gaukroger 1980, pp. 41–96.

———. 1986. "Cartesian Method as Mythic Speech: A Diachronic and Structural Analysis." In J. A. Schuster and R. Yeo, eds., *The Politics and Rhetoric of Scientific Method*, pp. 33–95. Dordrecht: D. Reidel.

Serrus, C. 1933. *La méthode de Descartes et son application à la métaphysique*. Paris: Librarie Félix Alcan.

Shapiro, Alan E. 1974. "Light, Pressure, and Rectilinear Propagation: Descartes' Celestial Optics and Newton's Hydrostatics." *Studies in History and Philosophy of Science* 5:239–96.

Shea, W. R. 1978. "Descartes as a Critic of Galileo." In R. E. Butts and J. C. Pitt, eds., *New Perspectives on Galileo*, pp. 139–59. Dordrecht: D. Reidel.

Sirven, J. 1928. *Les années d'apprentissage de Descartes (1596–1628)*. Albi: Imprimerie Coopérative du Sud-Ouest.

Smith, Norman Kemp. 1902. *Studies in the Cartesian Philosophy*. London: Macmillan.

Sorabji, Richard. 1982. "Atoms and Time Atoms." In Kretzmann 1982, pp. 37–86.

———. 1984. *Time, Creation, and the Continuum*. Ithaca: Cornell University Press.

Specht, Rainer. 1966. *Commercium mentis et corporis: Über Kausalvorstellungen im Cartesianismus*. Stuttgart-Bad Cannstatt: Friedrich Frommann Verlag.

Spink, J. S. 1960. *French Free-Thought from Gassendi to Voltaire*. London: Athlone Press.

Stephenson, Bruce. 1987. *Kepler's Physical Astronomy*. New York: Springer-Verlag.

Suarez, F. 1947. *On the Various Kinds of Distinctions (Disputationes Metaphysicae, Disputatio VII de variis distinctionum generibus)*. Trans. by C. Vollert. Milwaukee: Marquette University Press.

Sylla, Edith. 1979. "The A Posteriori Foundations of Natural Science: Some Medieval Commentaries on Aristotle's Physics, Book I, Chapters 1 and 2." *Synthese* 40:147–87.

Taliaferro, R. Catesby. 1964. *The Concept of Matter in Descartes and Leibniz*. Notre Dame Mathematical Lectures, number 9. Notre Dame: University of Notre Dame Press.

Thijssen-Schoute, C. Louise. 1950. "Le Cartésianisme aux Pays-Bas." In Dijksterhuis et al. 1950a, pp. 183–260.

Toletus, Franciscus. 1599. *Commentaria unà cum quaestionibus in Octo Libros Aristotelis de Physica Auscultatione*, 4th ed. Venice: Iuntas.

———. 1614. *Commentaria unà cum quaestionibus, in Universam Aristotelis Logicam*. Venice: Iuntas.

Van de Pitte, Frederick. 1979. "Descartes' *Mathesis Universalis*." *Archiv für Geschichte der Philosophie* 61:154–74.

Verbeek, Theo. 1988. *René Descartes et Martin Schoock: La querelle d'Utrecht*. Paris: Les impressions nouvelles.

———. 1991. *Descartes and the Dutch: Early Reactions to Cartesianism, 1637–1650*. Carbondale: Southern Illinois University Press.

Vigier, Jean. 1920. "Les idées de temps, de durée, et d'éternité dans Descartes." *Revue philosophique* 89:196–233; 321–48.

Waard, C. de. 1936. *L'expérience barométrique: ses antécédents et ses explications*. Thouars (Deux-Sèvres): Imprimerie Nouvelle.

———. 1939–53. *Journal tenu par Isaac Beeckman de 1604 à 1634*, 4 vols. The Hague: M. Nijhoff.

Wahl, Jean. 1920. *Du rôle de l'idée de l'instant dans la philosophie de Descartes*. Paris: Felix Alcan.

Wallace, William. 1981. *Prelude to Galileo*. Dordrecht: D. Reidel.

———. 1984. *Galileo and his Sources*. Princeton: Princeton University Press.

Weber, Jean-Paul. 1958. "Sur une certaine 'méthode officieuse' chez Descartes." *Revue de Métaphysique et de Morale* 63:246–50.

———. 1964. *La constitution du texte des Regulae*. Paris: Société d'édition d'enseignement supérieur.

———. 1972. "La méthode de Descartes d'après les *Regulae*." *Archives de Philosophie* 35:51–60.

Weier, Winfried. 1970. "Cartesianischer Aristotelismus im Siebzehnten Jahrhundert." *Salzburger Jahrbuch für Philosophie* 14:35–65.

Weisheipl, James. 1963. "The Concept of Matter in Fourteenth Century Science." In McMullin 1963, pp. 147–69.

———. 1965. "Classification of the Sciences in Medieval Thought." *Mediaeval Studies* 27:54–90.

———. 1978. "The Nature, Scope, and Classification of the Sciences." In Lindberg 1978, pp. 461–82.

———. 1982. "The Interpretation of Aristotle's *Physics* and the Science of Motion." In Kretzmann, Kenny, and Pinborg 1982, pp. 521–36.

———. 1985. *Nature and Motion in the Middle Ages.* Ed. by William E. Carroll. Washington: Catholic University of America Press.

Wells, Norman J. 1965. "Descartes and the Modal Distinction." *Modern Schoolman* 43:1–22.

Westfall, R. S. 1971. *Force in Newton's Physics.* New York: Neale Watson Academic Publications.

Wilson, Margaret. 1978. *Descartes.* London: Routledge & Kegan Paul.

Wippel, John F. 1982. "Essence and Existence." In Kretzmann, Kenny, and Pinborg 1982, pp. 385–410.

INDEX

A

Adam, Charles, 342–44
Adams, Marilyn McCord, 334
Aiton, E. J., 315, 349, 356
Albertus Magnus, 326
Alquié, Ferdinand, 361
Amesius, Guilielmus, 361
Amicus, Bartholomeus, 128
Aquinas, St. Thomas, 6, 68, 265–66, 274, 326, 329, 332, 361–62
Ariew, Roger, 339, 348
Aristotle, 103–4, 307–8, 329, 331, 335, 337, 340–41, 356; and atomism, 118, 123; in education, 5–9, 117; and the finitude of the universe, 149; and motion, 157–59, 194, 352, 363; and the persistence of motion, 225–26; and place, 135; and unity of knowledge, 37; and the vacuum, 127. *See also* scholastic philosophy
Armogathe, J.-R., 309, 315, 327
Arnauld, Antoine, 151–52
Arthur, Richard, 361
atomism: and indivisibility, 120–21; in comparison with Descartes' philosophy, chap. 5 passim; Descartes' critique of atoms, 120–26, 153; geometrical vs. physical, 123, 340–41; revival of, 117–18, 338
attribute, 66, 70, 362; force as an attribute of body, 296

B

Bacon, Francis, 118, 325, 336–37
Baillet, A., 13–14, 141–42, 312–14, 339, 343–44
Basso, Sebastian, 118

Battail, Jean-François, 330, 365
Beck, L. J., 317–18, 324
Beeckman, Isaac, 338; physics of, 10–11, 101, 118, 121, 129, 197, 312, 352, 363; relations with Descartes, 9–12, 14, 105, 118, 121, 129, 157, 197, 231, 309–13, 339, 352
Berkel, Klaas van, 312–13
Bérule, Cardinal de, 15, 314
Beyssade, Jean-Marie, 268–72, 361
Bizer, Ernst, 315, 336
Blackwell, Richard J., 186, 349, 358
Blum, P. R., 325
Boas, Marie, 338
body: and force, 293–99; essence of, 133–34, 151, 327–32, chap. 3 passim; existence of, 327–32, chap. 3 passim; hard, 357; individuation of, 160, 175–81; relation to space, 128, 131–36; some bodies actually divided indefinitely, 125–26; three Cartesian elements, 103, 106–7, 340. *See also* extension
Bon, François de, 23
Bouillier, Francisque, 315, 336
Boyle, Robert, 155, 346
Brockliss, L. W. B., 310, 337
Broughton, Janet, 332
Bruno, Giordano, 128–29
Buridan, 345
Burley, Walter, 326
Burns, Norman, 339
Buzon, F. de, 312

C

Campanella, 128–29
Carcavy, Pierre de, 142, 343–44

Carraud, V., 309
Carriero, John, 330
Caton, Hiram, 339
cause(s): causa secundum fieri vs. causa se-
cundum esse, 265–66, 295–96; direct
vs. mediate, 274–75, 283, 287; final,
119, 247, 253, 273–74, 292; modal vs.
substantial, 277–78; universal vs. partic-
ular, 200–202, 300, 353, 365
Chandoux, M. de, 15
Charlet, Father, 104
Charleton, W., 332
Charleton, Walter, 143, 340, 344
Cicero, 6, 338
Clagett, Marshall, 312, 325, 346, 356
Clarke, Desmond, 319, 324, 350, 358
Clauberg, Johannes, 300
Clavius, Christopher, 7–8, 311
Clerselier, Claude, 242, 300, 305, 365–66
Cohen, Gustave, 315, 336
Cohen, I. B., 352
Cohen, Lesley, 316, 318, 330
Coimbrian Commentaries, 6–7, 95–96,
128, 332, 334
common notions, 67, 174–75, 295
Condemnation of 1277, 127, 151, 154
Copernicanism, 7, 20–22, 229, 315, 324,
348–49; and Descartes' account of mo-
tion, 159, 162, 181–88. See also Galilei,
Galileo, condemnation of
Cordemoy, Gerauld de, 342, 345–46, 348,
366; and atomism, 155, 179; and body,
178–79; and occasionalism, 300, 305, 330
Cosentino, G., 311
Costabel, Pierre, 311, 318, 320, 343, 349,
358–59; on French translation of Principles, 248, 316, 357; on quantity of motion, 205, 252, 354
Cottingham, John, 313, 332, 362
creation, 314–15, 328–29, 338; and divine
sustenance, 264–65, 267, 271–73, 275,
280–92; everything follows
mechanistically from an initial chaos,
19, 119, 198, 228; space created together with body, 136
Cronin, T. J., 330
Curley, E. M., 54, 310, 322, 345

D

Dainville, François de, 310
Dalbiez, R., 330
Dear, Peter, 311, 314, 353–54

Debeaune, Florimond, 360
Democritus, 117–19, 155. See also atomism
Descartes, René: Annotations on the Principles, 153, 167, 185, 316–17, 345; Cartesius, 315; Compendium Musicae, 10,
312; controversy with Pascal, 138–44;
Democritica, 313; Description of the
Human Body, 28; Experimenta, 313; Geometry, 9, 15, 21–22, 45; Letter to Voëtius,
27, 105, 316; Notes against a Program, 27,
151, 365–66; Olympica, 13, 313;
Parnassus, 312–13; Passions of the Soul,
27–28; Praeambula, 14, 313; Primae
cogitationes circa generationem animalium,
317; Studium bonae mentis, 15, 313; three
dreams, 13–14, 313; Treatise on Man,
19, 43; youth and education, 5–12. See
also Dioptrics, Discourse on the Method,
Meditations, Meteors, Principles of Philosophy, Rules for the Direction of the Mind
Dijksterhuis, E. J., 312, 323, 342–43
Dinet, Father, 104
Diogenes Laertius, 118, 338–39, 340
Dioptrics: composition of, 21–22, 315;
hypotheses in, 45–46, 322; laws of motion in, 206, 214, 357; method in, 45–46
Discourse on the Method: as a biographical
source, 5–9, 12–20, 309–10; composition of, 21–22, 45, 49; and Copernicanism, 315; method in, 44–47; order of
knowledge in, 51–52
distinctions, theory of, 69–70, 168, 329,
346, 348
Donagan, Alan, 331
Drake, Stillman, 330, 356
Dubarle, Dominique, 357–58, 362
Dugas, René, 346, 349
Duhem, Pierre, 341
duration and time: continuity vs. discontinuity of time, 266–73, 287–88; duration inseparable from substance, 67,
174–75, 263, 272–73, 331; duration requires a sustaining cause, 263–66, 329;
instant and moment, 268–70, 285–88,
362–63; and motion, 172–75; successive vs. simultaneous duration, 328

E

elasticity, 357
Elisabeth of Palatine, Princess, 27, 144–
45, 301, 332

Epicurus and Epicureanism, 117–120, 123–25, 155, 338–341. *See also* atomism
essence: in Descartes' philosophy, 68–70, 80; vs. proper accident, 68–69, 79–80, 147–48, 151; in scholastic philosophy, 68
eternal truths, creation of, 325, 342, 345, 353–54; and the problem of vacuum, 153–54
Eustachius a Sancto Paulo, 24–25, 58–62 passim, 150–51, 194, 325–27, 329, 331–33, 335, 337, 341–42, 352–53, 356, 361, 365
extension: and divisibility, 121–26, 339–41; as essence of body, 63–64, 66–67, 69, 110, 133–34, 151; extension of body and space the same, 131–32; and impenetrability, 144–48. *See also* body

F

Fanton d'Andon, Jean-Pierre, 343
Feenstra, R., 309
Fitzpatrick, Edward A., 310
Fonesca, Pedro, 128
Fontialis, Jacobus, 334, 345
force: eliminaton of force in Descartes' physics, 247, 254, 297–99; force causing motion, 156, 160–62, 173, 208, 275–78, 283–84, 286–87, 302, 354–55, 363–65; and individuation of bodies, 179; Newtonian, 244; ontological status of, 293–99; for proceeding, 205, 207–8, 235–39, 243–44, 247, 294, 350, 358–59; and relativity of motion, 347; of resisting, 235–39, 243–44, 247, 358, 362; of rest, 240–44, 280, 294, 359
Forge, Louis de la, 299, 361, 365
François, Jean, 7–8, 311
Frankfurt, Harry, 345
Freddoso, Alfred, 365
Fromondus, Libertus, 108, 119, 336, 339–41
Funkenstein, Amos, 345

G

Gabbey, Alan, 193, 242, 312–13, 316, 337, 344–45, 349–50, 352, 354, 356–60, 363–65; on force, 244, 295–97
Gäbe, L., 327
Gadoffre, Gilbert, 310
Galen, 338
Galilei, Galileo, 7, 118, 307–8, 311, 325, 330; condemnation of, 18, 20, 181, 185; Descartes' critique of Galileo on

order of inquiry, 12, 367; and free fall, 209; and the principle of inertia, 228–30, 356
Garber, Daniel, 309, 316, 318–19, 321–24, 329–30, 333, 335–36, 339, 355, 361, 365
Gassendi, Pierre, 62, 125, 128, 136, 265, 325, 327, 330, 332, 338–42, 344, 356, 361; and atomism, 118, 123, 155; and the principle of inertia, 203; and the vacuum, 101, 143
Gaukroger, Stephen, 324
geometry, objects of, 68, 81–82: distinguished from physical objects, 293; have duration, 67, 175
Gervais de Montpellier, 105, 336
Geulincx, Arnoldus, 300
Gilbert, Neal, 313, 325
Gilson, Étienne, 310–11, 314, 323, 325–27, 330, 332–34, 337, 342, 352, 356, 361
Goclenius, Rudolph, 321, 329, 331, 333, 353
God, 54–55, 328; as cause of motion in bodies, 74, 198, 200, 202, 204, 227, 237, 273–92, 299–305; divine sustenance, 237, 263–66, 274–75, 277–78, 300–303; as ground of the laws of motion, 198–200, 215–17, 237, 273–74, 280–92; role in Descartes' refutation of atomism, 123–25; role in the proof for the existence of body, 71–73, 76; and space, 128–29, 136, 144–49; universal cause of everything, 300–304; and vacuum, 127, 133, 151–54
Gouhier, Henri, 309–10, 312–13, 330, 334, 352, 365
Gracia, J. J. E., 327, 329, 332
Grant, Edward, 312, 341, 344–45, 356
gravity, 11, 210, 312, 333, 355; scholastic conception of, 96, 98–99, 101
Gregory, Tulio, 338
Grene, Marjorie, 99
Grosholz, Emily, 318
Grosseteste, Robert, 326
Guenancia, Pierre, 343
Gueroult, Martial, 179, 183, 324, 329, 355, 361, 364; on discontinuity of time, 266–68, 362–64; on force, 295–97; on relativity of motion, 346–48
Gundissalinus, 326

H

Hatfield, Gary, 294, 297, 327, 363–65
Heereboord, Adriaan, 23, 105, 336

Henry, D. P., 329
Henry, John, 334, 338, 341
Heppe, Heinrich, 361
Hermes Trismegistus, 7
Hero of Alexandria, 338
Hill, Nicholas, 118
Hintikka, Jaakko, 318
Hobbes, Thomas, 62, 327, 344
Hoenen, P. H. J., 365
Hoffman, Paul, 332
Huygens, Christian, 203
Huygens, Constantijn, 24
hypotheses, Descartes' use of, 22, 42–43, 45–46, 51, 62, 109–10, 322–23, 338

I

Iltis, Carolyn, 354–55
impenetrability, 144–48
impetus theory, 10, 227, 312, 356
individuation: of bodies, 160, 175–81; physical vs. biological, 176–77; of substances, 175–76
inertia, 203, 224–25, 228–30, 253–54, 297. See also laws of motion, principle of inertia
infinite: infinite vs. indefinite, 339–40; some bodies actually divided indefinitely, 125–26, 340–41; unboundedness of universe, 118, 153, 328–29, 345

J

Jammer, Max, 342
Jardine, Lisa, 313
Jones, Howard, 338, 340
Joy, Lynn, 338, 340

K

Kargon, Robert, 338, 346
Kepler, Johannes, 101, 203, 253, 297, 333, 349, 360
Kilwardby, Robert, 326
Knight, David, 334
knowledge: and certainty, 31–33, 36, 59, 321, 325; clear and distinct perceptions, 56, 325; intuition and deduction, 31–33, 35–39, 43, 157, 325; moral certainty, 325; order of, chap. 2 passim; and scholastic philosophy, 58; unity of, 1–2, 12–15, 307–8, chap. 2 passim; validation of, 40–44, 54–55, 322–23
Knudsen, Ole, 349

Koyré, Alexandre, 173–74, 186, 312, 348–49, 355

L

La Flèche, 5–9, 128, 148, 154, 309–10
Laporte, Jean, 330, 332, 361
Lasswitz, Kurd, 338
laws of motion, chap. 7 passim, chap. 8 passim, chap. 9 passim; conservation principle, 11, 199–210, 232, 239, 251–52, 254, 281–84, 289, 312, 358; and experience, 358; impact contest model, 233–41, 243–45, 247, 249–51, 289–90; impact in the letter to Clerselier, 242–48, 260–62; impact law, 199, 201–3, 207–8, 288–92, 353, 356–60, chap. 8 passim; impact rules, 237–52, 358; laws of persistence, 10–11, 198–99, 201–3, 210–30, 251, 284–88, 312; oblique impact, 357–58; principle of inertia, 203, 224–25, 228–30, 356; Principle of Least Modal Change (PLMC), 245–47, 249–53, 292, 299, 359; in the Principles, 199–203, 211–12, 216–17, 234–42, 248–53, 290–91; in The World, 198–99, 211, 214–16, 231–34, 288–89 (see also inertia). See also motion
Lefèvre, Roger, 318
Le Grand, H. E., 334
Le Guern, Michel, 339
Leibniz, G. W., 284, 307, 313, 316, 337, 344–45, 348, 355; critique of Descartes' account of impact, 358; critique of Descartes on the individuation of body, 180–81; and Descartes' conservation principle, 209–10; and force, 363
Le Pailleur, 142
Le Tenneur, 140–41, 143, 343
Leucippus, 118
light, 20, 35, 121, 210–11, 218, 225, 355; and vacuum, 143
Lipstorp, Daniel, 311–12
Locke, John, 330–32, 344
Lucretius, 7, 117–19, 338–39, 341
Lukács, Ladislaus, S.J., 310

M

Magnen, Jean Chrysostom, 118
magnet, 316
Mahoney, Michael, 314
Maier, A., 312, 352, 356

Maimonides, 337
Malebranche, Nicholas, 62, 307, 331, 333, 361, 365, 367; and occasionalism, 300, 304–5
Marion, Jean-Luc, 317, 319–20, 326–27; on hypotheses, 322; on the notion of order in the *Rules,* 321
Marshall, David J., 348
Masi, Roberto, 346
mathematics: certainty of, 7–8; in Descartes' education, 7–8; its place in Descartes' system, 53, 324; its relation to physics, 7–8, 11, 311
mathesis universalis, 34–35, 319, 322–23
Mattern, Ruth, 332
McMullin, Ernan, 332
Meditations: composition of, 16–17, 23–24, 27–28; essence of body in, 64–65, 76–77, 80–89; existence of body in, 70–72; metaphysics in, 53–56; method in, 47–50; in relation to *Principles,* 65, 70, 73–74, 90–91, 328, 330; wax example, 77–78
Menn, Stephen, 353
Mercer, Christia, 334
Mersenne, Marin, 15, 143, 314, 343, 360
Mesnard, Pierre, 310
metaphysics, its place in Descartes' philosophy, 53–56, 60–61
Meteors: composition of, 21–22, 46, 324; hypotheses in, 45–46, 322; laws of motion in, 206; method in, 45–46
method, 14–16, 317–27 passim, chap. 2 passim; abandonment of, 46–48, 56–57; and mathematics, 318–20
Milhaud, Gaston, 319, 349–50
Miller, Fred D., 340–41
mode, 64, 66, 69–70, 329; motion as a mode of body, 161, 163, 172–75, 212–13
More, Henry, 27, 67, 181, 316, 339, 344, 346, 348, 362; on divine extension and vacuum, 144–46, 152; on impenetrability, 145–46; on motion, 161, 227
Morin, J.-B., 335
motion, chap. 6 passim; and action, 158–62, 279, 362; cause of distinguished from effect, 156, 161–62, 173, 200, 347; change of place vs. change of neighborhood, 158, 162–66; circular, 219–23, 225, 227–30, 285–87, 355; definition of, 157–72; determination, 171, 188–93, 208, 213–14, 219–21, 235–36, 245–47, 349–52, 356–57; and discontinuity of time, 267 (*see also* duration and time); of the Earth, 182–85, 229; in a fluid medium, 357, 360; free fall, 157, 210, 213, 224–25, 312, 355; geometrical conception of, 158, 173–74, 346; involves time, 172–75; local, 158, 160, 195; as mode of body, 161, 163, 172–75, 212–13; momentary parts of, 11, 285–88; quantity of, 206–7, 239, 284, 363; reciprocity of transference, 166–68, 346–48; rectilinear, 199, 201, 210–11, 213–14, 218–25, 227–28; relativity of, 162–72, 240–41; and rest, 158, 160–72, 216–18, 229–30, 235–37, 241, 247–48, 278–80, 362–63; scholastic conception of, 157–59, 164, 193–96, 216; speed, 171, 213, 235–37, 239–40, 245–47, 349–50; tendency, 10–11, 121, 219–23, 287, 355–56; tendency, centrifugal, 20, 210–11, 218–23, 225, 230. *See also* laws of motion
Mouy, Paul, 315, 319, 334, 350
Murdoch, John, 338, 352, 356
Mydorge, Claude, 360

N

Nadler, Steven, 330
Newton, Sir Isaac, 307–8, 348, 356; principle of inertia, 203, 228, 230; and relativity of motion, 170–71
Nicholas of Autrecourt, 344
Noël, Étienne, 5, 7, 9, 22, 128, 141–43, 310
Normore, Calvin, 330

O

occasionalism, 74–75, 202, 299–305, 365–66
O'Neill, Eileen, 330
Ong, Walter J., 311, 325
Osler, Margaret, 339

P

Pachtler, G.-M., 310
Pascal, Blaise, 310; and the vacuum, 136–44, 342–44
Pascal, Jacqueline, 138, 343
Patrizi, Francesco, 128
Pell, John, 311
Périer, Florin, 137, 343
Petersen, Kurt M., 349

Petit, Pierre, 136
Philoponus, John, 345
physics, 60; in Descartes' education, 7, 10–11; its relation to mathematics, 11, 292–93, 311, 326; terrestrial vs. celestial, 227–28
Picot, Abbé, 248
Popkin, Richard, 314
Prendergast, Thomas, 346
primitive notions: in Elisabeth letters, 90; in *Meditations*, 89–90; and the nature of body, 92; in the *Principles*, 90–92; and soul (mind) as cause of motion, 276–77
Principles of Philosophy: composition of, 24–28, 104, 316–17; essence of body in, 76, 78–80; existence of body in, 72–75; law of impact in, 234–42, 248–53, 290–91; laws of motion in, 199–203; laws of persistence in, 216–17; method in, 47–49; order of knowledge in, 52–53; in relation to *Meditations*, 65, 70, 73–74, 90–91, 328, 330; as a textbook, 65, 252, 307, 316; translation of, 28, 248
properties (propria) and accidents, 68–69; inseparable properties, 68–69, 79–80, 147–48, 151
Prost, Joseph, 330, 346, 365
Ptolemaic system, 182
Puy-de-Dôme experiment, 137–41, 343

R

rainbow, 46, 320, 323–23
Ramus, Peter, 59, 325
Ratio Studiorum, 7, 310
reality, objective, formal, and eminent, 76–77
refraction and reflection, 35, 121, 190–91, 210, 214, 224–25, 233, 240, 251, 357, 360
Regius, Henricus, 23, 27–28, 62, 151, 315–16, 327, 329, 334–36, 353
Reif, Patricia, 325, 334
Reneri, Henricus, 23
Revius, Jacob, 23, 105, 336
Rist, J. M., 338–41
Roberval, Gilles Personne de: and motion, 189, 355–56; and universal gravitation, 333; and the vacuum, 131–32, 140–43, 342
Rochemonteix, C. de, 309–10
Rochot, Bernard, 343, 356

Rodis-Lewis, Geneviève, 309–15, 323, 332, 348, 361
Rohault, Jacques, 62, 327, 342
Rosenfield, Leonora C., 326–27, 337
Roth, Leon, 327
Rubius, 6
Rules for the Direction of the Mind: anaclastic line example, 34–36; blacksmith example, 40–42; body in, 64; composition of, 15–16, 313, 317, 321–22; hypotheses in, 322; mechanical philosophy in, 105–6; method in, 31–44, 318; motion in, 157–58, 164, 346; order of knowledge in, 38–39; and the unity of knowledge, 14; vacuum in, 129, 153–54

S

Sabra, A. I., 193, 339, 351–52
Schmitt, Charles, 311, 337, 341
scholastic philosophy: Descartes' critique of hylomorphism, 103–16, 324, 336; Descartes' critique of scholastic logic, 49, 321, 336; Descartes' critique of the scholastic conception of motion, 157–59, 164, 194–96; and Descartes' education, 5–9; hylomorphism, 95–103, 274–77, 305, 332–34; hylomorphism derives from our conception of body and soul, 97–99, 111–12, 114, 276; hylomorphism derives from prejudice for the senses, 100–102; method in, 58–62; motion in, 157–59, 164, 193–96, 216; order of knowledge in, 58–62, 326–27; persistence of motion in, 225–30; quality, real, 96, 333; space and place in, 148–51; vacuum in, 126–28, 148–51
Schoock, Martin, 336
Schouls, Peter, 324
Schuster, John, 34, 312–14, 317–19, 322–23, 327, 356, 360
Sennert, Daniel, 118
sensory qualities, status of, 75–78, 80, 82–85, 91–92, 117–18, 212–13, 332, 334
Serrus, C., 318, 324
simple natures, 32–33, 35, 38, 43, 66–68, 157–58, 318, 320
Sirven, J., 312
Smith, Norman Kemp, 361
Sorabji, Richard, 340, 361
Sorbière, Samuel, 136, 342

soul (mind), 54–55; as cause of motion, 276–77, 302–3; Descartes' rejection of animal souls, 111–16; as distinct from body, 85–90, 331; and extension, 89–92, 144–48, 344; and hylomorphic conception of substance, 97–99, 111–12, 114, 276; substantial form of the body, 89, 99, 104, 276; united to body, 89–93
space and place: Descartes' views on, 128–29, 131–36; imaginary, 127–28; internal vs. external, 134–36, 342; in relation to God, 127–28; in scholastic philosophy, 127, 148–51; in sixteenth- and early seventeenth-century thought, 127–28
Specht, Rainer, 330, 365
Spink, J. S., 338
Spinoza, Baruch, 62, 307, 348
Stephenson, Bruce, 360
Suarez, Francisco, 69–70, 128, 327, 329, 341, 344
substance, 64–70, 327–28, 331, 362; individuation of, 175–76; in scholastic philosophy, 68–70
Sylla, Edith, 326, 352, 356

T

Taliaferro, R. Catesby, 349, 356
Telesio, 128–29
Tempier, Étienne, 127
Thijsson-Schoute, C. Louise, 334
Toletus, 6, 326, 332, 340–41, 344–45, 352; views on space, 149–51
Torricelli experiments, 136–44, 310
Tychonic system, 182–85, 347–48

U

Utrecht affair, 23, 27, 315, 333, 336

V

vacuum: and atomism, 126; Descartes' views on, 128–43, 145–46, 152–54, 336, 342, 345; in late scholastic philosophy, 126–28, 148–51; motion in, 129, 341; Pascal's views on, 136–44; in sixteenth- and early seventeenth-century thought, 127–28; vessel argument, 133, 145–46, 148–49, 152, 342
Van de Pitte, Frederick, 318
Verbeek, Theo, 315–16, 336
Vigier, Jean, 361
Voëtius, Gisbertus, 23, 25, 105, 108, 315, 333, 336
vortex theory, 19, 20, 26, 121, 210–11, 218, 227–28, 315, 349, 356

W

Waard, C. de, 342–43
Wahl, Jean, 361
Wallace, William, 310–12, 356
Weber, Jean-Paul, 313, 317, 319
Weier, Winifred, 334
Weisheipl, James, 325–26, 334, 341, 352, 356
Westfall, R. S., 346, 349
Wilson, Margaret, 337
World, The: body in, 64, 334; composition of, 18–20; conservation principle in, 204–5; and Copernicanism, 21, 315; and the creation of the eternal truths, 154, 325; its relation to the Principles, 26; law of impact in, 231–34, 288–89; laws of motion in, 198–99; laws of persistence in, 211, 214–16; mechanical philosophy in, 106–7; method in, 46–47; motion in, 158, 164; vacuum in, 130–31, 154